Harco Willems, Jan-Michael Dahms (eds.)
The Nile: Natural and Cultural Landscape in Egypt

Mainz Historical Cultural Sciences | Volume 36

Editorial

The **Mainzer Historische Kulturwissenschaften** [Mainz Historical Cultural Sciences] series publishes the results of research that develops methods and theories of cultural sciences in connection with empirical research. The central approach is a historical perspective on cultural sciences, whereby both epochs and regions can differ widely and be treated in an all-embracing manner from time to time. Amongst other, the series brings together research approaches in archaeology, art history, visual studies, literary studies, philosophy, and history, and is open for contributions on the history of knowledge, political culture, the history of perceptions, experiences and life-worlds, as well as other fields of research with a historical cultural scientific orientation.
The objective of the **Mainzer Historische Kulturwissenschaften** series is to become a platform for pioneering works and current discussions in the field of historical cultural sciences.

The series is edited by the Co-ordinating Committee of the Research Unit Historical Cultural Sciences (HKW) at the Johannes Gutenberg University Mainz.

Harco Willems, Jan-Michael Dahms (eds.)

The Nile: Natural and Cultural Landscape in Egypt

[transcript]

The print was sponsored by the Research Unit Historical Cultural Sciences (HKW).

Additional support in the printing costs was supplied by the Dayr al-Barsha project of KU Leuven.

An electronic version of this book is freely available, thanks to the support of libraries working with Knowledge Unlatched. KU is a collaborative initiative designed to make high quality books Open Access for the public good. The Open Access ISBN for this book is 978-3-8394-3615-8.

Bibliographic information published by the Deutsche Nationalbibliothek
The Deutsche Nationalbibliothek lists this publication in the Deutsche Nationalbibliografie; detailed bibliographic data are available in the Internet at http://dnb.d-nb.de

Cover layout: Kordula Röckenhaus, Bielefeld
Typeset by Mark-Sebastian Schneider, Bielefeld
Printed in Germany
Print-ISBN 978-3-8376-3615-4
PDF-ISBN 978-3-8394-3615-8

Contents

Preface 7
Harco Willems, Jan-Michael Dahms

Modelling the Nile Agricultural Floodplain in Eleventh and Tenth Century B.C. Middle Egypt
A study of the P. Wilbour and other Land Registers 15
Jean-Christophe Antoine

Harbours and Coastal Military Bases in Egypt in the Second Millennium B.C.
Avaris, Peru-nefer, Pi-Ramesse 53
Manfred Bietak

Development of the Memphite Floodplain
Landscape and Settlement Symbiosis in the Egyptian Capital Zone 71
Judith Bunbury, Ana Tavares, Benjamin Pennington, Pedro Gonçalves

Karnak's Quaysides
Evolution of the Embankments from the Eighteenth Dynasty to the Graeco-Roman Period 97
Mansour Boraik, Luc Gabolde, Angus Graham

Medamud and the Nile
Some Preliminary Reflections 145
Félix Relats Montserrat

The Nile in the Fayum
Strategies of Dominating and Using
the Water Resources of the River in the Oasis
in the Middle Kingdom and the Graeco-Roman Period 171
Cornelia Römer

Nilometers – or: Can You Measure Wealth? 193
Sandra Sandri

In Search of a Future Companion
Digital and Field Survey Methods in the Western Nile Delta 215
Joshua Trampier

The Dynamic Nature of the Transition from the Nile Floodplain to the Desert in Central Egypt since the Mid-Holocene 239
Gert Verstraeten, Ihab Mohamed, Bastiaan Notebaert, Harco Willems

The Analysis of Historical Maps as an Avenue to the Interpretation of Pre-Industrial Irrigation Practices in Egypt 255
Harco Willems, Hanne Creylman, Véronique De Laet, Gert Verstraeten

Landscapes of the Bashmur
Settlements and Monasteries in the Northern Egyptian Delta
from the Seventh to the Ninth Century 345
Penelope Wilson

Authors 369

Preface

HARCO WILLEMS, JAN-MICHAEL DAHMS

On 22 and 23 February 2013, the Forschungsschwerpunkt Historische Kulturwissenschaften at the Johannes-Gutenberg-Universität Mainz organised an international symposium on the Nile as a natural phenomenon, and on the impact of the river on Egyptian culture in the broad sense of the word.

That the Nile was of crucial importance to Egypt, a country surrounded by desert, is obvious to all, and has been realized already since Antiquity. Yet, the river as an environmental and cultural factor has been less intensively studied by archaeologists and Egyptologists than might be expected. One issue is that these scholars often work on the basis of an inadequate familiarity with the geomorphology of floodplains. For instance, texts and scenes concerning Nile deities or of religious customs related to the river are usually explained by referring to notions like 'ideal floodheights', which are never defined. Other scholars do take into account scientific evidence, but usually they base themselves upon a small amount of studies produced over a century ago, sometimes directly transferring the information provided by those accounts to the far more distant pharaonic past.

In this regard, the many publications by KARL BUTZER (most notably his *Early Hydraulic Civilization in Egypt* [1976]) mark a watershed. Outdated though many of his conclusions may now be, its lasting importance lies in showing that an integration of Egyptological evidence with data produced by the natural sciences *works*, and in for the first time pointing out the kinds of questions that can be approached in this way. The areas intensively dealt with by BUTZER were (settlement) archaeology, economic history, technology, and demography. However, one might also add religion, as the cycle of the Nile had such an impact on all aspects of life that it also co-determined for instance the religious calendar and the phasing of rituals. At roughly the same time as BUTZER published his seminal work, MANFRED BIETAK's *Tell el-Dab'a* II (1976) showed in an impressive fashion how the integration of geographic and archaeological data can contribute to the reconstruction of now-vanished landscapes.

More recently, STEPHAN SEIDLMAYER's book on historical and modern flood levels (2001) has created a new basis for understanding some of the effects of the Nile. It shows that we are facing a natural phenomenon, the study of which is fundamentally the domain of the natural sciences, but also that available evidence includes ancient and culturally biased material of a kind that lies far beyond the competence of most natural scientists. The problem in addressing the dispersed and incongruous sources of information is that an intensive interaction between numerous disciplines with little tradition of collaboration is needed.

Nowadays, significant progress is being achieved particularly in integrating earth sciences and Egyptian archaeology. One aim of the symposium was to enable natural scientists to compare the methods they deploy and the kinds of results they attain at the various sites. Another aim was to compare the results of regional interpretations from different parts of the country to address broader issues (like the size of the floodplain, the validity of hypotheses about the drift of the Nile bed, or the potential for economic and demographic analysis).

A further aim has been to assess ancient indigenous evidence testifying to how the Egyptians reacted to the environmental conditions imposed by the nilotic environment. For this, archaeological indications could be the spatial distribution of sites in relation to features of the floodplain landscape (e.g. settlement spread); the system of irrigation, or the date when certain changes in land form, land cover, or land use occurred. The importance of spatial data for modeling the modern and ancient landscape with the help of remote sensing and near surface geophysics has in the most recent years also come to the fore. But ancient Egyptian written and iconographic reflections on the landscape can be equally important. Specialists in these areas have been less prone to look at scientific evidence, and their work is often less accessible to the natural scientists.

Another issue is that Egyptology, while covering a time range that historians would consider very wide, is concerned with what to earth scientists is a very short period of time. All inhabitants have had to both control and exploit that ecological framework provided by the Nile. This was so in the pharaonic period, which is the domain of Egyptologists. However, the matter was no different in the Byzantine, medieval and post-medieval periods, and therefore historians interested in these more recent periods investigate material that is no less relevant.

The aim of the conference was to bring together a group of specialists from these diverse disciplines. The programme was as follows:

22 February 2013

Morning session: Chair: *Judith Bunbury*

Prof. Dr. Ulrich Försterman, Vizepräsident für Forschung (Johannes-Gutenberg-Universität Mainz): *Welcoming Speech*
Prof. Dr. Stefan Müller-Stach, GFK-Leitungsgremium ((Johannes-Gutenberg-Universität Mainz): *Welcoming Speech*
Harco Willems (Johannes-Gutenberg-Universität Mainz/KU Leuven): *The Analysis of Historical Maps as an Avenue to the Interpretation of Pre-Industrial Irrigation Practices in Egypt*
Luc Gabolde (CNRS Montpellier): *The Origins of Karnak – Geoarchaeological, Astronomical, Textual and Theological Sources*
Angus Graham (University College London): *The Origins of Karnak – Geoarchaeological and Geophysical Survey Results*
Gert Verstraeten (KU Leuven): *The Dynamic Nature of the Transition from the Nile Floodplain to the Desert in Central Egypt since the Mid-Holocene*
Harco Willems (Johannes-Gutenberg-Universität Mainz/KU Leuven): *The Hare Nome – from Physical Geography to Social Archaeology*

Afternoon Session – Chair: *Manfred Bietak*

Cornelia Römer (Deutsches Archäologisches Institute Abt. AI Kairo): *Irrigation Canals in the Fayum*
Dirk Blaschta (Universität Leipzig): *Geoarchaeological Investigations of the Area Surrounding the Dahshur Necropolis*
Judith Bunbury (University of Cambridge): *Migrating Memphis – The Development of a City in a River Floodplain*
Ana Tavares (Ancient Egypt Research Associates): *The White Walls – The Landscape of the Capital Zone*
Willem Toonen (RijksuniversiteitUtrecht): *Implications of the Holocene Palaeo-Environment on Cultural Dynamics in the Western Nile Delta*
Joshua Trampier (Oak Ridge Associated Universities): *Above, atop, and below – Integrated Methods for Reconstructing the Cultural and Natural Landscapes of the Western Nile Delta*
Manfred Bietak (Österreichische Akademie der Wissenschaften): *Harbours and Coastal Military Bases in Egypt in the 2ndMillenium: Avaris – Peru-Nefer-Piramesse* (keynote speech)

23 February 2013:

Morning Session: Chair: *Cornelia Römer*

Penelope Wilson (Durham University): *Isolation and Resistance in the Northern Nile Delta Landscape*

Rainer Nutz (Universität Basel): *Nile Gauge Readings and the Agrarian Potential in the Middle Kingdom*

Jean-Christophe Antoine (Université Jean Monnet Saint-Etienne): *Modeling the Nile Agricultural Floodplain from Papyrus Wilbour and Xth Century B.C. Land Registers*

Pierre Koemoth (Université de Liège): *Cultivable Land and Crocodiles – Ethology, Religion, and Economy in the Nile Floodplain in Roman Egypt*

Afternoon Session – Chair: *Harco Willems*

Jan Tattko (Eberhard-Karls-UniversitätTübingen): *Personifications of the Nile Flood according to Graeco-Roman Temple Inscriptions*

Sandra Sandri (Johannes-Gutenberg-Universität Mainz): *Nilometer – oder: Kann man Wohlstand messen?*

Stuart Borsch (Assumption College Worcester/MA): The Water Regime of Medieval Alexandria

The present, peer-reviewed volume offers the proceedings of this symposium. The result shows a wide range of topics, but the papers have in common that they show how the integration of evidence from different disciplines can change our perspective on ancient Egypt.

One series of papers published here are site- or regionally-specific studies integrating earth-scientific approaches with Egyptological and archaeological evidence. The papers address regions across Egypt, and demonstrate how the different types of environment in those regions impacted upon living conditions and, ultimately, social and historical processes there.

Penelope Wilson's article reports on the results of a recent archaeological survey of the surroundings of Lake Burullus at the northern fringe of the western Nile Delta. The work of her research group has led to the discovery of numerous hitherto unknown or potential archaeological sites there. She places their emergence in the context of long-term developments from the Graeco-Roman period until the late first millennium AD. Due to a decreasing sea level in the early part of this period, the area's agricultural potential increased, and many of the sites discovered, some of which are very prominent, go back to this period. However, the author notes evidence for the increasing development of swamps as of the fifth century A.D. Based on a discussion of five exemplary settlement sites, she shows how economic sustainability decreased in the early Islamic period and how archaeological remains help to understand the human response to these changing conditions.

Joshua Trampier's paper concerns work in an area immediately adjoining the one Wilson has been working in on the southwestern Delta fringe. The sites there date to the same period. Trampier discusses his work at one site – Kawm am-Qamḥa – in detail, so as to offer an overview of the various methods he has deployed (remote sensing, coring, field surveys, study of historical maps).

Manfred Bietak's article moves the perspective to the eastern Nile Delta around Tall al-Dabʿa. His discussion concerns the available evidence for the location of the New Kingdom harbor of Perunefer. In the past, this place was believed to have been located in the Memphite region, a point of view that was recently defended once again. Bietak's paper presents archaeological, geomorphological, and textual evidence demonstrating not only that the case for Memphis rests on a weak basis, but also that it is very likely that Perunefer referred to the harbor area of Avaris / Piramesse.

The Memphite region is among the earliest where archaeologists have shown an interest in contextualizing the archaeological record with floodplain research, but most of the relevant studies were published some twenty years ago. The article by Judith Bunbury, Ana Tavares, Benjamin Pennington, and Pedro Gonçalves offers an update on the current state of research. Their research addresses long term developments like climate change, the impact of sea level changes and wadi and Aeolian sand depositions on the nature of the floodplain. Their study adduces fresh indications that are interpreted as indicative of an eastward shift of the Nile bed. Moreover they argue that the head of the Nile Delta showed significant displacements over time. All these indications lead to a time series in which the archaeological record for settlement and cemetery use in the region is correlated with the evolving landscape.

Cornelia Römer studies the systems of water management deployed in the Fayyūm in the Middle Kingdom and the Graeco-Roman periods. Following Ball, she argues that the situation reportedly seen by Herodot, to the effect that water from inside the Fayyūm may have receded back into the Nile Valley after the flood, may have obtained also in the Middle Kingdom. She argues that the dams at the entrance of the Fayyūm may date back to this early period, and that they have played a role in controlling the movement of floodwater into and out of the Fayyūm, and from the Nile Valley to the north. The second part of her study argues that, as of the Ptolemaic period, the water level in the Fayyūm had shrunk considerably, and that this process went hand in hand with a more rational water distribution across the depression. Her account of this later period is particularly interesting by showing, through a combination of archaeological field surveys, geomagnetic research, and papyrological indications, how the flood regime worked in the Graeco-Roman period.

Two articles study the form of the landscape in Middle Egypt, roughly between where the borders of the Hare nome were in Antiquity (i.e. between Dairūṭ al-Sharīf in the south and the village of Itlīdim in the north). Verstraeten, Mohamed, Notebaert and Willems study the dynamics of the interface between floodplain and desert on both the eastern and western desert edges in this region. Using a similar array of methods as Trampier, they conclude that, on the east between al-Barshā and al-Dayr Abū Ḥinnis, the Nile has migrated from east to west since pharaonic days, and not in the opposite direction, as is often assumed. On the western desert edge, they were able to model the migration of dunes into the floodplain and their impact on the Baḥr Yūsif system in the course of the Holocene. The paper also demonstrates that interdune areas in that region were already used for agricultural purposes in the Old Kingdom. Willems, Creylman, De Laet and Verstraeten also investigate the western floodplain in the region of the Hare nome, this time primarily on the basis of the information of historical maps. Their paper demonstrates that, in the eightteenth century, the floodplain was still a mostly natural landscape, in which human intervention was restricted to the construction of dykes. The flood regime is reconstructed, the most important result being that much of the residual floodwater did not return to the Nile, but remained captured between the levees of Nile and of Baḥr Yūsif. Since the same situation prevailed further north, it is clear that the presence of a large, humid area in the centre of the western floodplain, as recorded in eighteenth century maps, must be taken seriously. The article studies the effects of this hydrological phenomenon on more northerly areas, and most notably on the Fayyūm, addressing several of the issues also dealt with by Cornelia Römer. Study of historical toponymy and archaeological evidence suggests that essential elements of this eightteenth century model were in force already in the pharaonic period.

Felix Relats Montserrat discusses the tribune and the quay that presumably accompanied it to the west of the temple of al-Madamūd. The question interesting him is whether it connected to a NE-SW canal allegedly linking the temple to Karnak North, or whether the Nile may in the First Intermediate Period and Middle Kingdom, when the temple emerged, have lain closer to the building. He concludes that only future geomorphological research can solve this problem, but argues that the latter possibility probably has the greater likelihood.

At the conference, Angus Graham and Luc Gabolde presented two fascinating, although partly contradictory papers on soil formation in the Karnak area. Since these papers will be published elsewhere, Gabolde, Graham and Mansour Boraik here present a new study, concerning the evolution of the quays at Karnak between the early New Kingdom and the Graeco-Roman period. They first discuss written evidence, showing that there were in antiquity many islands opposite Thebes. They also mention records referring to canals west of the temple, and perhaps also to its east. The rest of their study offers a fascinating overview of excavated embankments of different periods and of the results of geomorphological research in the same area. This leads to a series of partly hypothetical maps, arranged in a time series, which display how the development of the Karnak temple followed (and partly influenced) the changing landscape due to the westward migration of the Nile bed.

The remaining two papers discuss, not specific regions and landscapes, but cultural themes reflecting the ways how Egyptian culture reacted to the constraints and possibilities offered by the Nilotic landscape.

Late New Kingdom land registers like P. Wilbour and P. Reinhardt are among the most important administrative documents that have survived from ancient Egypt, and they have been intensively studied in the past. Jean-Christophe Antoine's contribution to this volume offers a new and highly interesting approach to study these documents as a source of information on land form, land use, and social structure. These documents offer information on thousands of plots of land, their location, the type of land concerned, the identity and profession of their owners, and so on. Deploying multiple correspondence analysis, multivariate logistic regression, and univariate analysis, Antoine demonstrates that hitherto unsuspected links existed between certain social categories and specific geographical zones, which he is able to link to hydrologically distinct areas in the landscape. Moreover, his statistics raise the interesting idea that the term *ḳ3i.t,* designating a type of 'high' land (perhaps on a levee), and the same term referring to a fiscal plot category, are actually conceptually entirely different things.

Finally, Sandra Sandri discusses the iconographic theme of the nilometer as it appears in the late Roman period across the Mediterranean. The earliest examples appear on second century Roman coins, and on reliefs from the Nilotic staircase on Elephantine island. Most of the other examples are much more recent, and many derive not from Egypt, but from other parts of the

Mediterranean. The article addresses the figures depicted on the nilometers, which are quite high. Although the few cases where the scenes indicate which nilometer is intended all agree that it concerns the one at Alexandria, the flood heights there must have been much lower than is indicated in the scenes, and the remains of the nilometer excavated at Alexandria by Alan Rowe look rather different than what the depictions show. Therefore, what is depicted is in many cases obviously an imaginary nilometer, which just has the aim of rendering an Egyptian scenery.

Last, but certainly not least, the editors have to express their gratitude to a number of persons and institutions at the Johannes-Gutenberg-Universiät Mainz. It is due to the generosity of the Johannes-Gutenberg-Forschungskolleg of this university that Harco Willems was able to spend a research period at the Forschungsschwerpunkt Historische Kulturwissenschaften (HKW) at the Johannes-Gutenberg-Universität Mainz between 1 April 2012 and 31 March 2013. In this context, Jan Dahms was moreover appointed as a scientific collaborator. This exceptional privilege not only gave us time to conduct research and to organize the symposium of which the proceedings are presented in this volume, we also have fond recollections of the friendly way we were received at the HKW and at the Ägyptologisches Institut at Mainz. At the HKW, we wish to express our thanks to Prof. Dr. Jörg Rogge and to Kristine Müller-Bongard for their support. But we owe a special debt of gratitude to Prof. Dr. Ursula Verhoeven and Prof. Dr. Tanja Pommerening. It was they who took the initiative to propose to the HKW to invite the first editor of this book as a Johannes-Gutenberg research fellow. Thank you very much for your friendship and support.

Modelling the Nile Agricultural Floodplain in Eleventh and Tenth Century B.C. Middle Egypt

A study of the P. Wilbour and other Land Registers

JEAN-CHRISTOPHE ANTOINE

1. Introduction

Although they contain a wealth of geographical information, land registers have not been fully exploited with the aim of reconstructing the Nile Valley landscape. However, compared to other written or pictorial sources, they have several advantages. 1) The geographical information they contain is based on onsite observation and therefore directly reflects the reality of the time at least from the point of view of the land assessors who surveyed the fields.[1] 2) Being administrative documents, we may expect them to be free from artistic, religious, political or ideological interferences which often make the interpretation of other ancient sources difficult. 3) For the Pharaonic Period we have a relatively high number of documents dating from the eleventh to the tenth century B.C. (Twentieth Dynasty to Twenty-second Dynasty) which concern Middle Egypt.[2] Unfortunately, several of these documents are very lacunous, devoid of useful geographical information, or still incompletely published so that, in the final analysis, only three texts can be fully exploited, namely P. Wilbour, the Louvre AF 6345-Griffith fragments and P. Reinhardt.[3]

---

1 On this particular point see ANTOINE, 2011, p. 12.

2 For a comprehensive list of these documents see VLEEMING, 1993, p. 78–80.

3 The Wilbour papyrus was published by GARDINER, 1941–1952, 4 volumes, especially volume II, Commentary and volume IV, Index by FAULKNER, 1952; for the

One reason that probably explains why scholars seldom used these documents is that they are difficult to interpret due to the pending uncertainty concerning the exact meaning of several words which designate features of the geographical landscape, different types of soil or administrative categories of fields. In addition, the number of data they encompass, which in P. Wilbour probably represents several hundred thousand bits of information, needs the use of powerful statistical methods to obtain a meaningful insight of their content. Here, using such methods, we propose a model of the agricultural landscape of the Nile Valley in Middle Egypt around 1000 B.C.

2. A brief description of the three documents used for the analysis

Due to its good state of preservation and the impressive amount of 3400 recorded plots, P. Wilbour is the most important of the three documents. It contains two texts, text A and text B, both dated to the middle reigns of the Twentieth Dynasty.[4] From a geographical point of view, plots form small clusters unevenly distributed over a region which almost corresponds to the province of Beni Suef at the time of the expedition of Napoleon Bonaparte to Egypt (see figure 4). The geography of text A, which is the easiest to reconstruct,[5] shows that fields were located on the west bank of the Nile, which at that time probably ran somewhere to the west of its present bed.[6] These fields are distributed over four different zones from north to south where they do not occupy the same part

Louvre AF 6345-Griffith Fragments Papyrus see GASSE, 1988, p. 3–73, and for the Reinhardt papyrus, VLEEMING, 1993.

4 Text A is dated to year 4 of Ramesses V and records a survey performed in the second half of July. Text B was based on a document anterior by a few years to text A, but it was written and modified after the redaction of text A, probably during the first years of the reign of Ramesses VI: GARDINER, 1948, p. 183–187; HARING, 1997, p. 316f; ANTOINE, 2010, p. 5–14.

5 GARDINER, 1948, p. 36–55; GOMAÀ et al., 1991, p. 105–166

6 BUTZER, 1976, p. 33–35; SAID, 1993, p. 61–63; LUTLEY/BUNBURY, 2008, p. 3–5. Changes also concerned Lake Karun, which probably had a larger extension at that time: HASSAN/TASSIE, 2006, p. 37–40.

of the floodplain (see figure 4).[7] The geography of text B is more complex and remains a puzzling problem. As a whole, the surveyed region corresponds to that of text A, but it probably extends to the north up to the level of Atfith and – at least in part – encompasses the east bank.[8]

P. Wilbour displays a very complex administrative and fiscal categorisation of land (see table 1) which has been the object of several studies.[9] However, this can be simplified for this analysis. Text B is devoted to *khato* (*ḫ3-t3*) land, a category of royal land held on temple domains while text A records fields belonging to temples or secular institutions including plots of *khato*-land. Two fiscal categories of land can be distinguished.

The first covers fields classified as *ḳ3yt*, *tni* or *nḫb*, which appear in both text A[10] and B, correspond to plots of relatively large size, are taxed in cereals and are held by high-ranking administrators or their subaltern middlemen.

The *ḳ3yt* category is by far the most frequent of the three, while the two others occur only occasionally. From a fiscal point of view, the value of the land decreases from *nḫb* and *tni* to *ḳ3yt*. In text A *nḫb* plots are taxed at a rate of 10 khars/aroura versus 7.5 for *tni* and 5 for *ḳ3yt* plots. These figures should be compared with the estimated average yield of cereals at the time which was probably between 6 to 10 khars/aroura,[11] which strongly suggests that the *nḫb* and *tni* categories correspond to the most productive lands.

The second group of plots, which appears only in text A, is held by a great number of mostly private smallholders with a variety of professions. The majority of these plots are measured in aroura and taxed in cereals on a reduced portion of their surface at a constant rate of 1 ½ khars/aroura (type I and IA

---

7 Gardiner, 1948, p. 36–55; Antoine, 2011, p. 9–27. The limits of the four zones are not completely similar in these two reconstructions, while Gomaà et al., 1991, p. 138–141, propose a different organization of the zone.

8 Ibid., p. 173–178.

9 Ibid.; Helck, 1960; Menu, 1970; Janssen, 1975, p. 127–185; Id., 1986, p. 352–366 (review article of Stuchevsky, 1982); Katary, 1989; Haring, 1997, p. 284–301. For a recent review on land organisation see Katary, 2013, p. 719–783.

10 In text A, *tni*, *nḫb* and *ḳ3yt* are not explicitly named, yet their existence is deduced from the tax rates applied to plots in normal domains, which follows the same proportional ratio as that of *tni*, *nḫb* and *ḳ3yt* in text B: Gardiner, 1948, p. 28–29 and 180. Although for the sake of convenience we shall use the term "tax" to designate these rates, it should be kept in mind that it is not clear whether they actually correspond to rents perceived by the landholding institutions or taxes.

11 Ibid., 1948, p. 71.

plots). Strikingly, another part is measured in land-cubits and left untaxed (type II and IIA plots). Some of these untaxed plots were apparently not planted while others were probably cultivated, although we ignore which crops were grown there. Despite an impressive number of recorded plots, text A certainly does not cover the totality of the fields held by landholding institutions in this region, but probably a category of them selected based on still unclear fiscal or administrative criteria.[12]

	TEXT B	TEXT A		
Date	Early Ramesses VI (?)	Year 4 of Ramesses V		
Owning institutions	temples	temples and secular institutions		
Types of domain	*khato*-land	normal domains	*pš* domains	
Administration/holding	officials	officials	small "private" holders	
Fiscal plot category	*ḳꜣyt-tni-nḫb*	*ḳꜣyt-tni-nḫb*	Type I/IA	Type II/IIA
Number of plots	602	540	1781	471
Unit of measure	aroura	aroura	aroura	land-cubit
Mean surface in aroura	30.4	13.9	6.3	0.50
Plantation	cereal	cereal	cereal	?
Tax/rent in grain	yes	yes	yes	no

Table 1: The organisation of fields in Text A and Text B of the Wilbour papyrus

The Louvre-Griffith fragments and P. Reinhardt are dated to the Twenty-first or Twenty-second Dynasty and record fields belonging to Theban temples and located in the Tenth Nome of Upper Egypt as the town of Tjebu is named in both

12 See KATARY, 1989, p. 23f and ANTOINE, 2011, p. 27. Based on the date of the record (end of July), Fairman once deduced that the assessment only concerned summer crops (FAIRMAN, 1953, p. 118–123) which is very unlikely owing to the distribution of plots across the floodplain (ANTOINE, 2011, p. 27).

texts.[13] The Louvre-Griffith fragments share several features with P. Wilbour. The recto concerns fields of a relatively large size that are fiscally classified as *k3yt* or *nḫb,* although the tax rate is clearly lower than in P. Wilbour.[14] The verso relates to small plots held by a variety of smallholders and taxed at a very low rate on a portion of their surface, while some are not taxed, all of this being reminiscent of the different category of plots of the *pš* domains of P. Wilbour.[15] Unfortunately, geographical information is preserved on the recto only.

P. Reinhardt as well is regrettably lacunar.[16] It records fields taxed in grain on a portion of their surface at the very high rate of 12 and sometimes 15 khars/aroura while the rest of the area is deducted from the assessment because the land was not suitable for cultivation or planted with other crops. Plots in this document fall into two administrative categories, namely cultivated fields (*iḥt*) and *corvée*-land (*bḥ* or *iḥt-bḥ*)[17] which are represented almost equally. Both are under the responsibility of officials or administrators. Among them are several water-chiefs (*ʿ3-n-mw*),[18] a rare title which, with the classification of the land as *corvée,* may indicate that the parcels covered by this document require specific administrative control concerning both their irrigation and exploitation.[19] Although the fiscal organisation underlying P. Reinhardt clearly departs from that of the two other documents, all three documents share a feature of importance for the present analysis, namely that some plots or a part of their surfaces are not taxed in grain because the land was not suitable for cultivation or because it was planted with another crop.

13 GASSE, 1988, p. 50 dates the Louvre-Griffith fragments to the end of the Twentieth Dynasty while VLEEMING, 1993, p. 79 dates them to between the Twenty-first and Twenty-second Dynasties on palaeographic considerations. On a discussion on this text see also VLEEMING, 1991 and HARING, 1997, p. 326–342.

14 1 khars/aroura for *k3yt* and 2 for *nḫb*.

15 On this opinion see JANSSEN, 1975, p. 149; HARING, 1997, p. 334; KATARY, 2005, p. 151f.

16 VLEEMING, 1993. On this document see also HARING, 1997, p. 326–342.

17 VLEEMING, 1993, p. 51–55.

18 Ibid, p. 56–57.

19 EYRE, 1994, p. 77.

3. Geographical information in land registers

The geographical information used in this analysis is contained in the way the scribes indicated the position of each plot in the floodplain by reference to one and, especially in P. Wilbour, two landmarks (hereafter designated as the main and secondary landmarks respectively).[20] These landmarks consist of either settlements including cities, temples, villages (*wḥyt*), houses (*ꜥt*), fortified villas (*bḫn*), keeps (*sgr*)[21] and various farm buildings, or remarkable landscape features such as islands (*ỉw*), mounds (*ỉ3t*), lakes (*š*), ponds (*brkt*) and groves together with different soil categories designated as *ḳ3yt* (high ground), *pꜥt*, *m3wt* (new land), *ỉw-n-m3wt* (island of new land) and *ỉdb* (for details see Annex 1).[22] On the whole, the land surveyors who assessed the fields preferred settlements or landscape features designated by or associated with a proper name[23] to indicate field location, probably because they are less ambiguous than anonymous features and made the use of a secondary landmark unnecessary. When plots were located in some remote area of the floodplain at a distance from inhabited or well-known places, a landscape feature was selected as an identifier, with a settlement often functioning as a secondary landmark. Such a method is the rule in P. Wilbour, but was used less frequently in the Louvre-Griffith fragments and P. Reinhardt, probably because the valley was narrower in the tenth nome of Upper Egypt, so that plot were only rarely located far from any well-known place.

From the geographical data, two sets of evidence can be derived. Firstly, by analysing the relative position of the main and secondary landmarks we can obtain information that helps to localise the different groups of landscape

20 ANTOINE, 2011, p. 9–27.

21 Following here the translation proposed by GARDINER, 1948, p. 35, but CERNY, 1958, p. 209f has proposed that *sgr* stands for *sg3* 'hill'.

22 On these words see mainly GARDINER, 1947, vol. I, p. 13; Id., 1948, p. 25–36; VLEEMING, 1993, p. 45–48; GRIESHABER, 2004. On *m3wt* see more specifically YOYOTTE, 2013, p. 231–237 and EYRE, 1994, p. 75–77.

23 Schematically, two categories of landmarks can be distinguished: those I have called "defined" (ANTOINE, 2011) and those that are "undefined" since they need a secondary landmark belonging to the first category to be identified. Undefined landmarks mostly correspond to landscape features while the defined landmarks are mostly localities using the town determinative, suggesting they were settlements. Although some situation may be ambiguous, the designation of defined localities can be considered as a proper name in most of the cases.

features in the Nile Valley.[24] Secondly, this being validated, the position of plots according to their main landmark may be used to reconstruct the distribution of the different agricultural categories of fields in the floodplain.

4. Statistical methods used for the analysis

Because of the large amount of data, P. Wilbour was the document used to construct a model of the Nile Valley landscape. In a second step, this model was checked against the two other documents. Handling a land register such as P. Wilbour requires statistical methods appropriate for identifying complex interrelationship and free from any preconceived point of view to obtain an unbiased insight of the structure. These methods must also provide a synthetic, but not simplistic understanding of complexity. Here we used two such complementary methods. The first is a multiple correspondence analysis (MCA).[25] This global exploratory and descriptive method projects the relationship between data (appearing as a point cloud) on a two-dimensional graphic space in an easily and intuitively understandable manner, the proximity of two points positively correlating to their interrelatedness. The second method, namely multivariate logistic regression, identifies and measures the strength of factors that specifically distinguish two variables.[26] The strength of the association is measured by the odds ratio (OR). An OR above 1 indicates a positive association, while an OR below 1 indicates a negative one; the greater the OR is above 1 or the lower below 1, the stronger the association. Univariate analysis was performed with the ANNOVA test for quantitative and with the chi-2 test for categorical variables. A p value <0.05 was considered significant. Statistics were performed with IBM-SPSS 20™ software.

24 Details concerning this method can be found in ANTOINE, 2011.

25 BENZECRI, 1973 and LE ROUX/ROUANET, 2010.

26 Each component of a multivariable category is transformed into a new category coded 0 or 1 depending on whether or not the condition is fulfilled. On this method: HOSMER/LEMESHOW, 1989.

5. Constructing the model from multiple correspondence analysis

The 3395 plots of texts A and B of P. Wilbour with all the information concerning their main landmark, land owning institution, administrative category of domain, administrator or holder's profession or occupation, fiscal category and unit of measure used, were entered into the analysis (see Annex 1 for details). The results are shown in figure 1.

Taking into account the respective effect of each variable and each plot on the entire population of plot, the analysis identifies two sub-clouds of points. Strikingly, landmarks are not evenly divided over these sub-clouds, thereby distinguishing two categories of landscape features. On one side are the soil categories *iw*, *iw-n-m3wt*, *m3wt*, *idb* and *pꜥt* which occur with plots measured in land-cubit or the *nḫb* and *tni* fiscal categories of land. Plots on these soils are closely associated with priests, overseers of cattle who are probably attached to the local temple staff and their middlemen including cultivators and administrators. Accordingly, local temples are the landowning institution in this sector. High-ranking state officials, mayors of local towns and high army officers behave differently and are linked to the *ḳ3yt* fiscal category of plots through the administration of *khato*-land. On the other side, we find settlements and landscape features closely attached to them, such as groves, mounds/kom, standing water areas (ponds and lakes) and high grounds (*ḳ3yt*). Plots in this sector are measured in aroura and belong to smallholders who mostly exert a profession linked with the army. Of note, although *ḳ3yt* is used to designate both a fiscal category of land and a landscape feature, the two meanings are quite distant, which indicates that they probably have little in common and warns us again the temptation of deducing the significance of fiscal terms from their geographical counterpart.

Thus, the MCA shows how the Nile Valley landscape in P. Wilbour was closely interconnected with the organisation of human settlements and marked by social, fiscal and economic considerations. However, this preliminary and global pattern requires further investigations and confirmation. I propose focusing on four aspects: 1) the organisation of the geographical features in the floodplain, 2) the specificity of *idb*, *pꜥt*, *iw*, *iw-n-m3wt*, *m3wt* and *ḳ3yt* categories of soil, 3) the irrigation system in the agricultural landscape and 4) the socio-geographical organisation of the land.

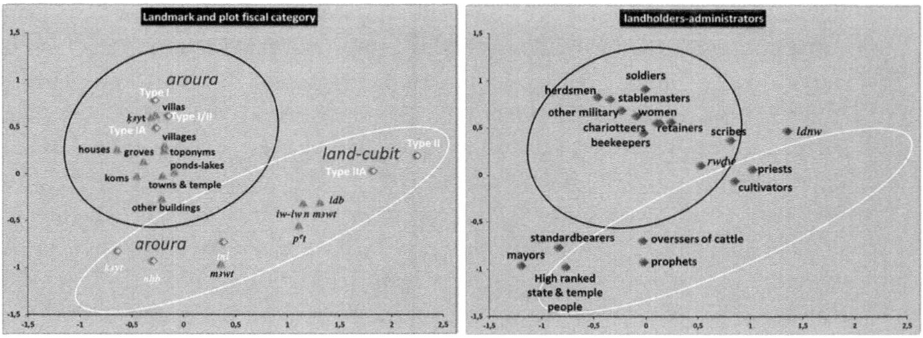

Figure 1. Results of the multiple correspondence analyses. For easier reading, a projection of the data is shown on two different screens. Ellipses encompass the two groups of data identified by the analysis. The scale of the two-dimensional space is arbitrary and was calculated by the software. The closer two points the more they are related and the greater their distance, the less they have in common.

6. Settling the organiation of the agricultural landscape

Comparing the relative position of the main and secondary landmarks in texts A and B affords the possibility of settling the landscape features in the Nile Valley in relation to one another. The analysis was performed by logistic regression and the results are summarised in figure 2, where arrows indicate the interrelations identified by the analysis. The ORs are not given here, as they have been published previously.[27] Making sense of this, at first glance, complex network requires the attempt to project the results on a typical profile of the valley at the level of Middle Egypt (see figure 5).[28]

Here, the Nile channel is gently meandering and probably ran at the west of its present bed in the eleventh century B.C. to the tenth century B.C., as previously mentioned. Typically, the floodplain near the river is delimited either by levees formed from deposits of fine sand and coarse silt during the inundation on the concaved deep side of the channel or by a sandy point-bar on the shallow convex side. Therefore, in a natural riverine system, levees form a discontinuous

27 Results of the analysis are detailed in Antoine, 2011, p. 9–27.

28 Butzer, 1976, p. 12–18; Said, 1993, p. 57–70; Hassan, 2010, p. 1–20.

line along the riversides as they alternate with low sandy bars on the same bank that are submerged during flood. Levees may span from several hundreds of meters to a few kilometres in width and are crisscrossed by undulations parallel to the main channel corresponding to oxbows or the ancient beds of the river. They rise to some meters above the valley flat but are usually covered by the flood. Only the highest grounds are likely to escape complete submersion. The floodplain corresponds to the low flat that lines the river channel beyond levees. In the region covered by P. Wilbour, the Nile River is not the only water course since the Bahr Yussef runs to the west on the desert fringe before entering the Hawara pass and debouching in the Fayyum depression. Compared to the Nile, this tributary probably meandered more than it is today, frequently abandoning old channels and oxbows.

If we now try to project the landscape features appearing in P. Wilbour on this Nile Valley profile, *ḳ3yt* (high grounds) are likely to be found on levees in keeping with the fact that they are associated with settlements.[29] Mounds (*i3t*) are also associated with settlements and high grounds. As modern koms, they probably result from the accumulation of debris generated during centuries of human occupation. Different varieties of groves are named in P. Wilbour. They are similarly associated with settlements suggesting that at least some of them may be artificial plantations. *Ỉw* can be securely identified with the islands of the Nile channel. Island formation is a continuous process resulting from the accumulation of sand followed by silt on point-bars. With time, the narrow channel separating the island from the bank can silt up, attaching the island to the riverside. *Ỉw-n-m3wt*, islands of new land, probably designate this phenomenon. The meandering course of the Bahr Yussef similarly allowed the formation of *iw* and *iw-n-m3wt*. *M3wt*, new land and *pʿt*, a rare word, probably designate different categories of new land formed over time by the meandering changes of the river or its tributary. They are likely to be found on the riverside since they connect with *iw* and *iw-n-m3wt*. *Ỉdb* is commonly rendered by "river banks".[30] Such a meaning is, however, unlikely since they are never associated with islands but appear to be located beyond *m3wt* to which they are associated. I have previously suggested that *idb* may designate the low flats of the floodplain[31], but it is more likely that they correspond to a subpart of them located in the meandering belt of

29 Interestingly, *ḳ3yt* were probably absent along the Bahr Yussef which was unable to develop prominent levees due to its low current (Antoine, 2011, p. 25) thereby confirming the identification of *k3yt* with levees.

30 On *idb* see Gardiner, 1948, p. 26–27 and Meeks, 1972, p. 149.

31 Antoine, 2011, p. 25.

the river. The last features to be considered in this landscape model are standing waters. They are designated by several words. *Brkt*, the ancestors of Arab *birket* or pond, is associated with settlements. *ḥn*, lakes or swampy lakes, probably correspond to back-swamps which developed in the lowest part of the Valley sustained both by the flood and ground water resurgences. The last category, *ẖnm/mẖnm,* is the most difficult to identify because of the scarcity of occurrence of this word which has been suggested to designate basins from which fields may be irrigated, as compared to other areas of standing water.[32]

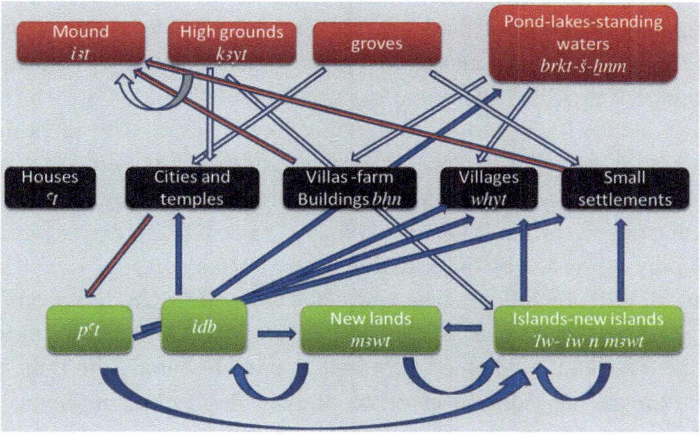

Figure 2. Schematic result of the logistic regression analysis of the relation between main and secondary landmarks. Arrows indicate the associations identified by the analysis between the different landscape features.

7. The specificity of *ỉw*, *ỉw-n-m3wt*, *m3wt*, *ỉdb*, *pʿt* and plots measured in land-cubits

As seen above, these words designate categories of soil we may suspect to lie within the river's meandering belt and for which the MCA indicates a specific pattern. To identify what characterises this pattern, multivariate logistic regression was performed comparing the plots in texts A and B located on these soils with those elsewhere in the floodplain in terms of fiscal categories of land

32 EYRE, 1994, p. 80, note 161.

and surface.[33] The proportion of plots on these soils is the same in the two texts and amounts to 16 %. The logistic regression identifies a strong association with *tni* and *nḫb* land (OR: 11.6; 6.6-20.5 95 % CI[34]) and untaxed plots measured in land-cubits (OR: 4.2, 2.3-7.8 95 % CI) while taxed plots measured in aroura and pertaining to private smallholders (Type I and IA plots) are conspicuously rare (OR: 0.16; 0.09-030 95 % CI).

Another important feature is that the area of plots on these soils is less than elsewhere in the valley, as shown in figure 3, regardless of the domain categories and of text A or text B. This is not only due to the very small size of plots measured in land-cubit, but also to the fact that *tni* and *nḫb* plots are significantly smaller than those fiscally classified as *ḳ3yt*. A sub-analysis with each of these soil categories provides the same results, except for *m3wt* which, although associated with *tni* and *nḫb,* does not specifically occur with plots measured in land-cubits. In contrast, plots situated on levees (*ḳ3yt*) follow a completely different pattern since the logistic regression retains that they are rarely of the *ḳ3yt* fiscal category of land (OR: 0.18; 0.09-0.35 95 % CI), thereby confirming the antinomy suggested by the MCA.

Thus, P. Wilbour tells us that lands planted with grain and promising the highest yield coexisted with plots measured in land-cubits on the meandering belt of the river and, as shall be seen below, probably along the Bahr Yussef.[35] But what are the agricultural characteristics of these plots measured in land-cubit? First, if 55 % of them lie on the soil categories considered above,[36] a significant proportion is found elsewhere, in particular near villas and small localities.[37] However, to interpret this point one should keep in mind that if

33 The analysis was adjusted for texts A and B to avoid bias due to a specific effect of one of the two texts.

34 CI indicates the 95 % confidence interval of the OR value. All the ORs specified in this study have a p value <0.05.

35 The co-occurrence of plots measured in land-cubit and those classified as *tni* and *nḫb* raises the question whether they were located in the same areas. This question can only be approached by a study of the individual landmarks in text A. Plots were considered to be in the same location when the orientation and the main and secondary landmarks were identical (Antoine, 2011). Compared to aroura measured plots, those in land-cubit do not specifically lie in the same localities as plots of normal domains.

36 241/440 land-cubit measured plots with 123 of 233 plots on *ỉdb*, 53 of 87 plots on *ỉw* and *ỉw-n-m3wt*, 40 of 84 plots on *pꜥt* and 25 of 52 plots on *m3wt*.

37 143 plots are near settlements (32.5 %).

settlements were preferentially used as landmarks because they provide precise information on plot localisation,[38] they do not provide information on the nature of the soils in their immediate surroundings, which may or may not be identical to the soil categories discussed here. P. Wilbour provides further important details on these small plots. Half of them are artificially presented as if part of their surfaces was liable to a tax not associated with grain, as a corn rate was not applied.[39] The other half was not cultivated for reasons that are clearly specified,[40] the most frequent being that the land was *wsf* (inactive or resting), which probably indicates a kind of fallow[41], and less frequently due to *wšr* (dryness). If Gardiner's reading and interpretation of this word is correct, this would mean that the field had not received water.[42] All of this suggests a method of irrigation which departs from that of lands naturally covered by the flood. P. Wilbour remains especially discrete on the nature of the crops raised here. Two plots are said to be cultivated with vegetables and five with flax, but we are probably close to reality when interpreting the scarcity of these indications as designating exceptional situations rather than a general rule. To obtain more information we need to turn to the Louvre-Griffith fragments and P. Reinhardt.

---

38 Compared to the previous plots, which are clearly said to be on (*m*) the soil categories studied here, the other plots are in the vicinity of settlements but at a distance which remains unknown. The usual absence of a secondary landmark with settlements prevents any conclusion on the nature of the soil in their vicinity. The only specification is their orientation toward the settlement used as landmark according to the cardinal points.

39 238/440 plots measured in land-cubit are expressed by figures a and b with a<b (type II plots). The total a+b follows the same distribution as the surface of untaxed plots (type IIA) which indicates that the actual surface of the plot is a+b. This pattern of expression by two figures is reminiscent of the way the taxed plots pertaining to small holders and measured in aroura (type I plots) are presented, but the relation between the two figures is different in type I plots, since b is clearly a part of a. On these plots see GARDINER, 1948, p. 93–94 and KATARY, 1989, p. 13.

40 These plots correspond to variety IIA of Gardiner's classification but, as showed by KATARY, 1989, p 82, a part of them is in fact measured in aroura but not assessed by the surveyors. This category is classified by Katary as variety I/II and must therefore be excluded from the analysis of Gardiner's type IIA category.

41 KATARY, 2005, p. 140f.

42 169/202 are *wsf* and 32 *wšr*. The reading of these annotations is difficult because of their very cursive script. On this see GARDINER, 1948, p. 93f.

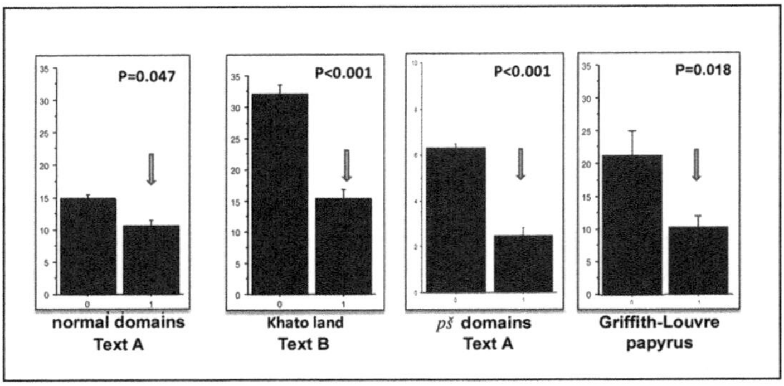

Figure 3. Comparison of the mean surface of plots on iw, iw-n-m3wt, m3wt, idb and pꜥt (arrow) with plots located elsewhere in the floodplain in the different documents. Bars indicate the standard error.

8. Validating and refining the model on the Louvre-Griffith fragments and P. Reinhardt

The Louvre-Griffith fragments provide information on 180 plots, all of the *ḳ3yt* and *nḫb* fiscal categories of land, *tni* being missing in what is preserved of the document. Unfortunately, because of numerous lacunae, only 84 plots are fully exploitable. As in P. Wilbour, the *ḳ3yt* fiscal category largely dominates, representing 92 % of the plots. Concerning landmarks, fields are mostly on high grounds (*ḳ3yt*) (80 % of cases), the others lying on *pꜥt*, *idb*, *iw* and in one case on *nḫb,* which therefore, as *ḳ3yt*, can be used to designate a soil category. According to what was encountered in P. Wilbour, these landscape features are mostly in the vicinity of settlements or hills.[43] Interestingly, plots located on *pꜥt*, *idb*, *iw* and *nḫb* are smaller than those on *ḳ3yt* with a mean surface of 11 arouras versus 22.8 ($p<0.05$) (see figure 3), which corroborates what appears in P. Wilbour, although this result should be interpreted with caution due to the high number of missing data. Departing from the situation prevailing in P. Wilbour is the strong association of the *ḳ3yt* category of land with their homonymous landscape

43 28 plots are near a locality designated by a proper name, 13 near a keep/hill (*sgr*/*sg3*), ten near a temple and five near a landscape feature situated in the countryside.

feature, which suggests that these plots were located on high flats probably covered by the flood along the river bed or near the desert fringe.[44]

P. Reinhardt provides further interesting information since all the analysable plots lie in the meandering river belt. Indeed, of the 13 preserved landmarks used to indicate plot localisation, seven are *iw*, four *iw-n-m3wt*, one *m3wt* and one *idb*, while *pꜥt* is absent.

Due to the lacunous state of the document most of the plots are missing a significant part of their information. However some interesting numerical data can be obtained. The median value, which cut the population of plot surfaces in two equal parts, is 2.4 aroura, varying from 0.125 to 22 arouras,[45] which is small if we compare these figures with the surface of fields located on *ḳ3yt* in P. Wilbour and the Louvre-Griffith fragments. Once more this confirms the small plot size in this part of the floodplain. Another very important argument for the model validation is the very high tax rate of 12 khars/aroura, probably the highest ever attested in pharaonic Egypt. This perfectly corroborates the model which predicts that the land in this part of the floodplain provides the best yield.

One characteristic of the fiscal organisation of fields in P. Reinhardt is that the plot surface is regularly reduced by iterative subtractions of an untaxed area.[46] The subtracted parts are tiny, with a median value of 0.63 arouras, which is very reminiscent of the surface of plots measured in land-cubits in P. Wilbour.[47] A very interesting feature for our purpose is that P. Reinhardt provides the reason as to why the reduced surface was excempted from the corn tax. The preserved examples are listed in table 2 with their translation as proposed by Vleeming.[48] Fifty seven % of the excemptions were due to the land being exposed to excessive moisture, as is indicated by the use of words such as *ẖr*, low land or fen, *mḥyt*, marsh, or *ẖ3t*, lagoon. In 19 % of the cases, the parcels were planted in cucurbits,

44 In this part of the valley, the profile is convex with the lowest part lying near the western desert fringe: Butzer, 1976, p. 15.

45 Calculated on 118 data.

46 Such an administrative method is also illustrated on the verso of the Louvre-Griffith fragments.

47 Calculated on 60 data. Range: 0.125-7.35 arouras. The median taxed surface calculated on 33 data is 2 arouras, ranging from 0.25 to 9. However there is no indication in the P. Wilbour that surfaces measured in land-cubit should be subtracted to an aroura-measured surface. Obviously the two documents use different methods of accountability. However, in both of them a series of small plots is not cultivated in cereals while grain assessment or taxation is the main goal of these texts.

48 Vleeming, 1993, p.65–69.

vegetables or grass for horse breeding. Thus, Reinhardt papyrus sheds an interesting light on the plots measured in land-cubit of P. Wilbour, confirming that a part of them was cultivated, and informing us about the nature of the cultivated plants. We have seen that the rare indications furnished by P. Wilbour in this domain probably reflect unconventional situations. It is thus tempting to consider that other, yet similar plants were cultivated in the land-cubit measured plots of P. Wilbour. The main difference lies in the reduced use of fallow (*wsf*) and the absence of dry fields (*wšr*) in P. Reinhardt. Therefore, P. Wilbour and P. Reinhardt probably reflect three limitations of cultivation on *iw*, *iw-n-m3wt*, *m3wt* and *pꜥt*, namely excess of water, dryness and land fatigue, which requires a fallow period to regenerate. All of this is probably connected with the irrigation system.

Land categories	Translation (VLEEMING)	Number of occurrences	Remarks
ẖr	Low land/fen	42	occurs with *tni*
mḥyt	Marshland	20	occurs with *b3nt*
ḥ3t	Lagoon	4	occurs with *b3nt*
šꜥ	Sand	2	
tni	Elevated land	13	occurs with *ẖr*
nẖb	Fresh land	1	
wsf	Fallow land	2	
w3ḏ	Vegetable land	2	
b3nt	Gourd land	10	occurs with *mḥyt* or *ḥ3t*
šꜥti	Mowing land	10	for the pharaoh's horses
TOTAL		115	

Table 2: Distribution of the different categories of land not liable to grain tax in the Reinhardt papyrus

9. The irrigation system in the land registers

For a long time it has been commonly admitted that the basin system of irrigation, which was for the first time fully analysed in the *Description de l'Egypte* at the end of the eighteenth century A.D.,[49] existed at least to some extend in pharaonic times.[50] This system depends on an adjustment of the natural flood basins by a series a dykes lining the Nile bed where levees are lacking, completed by transversal embankments extending from the river to the desert edges. Long feeder canals branching from the Nile are necessary to bring water to the desert's edge, while short canals feed levees near the riverbanks. This is completed by sluices controlling water movements between basins or within canals. A system such as this allows a single winter crop cultivated after the flood, followed by a summer fallow before the next inundation. However, artificial irrigation can be conducted during the dry period on restricted areas by means of water elevation devices near wells where the water table adjoins the surface or near canals, and during the flood on levees lining waterways.[51] However, this system was never fully developed and remained in an embryonic stage for centuries before the mid-nineteenth century A.D.[52]; in ancient times the available evidence indicates a limited use of artificial irrigation which was probably restricted to orchard plantations.[53]

This immediately raises the question regarding any evidence of the existence of an irrigation system in these documents. We have to admit that they are scant since words designating wells (*šdt*) and canals (*mr*) are absent, while dykes (*dnit*) are occasionally mentioned and in P. Wilbour only. However, localities incorporating *mr* in the formation of their name or using the canal sign  in their determinative are relatively frequent in P. Wilbour.[54] Figure 4 shows

49 Girard, 1824, p. 1–22. For a description of the classical basin system see Willcock/Craig, 1911; Barrois, 1904; Besançon, 1957, p. 199–200.

50 See for example Butzer, 1976, p. 41–46, Lloyd, 1983, p. 327 but also with a modulated opinion Kemp, 1989, p. 10.

51 Besançon, 1957, p. 89–90.

52 Alleaume,1992, p. 301–322; Michel, 2005, p. 253–276

53 Butzer, 1976, p. 41–51; Eyre, 1994, p. 57–80.

54 Gardiner, 1948, p. 29–30. In texts A and B, twelve localities incorporate *mr* in their name and 22 use the canal sign in their determinative. In text A they amount to nine and fifteen respectively, representing 20.7 % of all the toponyms in this text. The detail is as follows according to the four zones where the first figure gives the number of toponyms including *mr* in their composition and the second

their distribution among the four zones of text A.[55] A striking feature is their predominance in Zone 1 where 57 % of them occur versus 10 to 22 % in the other zones.[56] An analysis of the orientation of plots according to the cardinal points has suggested that fields in Zone 1 principally follow the Bahr Yussef, while another part lies on the left bank of the Nile.[57] In Zone 2 fields mostly adjoin the riverbank and in Zone 3 they are generally at a distance from watercourses, probably being located in the central part of the floodplain. Finally, in Zone 4, the situation is a mixture of these patterns since some plots are positioned along the river bed, while others are at some distance (see figure 4). Thus, the high frequency of toponyms referring to a canal in Zone 1 strongly suggests that the canal in question was the Bahr Yussef or its embranchments and probably not artificial constructions. One illustrative example of this may be the well-known town of Miwer (*Mr-wr*, "The Great Canal") that recent investigations have shown to have been located close to the Bahr Yussef in antiquity.[58] Whether these conclusions may be extended to the other zones remains speculative, but here also the natural branch of the Nile was possibly alluded to in toponyms referring to a canal.[59]

The absence of canals among landmarks in land registers is thus problematic. It is possible that the land surveyors considered that they did not provide reliable topographic information, but it is also noteworthy that *mr* is absent from the hydrological vocabulary in the Amenemope Onomasticon,[60] which could indicate that another word was used to designate canals while the usage of *mr*

---

the number of those using a canal as determinative (the first group being excluded): Zone 1: 4/8; Zone 2: 2/1; Zone 3: 2/5; Zone 4: 1/4.

55 On this map and the discussion of the localisation of fields in the four zones of text A, see ANTOINE, 2011, p. 17–21. For another interpretation of the data see GOMAÀ et al., 1991, p. 138–141.

56 Zone 2: 13 %; Zone 3: 10.2 %; Zone 4: 21.7 %.

57 Ibid., p. 18.

58 BUNBURY, 2012, p. 52–54.

59 A close analysis of these toponyms and of the plots adjoining them may therefore help to precise the limits of the different zones.

60 *mr* still occurs in official or literary compositions using an academic style such as P Harris I, I 50 (GRANDET, 1994, Vol. 3, p 74). In P. Turin Cat 1923 Rt1-8 (KITCHEN, 1983, p. 368), an administrative document, *mr* probably designates the canal joining the Ramesseum to the Nile, which may indicate a restricted meaning of this word. To the best of my knowledge, among land registers of the period a canal (*mr*) only appears in P. Berlin 23253 II, 53 and III, 21 (GASSE, 1988, p. 102–103).

was limited to the name of toponyms whose origin probably went back to a remote past. It has been proposed that *ẖnm*/*mẖnm* may designate feeder canals[61] but this word does not appear in the other land registers. In the Amenemope Onomasticon,[62] it follows swampy lake (*ḥnw*), lake (*š*), well (*ẖnmt*) and precedes a word which in the Miscellanies designates a water area associated with a villa (*bẖn*) and in which fish and hippopotamuses abound.[63] All of this suggests that *ẖnm* designates some kind of natural standing water area; all the more so since it never occurs in association with the soil categories located in the meandering belt of the river, the very region where we may expect feeder canals to originate.[64] Figure 4 provides further interesting findings since it shows that the distribution of plots located on *iw*, *iw-n-m3wt*, *m3wt*, *idb* and *pʿt* and that of toponyms referring to a canal in the four zones of text A are well correlated. Where reference to a canal is frequent, the proportion of fields located in the meandering belt of a water course is high, which suggests that a significant proportion of fields located on *iw*, *iw-n-m3wt*, *m3wt* and *pʿt* were probably along the Bahr Yussef.

Two dykes are mentioned in the P. Wilbour, namely the dyke of Spermeru in Zone 3 and that of Pi-Ihay ("The Byre") in Zone 4. Schenkel has proposed to identify them with two of the transversal dykes that crossed the valley in the nineteenth century and appear on the map of Linant de Bellefonds, but this remains hypothetical.[65] The exact localisation of Spermeru and Pi-Ihay is unknown. The former probably needs to be searched in the northern half of Zone 3[66] on the western fringe of the valley, while the latter is associated with fields belonging to the sanctuaries of Sako at the eastern limit of Zone 3 and 4. An *idb* is said to adjoin this dyke, which possibly indicates a location adjacent to the Nile near Zone 3. What is relevant for this study is that dykes occur in a region

61 EYRE, 1994, p. 80, n. 161.

62 GARDINER, 1947, Vol.I, p. 7f.

63 *ẖnini*: P. Lansing 12, 10; P. Anastasi IV 1b5 (GARDINER, 1937, p. 35 and 111); *ẖnm* appears in the Stela of Shoshenk (line 12) where a field is said to be fed (with water) from a *ẖnm* (BLACKMAN, 1941, p.83–95).

64 A further argument against the hypothesis that *ẖnm* could have succeeded to an obsolescent *mr* comes from the analysis of the distribution of this word in the four zones of Text A. Indeed among the 11 *ẖnm* named in this text, 5 are in Zone 4 and 2 in each of the other zones which clearly departs from the distribution of toponyms referring to a *mr*, which largely predominate in Zone 1.

65 SCHENKEL, 1994, p. 29f. On ancient dykes see also GOMAÀ et al., 1991, p. 33–72.

66 Spermeru has been identified with Safaniya by GOMAÀ et al., 1991, p. 78.

where the assessed fields probably lay in the central part of the floodplain, which suggests that they are associated with the more common categories of land, namely those located in natural inundation basins.

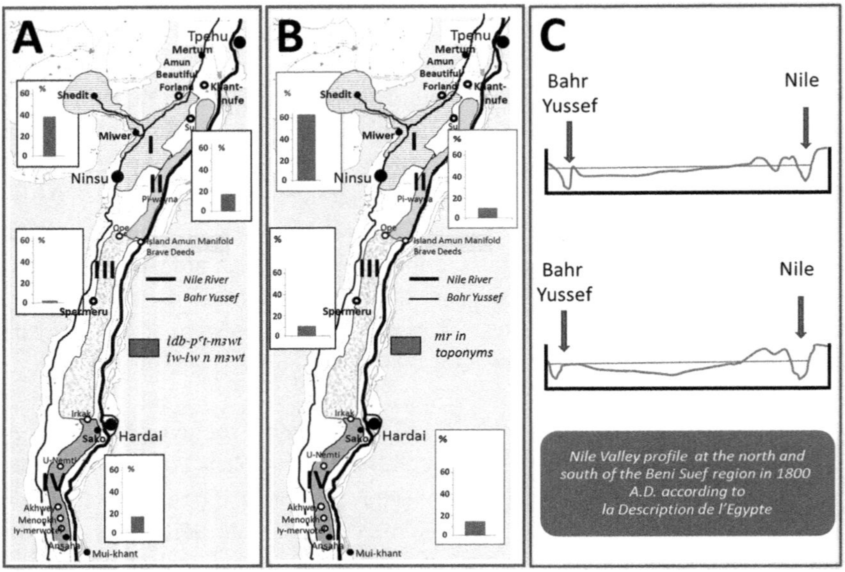

Figure 4. Proportion of fields on idb, pꜥt, iw, iw-n-m3wt and m3wt (A) and of toponyms incorporating the word mr (canal) or using the canal sign as determinative (B) in the four zones of Text A of P. Wilbour. Roman numerals indicate the zone number. The possible distribution of plots in the floodplain in the four zones is from ANTOINE*, 2011. The Nile River and Bahr Yussef courses are hypothetical. In C is the schematic Nile valley profile according to the* Description de l'Egypte *at the level of Zone 1-2 (upper row) and Zone 3 (lower row).*

10. The socio-geographical pattern of the agricultural landscape

Multiple correspondence analyses have shown the importance of social factors in the organisation of land in P. Wilbour. A detailed analysis of plots in the *pš* domains of text A was therefore conducted to specifically assess this point using multivariate logistic regression with the primary landmarks categories as dependent variables and professions or occupations as explanatory variables.

The results are summarised in table 3. They confirm what was suggested by the MCA. Two groups of association clearly occur. On the one hand, plots on soil categories lying in the meandering belt of waterways, namely on *iw*, *iw-n-m3wt*, *m3wt*, *idb* and *pꜥt*, which are associated with local temple staff including priests (*w3b*) and prophets, overseers of cattle, scribes and high ranking officials but also administrators and cultivators who for the most part were the middlemen of these people in the administration of normal domains.[67] On the other hand, plots near settlements and their neighbourhood, such as groves, mounds and high grounds (*ḳ3yt*), are associated with professions mostly linked with the army. Some of these associations provide interesting clues to the economic activity conducted in these categories of settlements.

This is particularly the case for the strong relationship of herdsmen with *ꜥt* (OR: 13.15; CI 3.05-56.8) which highly suggests that these settlements where closely linked with cattle farming. The word *ꜥt* is usually translated as "house" which is here inadequate since it clearly has a professional connotation and something like farmstead is probably more accurate.[68] This kind of settlement is especially abundant in Zone 3 where military professions and particularly simple soldiers predominate and where Spermeru is the only centre of some importance. Another interesting association is that of stable masters (*ḥry-iḥw*), a category of support personnel for the army[69] (OR: 2.0; CI 1.45-2.77), and charioteers (*kṯ*) (OR: 3.62; CI: 1.86-7.06) with *bẖn*.[70] In literary sources[71] *bẖn* is depicted as a large and idyllic farm villa, possibly fortified,[72] reflecting the wealth

67 On *iḥwtyw* see MENU, 1970, p. 135–145; MORENO GARCIA, 2009–2010; JANSSEN 1986, p. 356f, reporting the translation of "agent of the fisc" proposed by Stuchevsky. On *rwḏw*: KRUCHTEN, 1979. On the organization of temple land holding see mainly JANSSEN, 1986; HARING, 1997; Id., 2009; KATARY, 2013.

68 Similarly at Deir el-Medina, *ꜥt* designates a kind of workshop while house (the dwelling place) is designated by *pr*. DEMARÉE, 2006, p. 57–66 proposes "workplace" as the best translation in the theban context. On *ꜥt* designating a farm building isolated in the fields see MEEKS, 1972, p. 59 with further references. See also WILLEMS, 2007, p. 47.

69 SCHULMAN, 1964, p. 81–86; CHEVEREAU, 1994, p. 45; GNIRS, 1996, p. 78–79.

70 Of 52 individuals holding plots near a castle (castle of Piot, of Iot and of Meryset), 26 are stablemasters and four charioteers.

71 P. Lansing 12, 1 and P. Anastasi III 3, 7.

72 GARDINER, 1948, p. 34 translates *bẖn* by "castle" but O'CONNOR, 1972, p. 693 rightly proposes "villas" which is more appropriate.

of its high-ranked owner.[73] The specific presence of plots around *bḫn* associated with professions linked with the use of horses for military purposes suggests that at least some of these settlements were specialised in horse breeding.[74]

	lady	soldier	sherden	charioteer	stablemaster	herdsman	priest	prophet	scribe	officials & over-seer of cattle	rwdw	cultivator
Villages (*wḥyt*)	+	+	+	+	+			+	+		+	+
Houses (*ʿt*)						+						
Villas (*bḫn*)				+	+							
Small settle-ments		+	+			+						
Cities												
Mounds (*i3t*)	+											
Ḳ3yt	+											
Groves		+	+									
Idb								+	+	+		+
Pʿt							+				+	+
m3wt								+			+	+
jw/iw-n-m3wt									+	+	+	+

Table 3: Logistic regression of the 2253 plots of pš domains of Text A of P. Wilbour. The main landmarks are used as dependent variables and professions and occupations as explanatory variables. + indicates a statistically significant positive association. Empty boxes indicate absence of association. For the sake of simplicity ORs are not given.

73 In P. Wilbour the adjunction of the town determinative suggests that *bḫn* was then perceived as a toponym.

74 Of note, in the aforementioned passage of P. Lansing, the *bḫn* harbours a stable and chariots are built there from wood.

Finally, the only settlement category which escapes this dichotomous socio-geographical organisation is *wḥyt*, since plots around them pertain to a wide variety of professions including the military and their associated professions and members of temple staff. In several sources *wḥyt* contrasts with *dmi* ("town") and it is frequently translated as "village".[75] However, the word probably had a more specific meaning. Names such as "The Village of the army/troop" suggest that military colonies settled here at an undetermined point in time. However, in many instances there is no specific clue as to the origin of these localities. On the whole, ladies, soldiers and Sherden are over-represented near *wḥyt*.[76] The presence of Sherden may be a remnant of foreign troops who once were installed in specific regions of Egypt.[77] The association of some of the *wḥyt* mentioned in P. Wilbour with soldiers is in keeping with the frequent military connotation of the word during the New Kingdom and its association with people of foreign origin may refer to the notion of tribes also conveyed by the word.[78] The presence of scribes and priesthood members may be explained by the fact that several of these villages harboured a small temple[79] and needed administrative staff, but it may also indicate an organisational and economic dependence on the distant sanctuaries of the regional metropolis, since the personnel of metropolis temples probably were the main holders of plots on the meandering belt of waterways.

11. Discussion

The model developed here was obtained from statistical methods appropriate for the purpose and commonly used in the social or biological sciences. The main interest of these methods is that the resulting model is independent from

75 On *wḥyt* see SPALINGER, 2008, p. 154–162; WILLEMS, 2007, p. 47–48.

76 Among 132 individuals holding plots near villages, 24 are women, 21 soldiers and 8 Sherden. Most of these women were probably the spouses or widows of members of the army: KATARY, 2001, p 61–82; ANTOINE, 2010, p. 5–14.

77 See P. Harris I, 77, 4-6, rhetorical stelae from Deir el-Medina (KITCHEN, 1983, Vol. 5, p. 91), and P. Amiens-Baldwin, Rt. A V, 4. (JANSSEN, 2004).

78 Familial lineage, even expanded, may be an important factor of the structure of *wḥyt*, which also applies to settlement founded by an individual as is clear from Mose's inscription: SPALINGER, 2008, p. 154–162 and WILLEMS, 2007, p. 47–48.

79 See for example the chapel (*ḫnw*) of Montu in the Village of Inroyshes (A 29, 18). Villas may also have some place of worship as the Tabernacle (*sšmw*) of Pre in the Villa (*bḫn*) of Meryre (A 34, 21).

any preconceived point of view and affords a synthetic and holistic vision of the organisation of the ancient Egyptian landscape around 1000 B.C. both in terms of geography and sociology. As seen above and as will be discussed here, this model is coherent, in a large number of cases, it is able to predict the expected profile of a plot or an individual exercising a given occupation; although none of them evidently follows this profile in 100 % of the cases.[80] The fact that this model was obtained from the most informative document, P. Wilbour, unavoidably raises the question of the general value of this document, as it covers only a limited part of the Nile Valley.[81] Interestingly, two other land registers, dated one or two centuries later and covering another part of the valley, not only confirm what the model predicts, but which, despite their lacunous state, also shed further light on what remains unclear in P. Wilbour. Other land registers, especially P. Prachov, would also help to improve and test the model but, unfortunately, they still await full publication.[82] Archaeological and botanical methods may also contribute to the model but, for evident reasons, they focus on settlements and their immediate surroundings while the countryside proper still escapes exploration.

Text A and text B of P. Wilbour in their present state cover 20,100 and 16,255 arouras respectively.[83] After elimination of plots counted twice in text A[84] and estimation of missing data,[85] the total amount of land is 22,700 arouras in text

80 Using ROC curves and scoring of the variables retained by logistic regression as characterising a specific category it is possible to demonstrate that, on the whole, 65 % of landmarks follow the model. The proportion is 71-91 % for *idb*, *pꜥt*, *mꜣwt*, *iw*, *iw-n-mꜣwt*, groves, and villages and 55-64 % for the other settlements and associated landscape features. Concerning the localisation of plots according to profession, 50 % of plots, on the whole, follow the model. The proportion is 65-68 % for the military and 35-45 % for priests and their middlemen because their plots were not restricted to the meandering belt of the river. However, 76 % of plots on *idb*, *pꜥt*, *mꜣwt*, *iw* and *jw-n-mꜣwt* follow the social categories predicted by the model.

81 FAIRMAN, 1953, p. 118–123.

82 The P. Prachovs (TURAYEV, 1927) may be particularly promising since it probably records several hundreds and possibly more than one thousand plots. At present it has only be partially published by GASSE, 1988, p. 123–138.

83 Some plots of *khato*-land appear in both texts but the total amount is negligible.

84 In *pš* A and B entries. For this see GARDINER, 1948, p. 72f.

85 This, of course, includes plots for which the surface is lost but also in text A plots pertaining to Theban and Heliopolitan temples of Zone 1, that once were recorded on another papyrus roll, and the erased section of *khato*-land in Zone 4.

A and 20,000 arouras in text B, totalling approximately 43,000 arouras. This represents only a limited proportion of the cultivated surface of this part of the Nile Valley during the Ramesside Period. However, these figures may help to obtain a rough estimation of the actual arable surface. *Khato*-land in text A represents 5.45 % of the total estimated surface and text B in all certainty contains a complete record of *khato* land in approximately the same region and at the same time. Assuming that the distribution of fields among the main landholding institutions in text A[86] is representative of the general situation in Middle Egypt at that time, we may estimate that the total cultivated area was approximately 367,000 arouras,[87] meaning that P. Wilbour may represent 12 % of the cultivated surface, which is an acceptable and representative sample. For comparison, the surface cultivated in 1800 A.D. can be estimated by planimetry from the map of the *Description de l'Egypte*[88] using ImageJ software. This gives an estimate of 2061 km^2 making some 841,000 arouras, more than twice the surface obtained for the Ramesside Period.[89] Another illustrative comparison can be made with the 292,500 aroura of arable land ascribed to the Oxyrhynchite Nome in the fourth century A.D.[90] Taking into account that the nome covered about 70-80 % of the region surveyed in the P. Wilbour at that time, the order of magnitude is very similar to that of the Ramesside Period. Thus, the assumption used here to estimate the cultivated surface of P. Wilbour at the time appears to be quite plausible. Even if we add landed property of institutions not named in the document, comparison with the arable surface at the time maps of the

86 Namely Theban, Heliopolitan, Memphite, and local temples and secular institutions provided there were no other categories of landholding, particularly completely private holdings, which escaped administrative recordings.

87 Calculated as follows: (43,000 arouras/5.45) x100.

88 Using maps 15–19 of the *Description de l'Egypte* Volume V, Paris 1818. Although of fairly good quality, these maps show a slight distortion of reality when attempting to overlap the position of BeniSuef, Medinet al Fayyum and Samalut with satellite images; this leads to an overestimation of surface of about 10.4 %. This was taken into account in the figure given here.

89 The same methodology can be applied with modern surfaces by using satellite image. This results in 8900 km^2, about 3.26 million arouras, i.e. more than ten times the area estimated for the P. Wilbour.

90 ROWLANDSON, 1996, p. 17. This estimation relies on a fourth century text which indicates the total amount of land cultivated in cereals to which is added the estimated surface of vineyards, gardens, inhabited places, dykes and canals, totalling about 800 km^2.

Description de l'Egypte were produced raises the suspicion that cultivated zones probably predominating around settlements, alternated with regions where the floodplain was left to its natural regime.

An important aspect revealed by the model is the almost dichotomous organisation of the landscape opposing *iw*, *iw-n-m3wt*, *m3wt*, *idb* and *pꜥt*, which probably lay in the meandering belt of the River Nile and the Bahr Yussef to landscape features lying near settlements or in the floodplain. The former represent 11 % of the total surface of texts A and B,[91] which compares with the modern proportion of land situated in the same area of the valley in the region of Beni Suef (15.3 %) and Beni Mazar (4.1 %) as estimated from satellite imaging – that is to say in the north and south part respectively of the area covered by P. Wilbour.[92] It is particularly interesting that the modern repartition of land follows what is observed in text A with a higher proportion of land in the meandering belt near Beni Suef, next to Zone 1 and 2 and a lower one near Beni Mazar, which corresponds to Zone 3, thereby confirming that P. Wilbour may be fairly representative of the general situation at the time.

The *Description de l'Egypte*[93] provides further interesting information of the Nile Valley profile in this part of Middle Egypt (see Figure 4). In the north, at the level of Beni Suef, it mentions a 2 km wide levee lining the river followed in the east by a sloping floodplain gradually descending to the Bahr Yussef and the desert's edge. A different situation prevails in the south where the floodplain is marked by a longitudinal central depression called the Bahr Bathen, "the inner river",[94] where the state of flood persists longer than elsewhere. The *Description de l'Egypte* adds that these profiles were associated with different irrigation systems. Due to the westward slope, a natural irrigation was possible all over the floodplain in the north; artificial irrigation was necessary on the large meandering belt that lines the river while natural irrigation was centred on the middle of the floodplain in the south and artificial irrigation was restricted to

91 Calculated on the preserved data. To this, the surface of plots on high grounds (*ḳ3yt*) which also lie in the meandering belt of the Nile should be added, which brings the proportion to 12.2 % of the total surface.

92 As measured by GIS analysis from satellite data: Afify, 2010, p. 6 and Wahab/El Semary, 2012, p. 5685.

93 Martin, 1825, p. 6–15. For an analysis of the geographical information provided by the *Description de l'Egypte* see more particularly Gomaà et al., 1991. On the presence of a central channel probably temporarily flooded in this region of Middle Egypt see Subias et al., 2013, p. 27–44.

94 Or the "the river belly".

the very banks of the Nile and the BahrYussef.[95] These data shed light on what occurs in Zone 3 and 4 of P. Wilbour where fields were probably mainly located in the central part of the valley, suggesting that their irrigation depended on inundation basins in keeping with the presence of dykes in this region.[96]

As already discussed, the fact that land registers seldom, if ever, mention canals is a puzzling problem.[97] In the P. Wilbour, toponyms referring to a canal perhaps allude to the Bahr Yussef. However, the existence of irrigation canals is attested in inscriptions or representations since the origin of Pharaonic Egypt[98] but usually in a religious setting or in a context of propaganda to the benefit of a ruler or a nomarch. This raises the question of the real development of this irrigation system and of the surface that was actually irrigated by means of dykes and canals. Land registers show that – at least in the region of Middle Egypt studied here and in the eleventh to tenth century B.C. – there was probably only limited modification of the riverine system. Local short dykes leaning on natural heights in the valley may have sufficed to increase the duration of water stagnation in inundation basins, while adjustments of the short temporary channels branching on the riverbanks[99] may improve irrigation in the meandering belt by facilitating water access and drainage.[100] Butzer has estimated that "an average flood would allow a single crop season over perhaps two-thirds of the alluvial surface"[101] with an unmodified water regime, which largely covers the cultivated area estimated from P. Wilbour. In this system, levees are formed during the inundation season when the floodwater tops the bank, rapidly depositing its load of fine sand and coarse silt. As floodwater flows away from the channel, water velocity diminishes and sediments accumulate with greater thickness closer to the channel and finer silt accumulates in depressions, provided water remains for a sufficient time.[102] If we project these

95 GIS analysis shows that the valley is occupied by a series of inundation or discharge basins arranged in a north south manner in the region of Beni Mazar: Wahab et al., 2012, p. 5685.

96 Subias et al., 2013, p. 33–35.

97 Among the eleventh to tenth century B.C. land registers, a canal is only named in P. Berlin 23253 II, 53 and III, 21. On canals in general in the Nile Valley landscape see Michel, 2005, p. 257f.

98 Schenkel, 1978.

99 These channels are well illustrated on the maps of the *Description de l'Egypte*.

100 Willems, 2013, p. 347–349.

101 Butzer, 1976, p. 20.

102 Hassan, 2010, p. 5–6.

data on the typical Nile valley profile of figure 5, it appears that the maximum of silt deposit, and thus of lands able to provide the best yield, lies in the low flats lining the Nile or the Bahr Yussef and in the central depression of the valley. These regions may be exposed to moisture in cases of excessive flood or, if waters stay longer or if drainage becomes deficient while, with a too brief stage of the flood or an insufficient inundation, they may be exposed to dryness.

These elements help to understand the agricultural specificity of plots localised on *ỉw*, *ỉw-n-m3wt*, *m3wt*, *idb* and *pꜥt*, which provide the best yield when planted with grain and can also be cultivated with vegetables, fodder and cucurbits on small parcels. Land with a high grain yield or under an alternative culture was probably not restricted to this part of the floodplain;[103] however, they both occur on the same category of soils, although not necessarily in the same fields at any given time.[104] Furthermore, cultivation on these lands needs adaptation and is not a straightforward process as it is exposed to particular risks such as dryness or excessive moisture, and may need periods of fallow if the interpretation of *wsf* is correct. Constant high grain yield over the years is not secure as is shown by fluctuations of the classification of a given plot as *ḳ3yt*, *tni* or *nẖb* in text B. All of this probably accounts for the designation of half of the fields as *corvée*-land in P. Reinhardt, that is to say as a category of land which needs compulsory and administratively controlled exploitation. The high proportion of fields held by a water-chief in this document is certainly an indication that water control plays an important role in their management and needs specific skill.[105] The limitation of the available surface in this part of the floodplain is another important factor and probably explains why plots were smaller here than elsewhere. Alternatively, the reduced size of the plots and the need for compulsory labour may be due to the necessity of developing more work and energy to obtain the expected yield, which could only be achieved on smaller surfaces.

A system of fallow (*wsf*) restricted to very small plots measured in land-cubits regularly occurs in P. Wilbour where it represents 38 % of the fields measured in this unit, in contrast with other documents, including the verso of the Louvre-Griffith fragments, P. Reinhardt and P. Berlin 23253, where

103 44 % of plots measured in land-cubits in the P. Wilbour are located elsewhere.

104 The situation is apparently different in the P. Wilbour and the P. Reinhard. In the former, the coexistence in the same place of both types of culture is uncommon while it is the rule in the latter.

105 VLEEMING, 1993, p. 56–67; ENDESFELDER, 1979, p. 37–51 compared this title with *ḥry-mw*, *ꜥ3-bꜥḥ*, *mr-bꜥḥ* and *ḥry-bꜥḥ*. On *bꜥḥ* designating a garden land possibly artificially irrigated see EYRE, 1994, p. 70–71.

they are mentioned only occasionally. This may indicate regional differences in agricultural practice, the consequence of local variations of soil quality, or climatic changes between the eleventh and tenth centuries B.C.[106] Katary recently proposed that *wsf* plots may be subject to a rotation of culture alternating grain, vegetables and fallow in a way prefiguring what is illustrated in the Fayyum in the Ptolemaic period,[107] i.e. in a region which entirely depends on artificial irrigation as it is not reached by the flood.[108] This remains speculative, however, it is clear that the soil was not suitable for cultivation for a period of time and had to rest.[109] This may be explained by the fact that plants cultivated on these fields needed more nutrients than those supplied by the flood. This raises the question of whether some of these plots may have been artificially irrigated or whether they produced a second crop during the dry season, which would expose the land to exhaustion and would explain the necessity of a fallow period if this was repeated over several consecutive years. As we have seen, land-cubit measured plots may have predominated along the Bahr Yussef, which has only low levees, so that a mechanical or manual water-lifting may be easier here than on the Nile banks.[110] Artificial irrigation and additional work demand may also explain why these plots were so small. In P. Reinhardt, vegetables, cucurbits and grass were cultivated in regions exposed to an excess of water, probably in low flats. In P. Wilbour vegetables were probably not cultivated on plots measured in land-cubits, but other crops such as cucurbits, lentils or leguminous may have been planted there. A second crop of cereals may also explain why tax was higher on *nḫb* and *tni* lands if it is assumed that tax collection took into account two harvests a year.[111] P. Berlin 8523, which dates to this period,[112] illustrates several of these points and shows another way of irrigating this type of fields, which is not illustrated in the land registers. Here, a small parcel (one aroura)

106 There are indications that the Second Intermediate Period and especially the tenth century B.C. were particularly wet: SAID, 1993, p. 152.

107 KATARY, 2005, p. 145–147.

108 MANNING, 2003, p. 30.

109 As *wsf* plots represent about 1/3 of plots measured in land-cubit, we may speculate that fallow occurs for one year every two years.

110 Levees are about one meter above the flood plain along the Bahr Yussef agains 2 to 3 meters along the Nile (SUBIAS et al., 2013, p. 38). See also WILLEMS, 2007.

111 On the hypothesis that *tni* land which Vleeming translates as "elevated land" were possibly artificially irrigated and may give two crops a year, see VLEEMING, 1993, p. 68.

112 ALLAM, 1973, p. 275–276 and Id., 1994, p. 1–7.

of a landholding is planted with vegetables and depends on a well (*šdt*) for its irrigation. Interestingly, the landholding is composed of *nḫb* and *ꜥmꜥmt* land, the latter term designating a mud flat.[113] This text corroborates the hypothesis that *nḫb*-land may have been artificially irrigated to achieve a non-cereal crop or lay near over-watered soils, probably in low parts of the floodplain just above the water-table, where a well may be easily dug and water may stay a long time, thus transforming the soil into mud.

Whatever the uncertainties still pending regarding the nature of these lands, and despite the difficulties involved in their exploitation, lands located on the meandering levee belt of the river clearly had significant economic value, as is shown in P. Wilbour by the contrasting sociological pattern of their holders, who are mostly local temple staff and high-ranking officials. The weight of social factors in the organisation of the Nile valley landscape is another important result of this analysis and deserves more specific comment. It is important to underscore here how social factors may help to clarify economic activity in or around some categories of settlements.

12. Conclusion

The study of landmarks used in land registers to describe plot localisation allows the reconstruction of the ancient Nile Valley landscape with the help of statistical methods to analyse the complex structure underlying such texts as P. Wilbour. The model deduced from this text is coherent and is validated by documents of the tenth century B.C. in another part of Middle Egypt, and by confrontation of the results with information on the modern and pre-modern situation, despite obvious differences. This model improves our understanding of the terminolopgy used to describe landscape features, whether inhabited or not. It suggests minimal adjustment of the natural riverine system and a specific agricultural exploitation of the meandering belt of both the Nile and the Bahr Yussef. Besides regional geomorphological factors, social and economic aspects significantly contributed to the organisation of this landscape.

---

113 On this word see GARDINER, 1947, p. 10–12, where it is interestingly associated with *ḳꜣyt*, *tni* and *nḫb*. The word occurs as a landmark in the Louvre-Griffith fragments and in text B of the P. Wilbour (18, 8) where it enters in a compound name: *š-ꜥmꜥmt*, Lake of Mud, associating here also a mudflat with a standing water area.

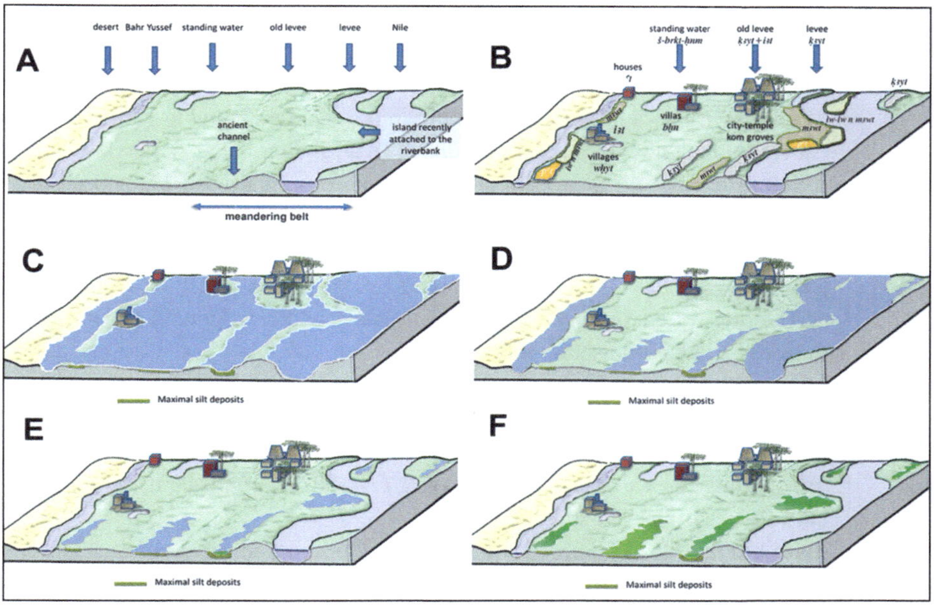

Figure 5. Typical profile of the Nile Valley in the region covered by the P. Wilbour. A: Natural profile with landscape features. B: Projection of landscape features named in the papyrus according to the results of logistic regression analysis of the relationship between main and secondary landmarks. C-F. Hypothesised flood regime with no or minimal artificial adjustment of the floodplain, showing the expected distribution of maximum silt deposits and land of the best quality (green).

Annex 1

Variables and their categorisation entered in the analysis. One entry corresponds to one plot to which the variables listed below were attached as appropriate.

- Text A or text B.
- Main landmark: landmark categories describing plot localisation pertain to three groups.[114] The first refers to elements of the agricultural landscape. These include *idb*, *pꜥt*, *ḳ3yt*, island and new island (*iw* and *iw-n-m3wt*) and new land

114 Gardiner, 1948, p. 25–36; Antoine, 2011, p. 9–27.

(*m3wt*) which were considered as many categories; the second group refers to human settlements and includes the following categories: well-known cities[115] and their temples with their pylons (*pr*, *ḥwt*, *bḫn*), villages (*wḥyt*), houses (*ʿt*), villas (*bḫn*) and agricultural buildings including byres (*iḥ3y*), granaries (*šnwt*), and stables (*iḥw*), other buildings (*isbt*, shelter; *sgr*/*sg3*, keep/hill; *wḏ3*, storehouse; *ẖr*, tomb), and finally the numerous otherwise unknown small localities (other localities) which the town determinative suggests to have been inhabited places. The last group comprises the following categories: mound (*i3t*), grove (*nh3t*, sycamore, *št3*, copse; *ḥḏ*, clearing (?), and other plantations), and resting waters (*ẖnm*, *m-ẖn*: basins, *š*: lakes, and *brkt*: ponds) which appear to be associated with settlements.

- Land holding institution: secular,[116] local, regional,[117] Theban, Memphite, Heliopolitan temples.
- Domain category: *rmnyt*, *rmnyt pš*, *šmw pš*, field for the white goats (*mkib ḥḏ*), herbage (*smw*), field of Pharaoh (*3hwt Pr-ʿ3*), *khato*-land (*ḫ3-t3*)/*mine*-land (*mint*), other, not specified/lost.[118]
- Land holders: P. Wilbour lists more than one hundred different occupations which were assigned to 29 socio-professional categories.[119] Women (*ʿnḫt-nw-niwt*), cultivators (*iḥwty*), herdsmen (*mniw*), beekeepers (*bity*), representatives (*rwḏw*), deputies (*idnw*), and godfathers (*it-nṯr*) were easily classified since they bear only one title. All individuals bearing the title soldier (*wʿw*), charioteer (*kṯ*), stable master (*ḥry-iḥw*), priest (*wʿb*), prophet (*ḥm-nṯr*), scribe (*sš*), whether with or without extension, were entered into as many categories. We considered the Sherden mercenaries (*šʿrdn*) proper as a category, while their retainers (*šmsw*) and standard-bearers (*ṯ3ysryt*) were grouped with individuals of the same rank. High ranking persons included a King's son, the Vizier, the High Priests of Thebes and Heliopolis, the Steward of the House of Amun, and overseers of cattle and local temples and mayors of regional towns. Other military included shield bearers, retainers, quartermasters and others. The remaining titles were summed up under the category "other" while the "lost" or "unspecified" categories encompassed persons for whom the title

115 GARDINER, 1947, Vol. II and Id., 1948, p. 196, table III.

116 Harems, landing places of Pharaoh, and *mine* and *khato*-lands

117 Regional sanctuaries designate temples not located in the surveyed region or in Thebes, Heliopolis, or Memphis.

118 GARDINER, 1948, p. 23–25 and 169f. For fields of Pharaoh see also HARING, 1997, p. 321f.

119 GARDINER, 1948, p. 79–84.

was lost or not given. For normal domains of text A, the profession of the main responsible administrator (the one introduced by *r-ḫt*) was kept when a subordinate administrator introduced *bym-ḏrt* was also named.[120]

- Fiscal categories of plot: *ḳ3yt*, *tni*, *nḫb*, type I, type IA, type I/II, type II, type IIA, lost.
- Surface unit: aroura, land-cubit.
- Plot surface in aroura: small plot: <10 arouras in text B and in normal domains of text A and <5 arouras in *pš* domains of text A; large plot: >10 arouras in text B and in normal domains of text A and >5 arouras in *pš* domains of text A; lost.

Bibliography

AFIFY, AFIFY/ARAFAT, SAYED/ABOEL GHAR, MOHAMED/KHADER, MAGDI, Physiographic soil map delineation for the Nile alluvium and desert outskirts in the middle north of Egypt using remote sensing data of Egyptsat-1, in: US-Egypt Workshop on Space Technology and Geo-information for Sustainable Development, Cairo 2010. http://www.aag.org/cs/projects_and_programs/geoinformation_for_sustainable_urban_ management/2010cairoworkshop, 14.6.2010.

ALLAM, SCHAFIK, Hieratische Ostraka und Papyri aus der Ramessidenzeit (Urkunden zum Rechtsleben im alten Ägypten 1), Vol. I, Tübingen 1973, p. 274-275.

ID., Implications in the hieratic P Berlin 8523 (registration of land-holding), in: Essays in Egyptology in honor of Hans Goedicke, ed. by BETSY MORREL BRYAN/DAVID LORTON, San Antonio1994, p. 1-7.

ALLEAUME, GHISLAINE, Les systèmes hydrauliques de l'Egypte pré-moderne. Essai d'histoire du paysage, in: Itinéraires d'Egypte, Mélanges offerts au père Maurice Martin, ed. by CHRISTIAN DÉCOBERT, BdE 107 (1992), p. 301-322.

ANTOINE, JEAN-CHRISTOPHE, Dead people in P Wilbour. What can we learn from them?, in: Göttinger Miszellen 225 (2010), p. 5-14.

ID., The Wilbour papyrus revisited. The land and its localisation. An analysis of the places of measurement, in: Studien zur altägyptischen Kultur 40 (2011), p. 9-27.

120 Ibid, p. 65–70 and MENU, 1970, p. 44–53.

ID., Social position and the organisation of landholding in Ramesside Egypt. An analysis of the Wilbour Papyrus, in: Studien zur altägyptischen Kultur 43 (2014), p. 16-46.

BARROIS, JULIEN, Les Irrigations en Egypte, Paris 1904.

BENZCÉRI, JEAN-PAUL, L'analyse des Données, Vol. 2. L'analyse des correspondences, Paris 1973.

BESANÇON, JULIEN, L'Homme et le Nil, Paris 1957.

BLACKMAN, AYLWARD, The stela of Shoshenk, Great Chief of the Meshweh, in: Journal of Egyptian Archaelogy 27 (1941), p. 83-95.

BUNBURY, JUDITH, Geoarcheology in: SHAW, IAN, The Gurob Harem Palace Project, Spring 2012, in: Journal of Egyptian Archaeology 98 (2012), p. 52-54.

BUTZER, KARL, Early hydraulic civilization in Egypt. A study in cultural ecology, Chicago 1976.

CERNY, JAROSLAV, Some Coptic Etymologies III, in: Bulletin de l'institut français d'archéologie orientale 57 (1958), p. 203-213.

CHEVEREAU, PIERRE-MARIE, Prosopographie des cadres militaires égyptiens du Nouvel Empire, Antony 1994.

DEMARÉE, ROBERT, A house is not a Home. What exactly is a hut?, in: Living and Writing in Deir el-Medine, ed. by ANDREAS DORN/TOBIAS HOFMANN, Basel 2006, p. 57-66.

ENDESFELDER, ERIKA, Zur Frage der Bewässerung im pharaonischen Ägypten, in: Zeitschrift für ägyptische Sprache und Altertumskunde 106 (1979), p. 37-51.

EYRE, CHRISTOPHER, The water regime for orchards and plantations in Pharaonic Egypt, in: Journal of Egyptian Archaeology 80 (1994), p. 57-80.

FAIRMAN, HERBERT, Review work: the Wilbour Papyrus by Alan H. Gardiner, in: Journal of Egyptian Archaeology 39 (1953), p. 118-123.

FAULKNER, RAYMOND, The Wilbour Papyrus, Vol. 4, Index, Oxford 1952.

GARDINER, ALAN, Late-Egyptian Miscellanies, Brussels 1937.

ID., Ancien Egyptian Onomastica, Oxford 1947.

ID., TheWilbour Papyrus, Vol. 2, Commentary, Oxford 1948.

GASSE, ANNIE, Données nouvelles administratives et sacerdotales sur l'organisation du domaine d'Amon. XXe-XXIe dynasties, Bibliothèque d'études 104 (1988).

GIRARD, PIERRE-SIMON, L'agriculture et l'industrie, in: Description de l'Egypte, Vol 17, Etat moderne, Paris 1824.

GNIRS, ANDREA M., Miltär und Gesselschaft. Ein Beitrag zur Sozialgeschichte des Neuen Reiches, Studien zur Archäologie und Geschichte Altägyptens 17, Heidelberg 1996.

GOMAÁ, FAROUK/MÜLLER-WOLLERMANN, RENATE/SCHENKEL, WOLFGANG, Mittelägypten zwischen Samalut und dem Gabal Abu Sir, Beihefte zu Tübinger Atlas des Vorderen Orients B/69, Wiesbaden 1991.

GRANDET, PIERRE, Le papyrus Harris I (BM 9999), Bibliothèque d'études 109 (1994).

GRIESHABER, FRANK, Lexikographie einer Landschaft. Beiträge zur historischen Topographie Oberägyptens zwischen Theben und Gabal as-Silsila anhand demotischer und griechischer Quellen, Göttinger Orientforschungen IV/45, Wiesbaden 2004.

HASSAN, FEKRI, Climate change, Nile floods and riparia, in: Riparia Dans l'Empire Romain, ed. by ELLA ERMON, Oxford 2010.

HASSAN, FEKRI/TASSIE, GEOFFREY, Modeling environmental and settlement change in the Fayum, in: Egyptian Antiquities 29 (2006), p. 37-40.

HARING, BEN, Divine households. Administrative and economic aspects of the New Kingdom royal memorial temples in Western Thebes, (Egyptologische Uitgaven 12), Leiden 1997.

ID., Economy, in: UCLA Encyclopedia of Egyptology, ed. by ELIZABETH FROOD/ WILLEKE WENDRICH, Los Angeles 2009, http://repositories.cdlib.org/nelc/ uee/1028.

HOSMER, DAVID/LEMESHOW, STANLEY, Applied logistic regression, New York 1989.

HELCK, WOLFGANG, Materialien zurWirtschaftsgeschichte des Neuen Reiches, Vol. 2, Mainz 1960.

JANSSEN, JACOBUS, Prolegomena to the study of Egypt's economic history during the New Kingdom, in: Studien zur altägyptischen Kunst 3 (1975), p. 127-185.

ID., Agrarian administration in Egypt during the Twentieth Dynasty, in: Bibliotheca Orientalis 43 (1986), p. 352-366.

ID., Grain Transport in the Ramesside Period: Papyrus Baldwin (BM EA 10061) and Papyrus Amiens, (Hieratic Papyri in the British Museum 8), London 2004.

KATARY, SALLY, Land tenure in the Ramesside Period, London 1989.

ID., Land tenure in the New Kingdom. The role of women smallholders and the military, in: Agriculture in Egypt. From pharaonic to modern times, ed. by ALAN BOWMAN/EUGENE ROGAN, Oxford 2001.

ID., The administration of institutional agriculture in the New Kingdom, in: Ancient Egyptian administration, ed. by JUAN CARLOS MORENO GARCIA, Leiden 2013, p. 719-783.

Id., The *wsf* plots in the Wilbour Papyrus and related documents. A speculative interpretation, in: Cahiers de recherches de l'Institut de papyrologie et égyptologie de Lille 25 (2005), p. 137-155.

Kitchen, Kenneth, Ramesside inscriptions, Vol 5 and 6, Oxford 1983.

Kemp, Barry, Ancient Egypt. Anatomy of a civilization, London 2006.

Kruchten, Jean-Marie, L'évolution de la gestion domaniale sous le Nouvel Empire Égyptien, in: State and temple economy in the Ancient Near East, Vol 2, (Orientalia lovaniensia analecta 6), ed. by Edward Lipinski, Leuven 1979, p. 517-525.

Lloyd, Alan, The Late Period, 664-323 B.C., in: Ancient Egypt, a social history, ed. by Bruce G. Trigger et al., Cambridge 1983, p. 279-348.

Lutley, Katy/Bunbury, Judith, the Nile on the move, in: Egyptian Antiquities 32 (2008), p. 3-5.

Le Roux, Brigitte/Rouanet, Henri, Multiple correspondence analysis, London 2010.

Meeks, Dimitri, Le grand texte des donations au temple d'Edfou, Bibliothèque d'études 59 (1972).

Martin, Pierre Dominique, Des Provinces de Beny-Soueyf et du Fayoum, in: Description de l'Egypte, Vol XVI, Etat moderne, Paris 1825, p. 6-15.

Manning, Joseph, Land and power in Ptolemaic Egypt, Cambridge 2003.

Menu, Bernadette, Le régime juridique des terres et du personnel attaché à la terre dans le Papyrus Wilbour, Lille 1970.

Michel, Nicolas, Travaux aux digues dans la vallée du Nil aux époques papyrologique et ottomane. Une comparaison, in: L'agriculture institutionnelle en Égypte ancienne. État de la question et approches interdisciplinaires, ed. by Juan Carlos Moreno Garcia, Cahiers de recherches de l'Institut de papyrologie et égyptologie de Lille 25 (2005), p. 253-276.

Moreno Garcia, Juan Carlos, Les *iḥwtjw* et leur rôle socio-économique du III^e^au Ier millénaire avant J.-C., Cahiers de recherches de l'Institut de papyrologie et égyptologie de Lille 20 (2009-2010), p. 321-351.

O'Connor, David, The geography of settlement in Ancient Egypt, in: Man settlement and urbanism, ed. by Peter Ucko et al., London 1972, p. 681-698.

Rowlandson, Jane, Landowners and tenants in Roman Egypt. The social relations of agriculture in the Oxyrhynchite Nome, Oxford 1996.

Said, Rushdi, The River Nile, geology, hydrology and utilization, Amsterdam 1993.

Schulman, Alan, Military rank, title, and organization in the Egyptian New Kingdom, (Münchner Ägyptologische Studien 6), Berlin 1964.

SCHENKEL, WOLFGANG, Die Bewässerungsrevolution im Alten Ägypten, Mainz 1978.
ID., Les systèmes d'irrigation dans l'Egypte Ancienne et leur génèse, in: Archéo-Nil 4 (1994), p. 27-35.
SPALINGER, ANTONY, A garland of determinatives, in: Journal of Egyptian Archaeology 94 (2008), p. 154-162.
STUCHEVSKY, IOSIF, Zmledel'tsy gosudarstvennogo chozyaïstva drevnego Egipta epoki Ramessidov, Moscow 1982.
SUBIAS, EVA/FIZ, IGNACIO/CUESTA ROSA. The Middle Nile Valley. Elements in an approach to the structuring of the landscape from the Greco-Roman era to the nineteenth century, in: Quaternary International 312 (2013), p. 27-44.
TURAYEV, Boris, Papyrus Prachov (Sobranya B.A. Turayeva), Leningrad 1927.
VLEEMING, SVEN, Papyrus Reinhardt. An Egyptian land list from the tenth century B.C, Leiden 1993.
ID., A review of GASSE, 1998, in: Enchoria 18 (1991), p. 217-227.
WAHAB, MOHAMED AHMED/EL SEMARY MAHMOUD, The correlation between physiography and soils west of the Nile Valley. A case study from Egypt, Biba-Bani Mazar area, in: Journal of Applied Sciences Research 8 (2012), p. 5682-5689.
WILLEMS, HARCO, Dayr al-Barshā. Volume 1: the rock tombs of Djehutinakht (No. 17K74/1), Khnumnakht (No. 17K74/2) and Iha (No. 17K74/3); with an essay on the history and nature of nomarchal rule in the early Middle Kingdom, Orientalia Lovaniensia Analecta 155, Leuven 2007.
ID., Nomarchs and Local Potentates. The Provincial Administration in the Middle Kingdom, in Ancient Egyptian Administration, ed. by JUAN CARLOS MORENO GARCIA, Leiden/Boston 2013, p. 341-392.
WILLCOCKS, WILLIAM/CRAIG, JAMES, Egyptian Irrigation, 3rd ed., London 1911.
YOYOTTE, JEAN, Histoire, géographie et religion de l'Egypte Ancienne, Opera selecta, Leuven 2013.

Harbours and Coastal Military Bases in Egypt in the Second Millennium B.C.

Avaris, Peru-nefer, Pi-Ramesse

MANFRED BIETAK

1. Introduction

During the Middle Kingdom, it seems that settlement in Lower Egypt was concentrated in the eastern part and at the extreme western edge of the Delta. Thus far, however, the greater part of the western half of the deltaic landscape has not yielded any sites of this period. It seems that this situation continued until the time of the New Kingdom. I have argued that the concentration of sites of the Second Intermediate period in the eastern Delta was the result of the political situation at that time and the development of a kind of homeland for an immigrated Near Eastern population which later caused the Hyksos rule in Egypt. But it now seems that this concentration of settlements in the eastern Delta had its roots in the physical geography of the Delta already in the Middle Kingdom. One has to face the fact that the western part of the Delta was void of habitation sites. It seems that only the cults at sacred places such as Sais and Buto were kept going, but we have no evidence of settlement there during this period. This is the result of extensive archaeological surveys conducted by the EES, the German Archaeological Institute, the University of Amsterdam and University of Liverpool (see Figure 1).[1] Explanations for this situation have

1 See the survey of the EES: http://www.deltasurvey.ees.ac.uk/ds-home.html. At this homepage one finds a full bibliography on the survey activities in the Nile Delta. I am indebted to Alan Jeffrey Spencer for more information on this survey. For the survey of the University of Amsterdam see: VAN DEN BRINK 1987; 1988. For the Survey of the University of Liverpool s. SNAPE 1986. For the survey

thus far not been put forward. However, it would seem that there are not many options that might explain this apparent lack of sites. One possibility is that sediment accumulation has caused sites to disappear under substantial layers of Nile mud.[2] This is not very likely, however, as the subsidence rates in most parts of the western Delta are moderate,[3] while sites older than the Middle Kingdom, dating to the fourth and third millennium B.C., have been found in this region.[4] One has to add that already existing settlement mounds (Tells), which jut out from the floodplain and provide secure settling ground during the flood season, were usually chosen as habitation sites.

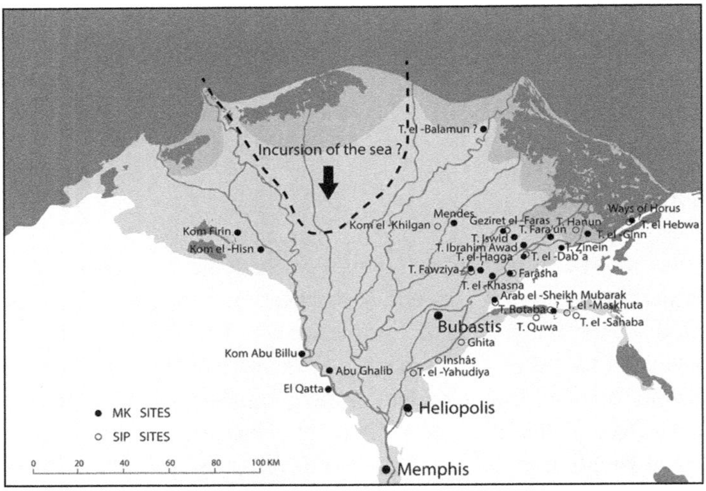

Figure 1. The Delta with sites of the Middle Kingdom and the Second Intermediate Period (graphic work Marian Negrete-Martinez).

As an alternative hypothesis to explain the void of Middle Kingdom and Second Intermediate Period sites in this area one could also propose that sediment accumulation rates may not have sufficiently compensated for the eustatic rise of the Mediterranean Sea level, a situation which would have led to a long-

of the German Archaeological Institute s. Schiestl, https://dainst.academia.edu/RobertSchiestl.

2 Stanley/Toscano, 2009, p. 161-167.

3 According to Wunderlich/Andres, 1991, p. 115-118.

4 See note 2.

term submersion, rendering large parts of the western Delta uninhabitable. It seems that even in the Late Period the western central Delta was less inhabited than the eastern part.[5] According to the medieval chroniquer Abu el-Hassan el-Makhzoumi there was a substantial sea incursion at 961 AD[6] causing the creation of the present coastal Nile Delta lakes which show remains until today in shore sediments south of the present inshore lakes.[7] It is possible that the central western Delta had a repetitive weakness in respect to sea incursions. There is also the possibility of an additional effect of tsunamis after volcanic or tectonic events in the Aegean or in Asia-minor.[8] With the wetlands inundated and only the levees and mounds emerging, the land would have been deprived of agriculture and flock-keeping and thus of sustenance. The reconstruction map of Stanley and Warne[9] for 4000 non-calibrated years BP, based on numerous cores, shows extended wetlands reaching far south into the Delta, west of the so-called Mendesian branch of the Nile, whereas eastwards, the floodplain suitable for agriculture and pasture reached as far north as the region of what is today San el-Hagar (see Figure 2). The eastern Delta with its substantial Pleistocene sand substratum and its numerous turtlebacks (sand geziras) differs from the western Delta by offering an ideally stable settling ground.[10] Additionally it seems that sediment accumulation rates were higher there than in the central and western Delta.[11] The extreme western edge of the Delta must also have been a more stable place of settlement in the Middle Kingdom. Therefore, the western – and even more so the easternmost – Nile branches must have been the most suitable water courses also for harbours.

5 Wilson and Grigoropoulos 2009.

6 I am grateful to Penelope Wilson of Durham University of informing me about this event.

7 Shafei, 1962; Frihy, 1992, p. 392.

8 Some information can be gathered from Papyrus Hearst of the early Eighteenth Dynasty, which renders a magic spell about the god Seth who seems to have stopped an incursion of the Mediterranean – an event which has to be tied because of the invocation of god Seth to the eastern Delta: "Just as Seth has banned the Mediterranean Sea Seth will ban you likewise..." (Goedicke, 1984, p. 46).

9 Stanley/Warne, 1998, fig. 8D.

10 Andres/Wunderlich, 1992.

11 Stanley et al., 1992, p. 30-39, figs. 10, 12.

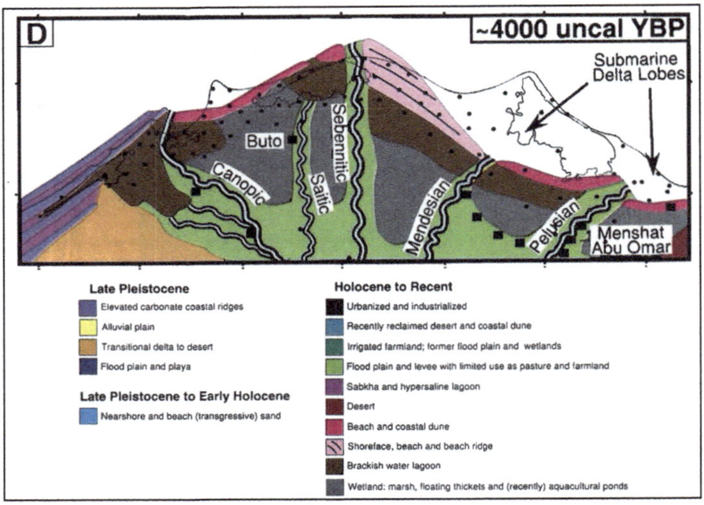

Figure 2. Reconstruction of the Delta in the time of 4000 BP (Stanley/Warne, 1998, fig. 8D).

2. The Nile Regime, the harbour situation and the identification of Avaris, Peru-nefer, Pi-Ramesse

Before the construction of the barrages in the nineteenth century A.D., perennial harbours for seagoing ships had to be situated within the reach of the sea.[12] The period of drought between March and early July, when the Nile shrank to one fifth of its average volume, made river navigation very difficult.[13] During this period, seawaters penetrated the nearly empty channels of the Delta and made perennial navigation within the reach of the sea possible. For this reason we find deltaic harbours such as Rashid (Rosette) and Dumiat (Damiette) in locations protected against winter gales, yet within a relatively short distance to the sea until the nineteenth century. Also Tanis served as a harbour for seagoing traffic at least until the fifth century A.D.[14] Harbours further upstream, beyond the reach

12 Bietak, 2010a, p. 165-169; 2010b, p. 19-21.

13 Le Père, 1822, p. 140-141; Clot, 1840; Cooper, 2012.

14 A manuscript in the monastery of Arezzo gives an account of a pilgrimage by abbess Aetheria to the Holy Land. The ship which she uses for her transfer to Egypt landed at Tanis from where the travel continued overland from the Nile Delta to

of the sea during the months of low river levels, could not provide perennial navigation for sea-bound traffic and had difficulties to reach the Mediterranean for nearly half of the year. On top of such limitations we know from Aramaic custom duty papyri with lists of incoming and outgoing ships from the Persian Period from Elephantine that there was no sea traffic at all during the months Thoth and Paophi (approx. January and February),[15] presumably because of usual fogs and winter gales during this time.

Based on these environmental conditions we can once and for all exclude Memphis as a candidate for the identification with the famous New Kingdom naval base of Peru-nefer[16]. It is unthinkable that the major naval base targeting the Near East was only operational from the second half of July until the end of the year during times of increased Egyptian warfare; and even then, the long distance from the Mediterranean would prolong the reaction of the Egyptian crown to any happenings in the Levant or would render a necessary mission impossible for six months. Therefore, we have to look for an alternative candidate, which was found in a huge harbour basin at Tell el-Dab'a connected with entry and exit channels to the Nile system within reach of the sea in the second millennium

the Sinai, to the Holy Land and back. See BROX, 1995; RÖWEKAMP, 1995; VRETSKA, 1958.

15 PORTEN/YARDENI, 1993, C3.7; YARDENI, 1994, p. 69. The name of the harbour is not mentioned, but it is clearly a harbour for seagoing ships (YARDENI, 1994, p.77, n. 13).

16 BADAWI, 1948, p. 34-36, 55-63, 137-139; GLANVILLE, 1931, p. 109; ID., 1932; HELCK, 1939, p. 49-50; ID., 1971, p. 160, 166, 356-357, 447-448, 454-456, 460, 471, 473, 501; JEFFREYS/SMITH, 1988, p. 61; EDEL, 1977, p. 155; KAMISH, 1985; EAD., 1986; DER MANUELIAN, 1987, p. 159; SÄVE-SÖDERBERGH, 1946, p. 37–39; STADELMANN, 1967, p. 32-35; ZIVIE, 1988, p. 107. Recently FORSTNER-MÜLLER, 2014, argued again in favour of locating Peru-nefer at Memphis and suggested that this town was reachable during the time of inundation, which lasted only a few months and is therefore unfeasible. The fact that she did not find any New Kingdom remains before the Ramesside Period in the very restricted area she excavated at the northern edge of the big harbour basin at Tell el-Dab'a is not a valid argument as we don't know yet where the Eighteenth Dynasty installations were positioned. On the other hand the presence of one of the biggest palaces in Egypt, dating to the Thutmosid Period, situated not far from the harbour basin, is begging for an explanation, which can only be offered by identifying this site with Peru-nefer.

B.C.[17] It was also situated at the easternmost Nile branch which, besides the so-called Western River, was the major water way during the second millennium B.C. This considerably narrows the possibilities of location of the major naval base, besides the impressive presence of the Thutmosid Period (see below). The direction of the reconstructed Nile channel to the north-east was ideal, as one could sail and return half-winds under the prevailing northerly winds.[18]

This harbour should be identified as the harbour of Avaris, which could accommodate hundreds of ships according to the second stela of Kamose.[19] Military installations and especially a 13 acre palace precinct of the Thutmosid Period,[20] embellished with Minoan wall paintings, make it highly likely that this had also been the famous harbour of Peru-nefer where Keftiu ships were moored[21] and which was a resort where Amenhotep II spent much time as a crown prince and as a king.[22]

The harbour and its channels were already assessed by my personal surface survey 1969 and 1973 and since by core drillings performed by Josef Dorner.[23]

---

17 Already Spiegelberg, 1927 and Daressy, 1927/28 thought that Peru-nefer was situated in the Delta. Habachi, 1972, Id., 2001, p. 9, p. 106–07, p. 121, thought that it had to be located at Avaris and the later city of Piramesse. See also Collombert/Coulon, 2000, p. 217; Pumpenmeier, 1998, p. 89-93; Röhrig, 1990, 126-127. Bietak, 2005a; Id., 2005b; Id., 2009a; Id., 2009b; Id., 2010a, p. 165-169; Id., 2010b, p. 19-21, was able to provide more solid evidence with the discovery of the Thutmosid palace of considerable size and the harbour basin. See especially the results of core drilling by Dorner, 1999, Plan 1, and the results of the geomagnetic survey of the Austrian Archaeological Institute (see notes 24 and 25).

18 This suggestion was made to me by the president of the Cairo Yachting Club, Mr. Yussef Mazhar.

19 Habachi, 1972, p. 36-37.

20 Bietak, 2005a; Id., 2005b; Bietak/Dorner/Jánosi, 2001; Bietak/Forstner-Müller, 2003; Iid., 2005; Bietak et al., 2007, p. 13-43.

21 Glanville, 1931, p. 121; Id., 1932, p.30, 36. The author proposes a foreign type of ship but also thinks it would be possible to suggest a meaning about the destination of the ship such as "East India-Man". Against this idea speaks that the Kfty-ship is only mentioned in the time of Thutmose III, neither before nor afterwards. The Minoan paintings in the enormous palace compound of the Thutmosid Period at Tell el-Dab'a is a strong indication that either real *Kfty*-ships are involved or that the Egyptians copied them in their dockyards.

22 Der Manuelian, 1987.

23 Dorner 1999, plan1.

Geomagnetic surveying was carried out by the Austrian Archaeological Institute and clarified the topography of Avaris (see figure 3).[24] The harbour of Avaris, known by the second stela of king Kamose to have accommodated hundreds of ships, was subject to recent paleogeographic studies by a team of the University of Lyon 2 under Hervé Tronchère and Jean-Philippe Goiran, who were able to identify the harbour function according to sediments with precision and to date the activity of the different river branches and the harbour basin itself.[25]

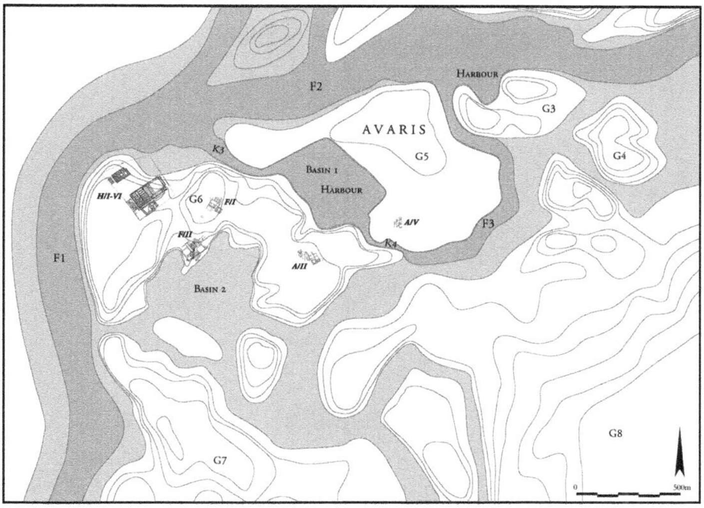

Figure 3. The paleogeography of the harbour of Avaris/Peru-nefer (graphic work Nicola Math).

Still more precision in tracing the outlines of the harbour can be obtained by studying surface features and the present use of land. The edges of the enormous trapezoid harbour basin of approx. 400 x 450 m can be identified by the present track from the Husseiniya road to 'Ezbet Rushdies-saghira, which is orientated exactly tangentially to the north-eastern edge of the basin (see figure 4). The south-western edge of the basin is not only apparent from the results of the geomagnetic survey, but is also marked by the position and direction of the eastern wall of the temple of Seth of Horemheb and the Nineteenth Dynasty (see figure 4). The orientation of the basin is likewise followed by the big temple precinct which was constructed just before, and remained intact throughout,

24 Forstner-Müller et al., 2007; Ead., 2009; Ead. 2010.

25 Tronchère et al., 2008; Tronchère, 2010.

the Hyksos Period (see figure 5)[26] The similarity of alignment of the basin edges with important archeological features is evidence that the orientation of the architecture must have been influenced by the presence of the harbour what should be considered as chronological evidence. Recent excavations by the Austrian Archaeological Institute and geophysical surveying showed that the extreme northern part of the basin was in the process of being filled by sediments during the Hyksos Period so that houses with tombs started to invade the former harbour space.[27] According to these investigations this process was more advanced during Ramesside times (see below).

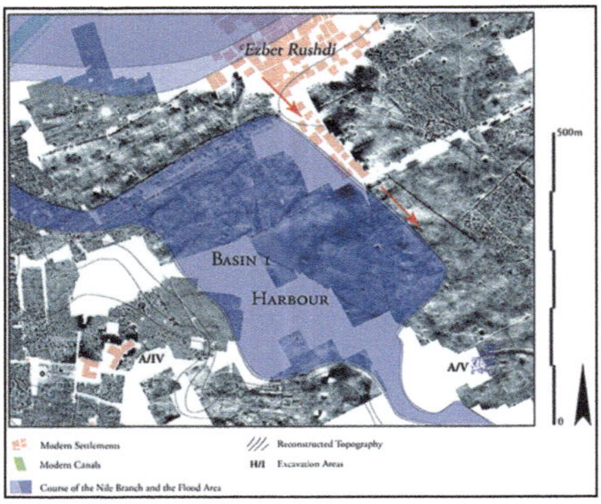

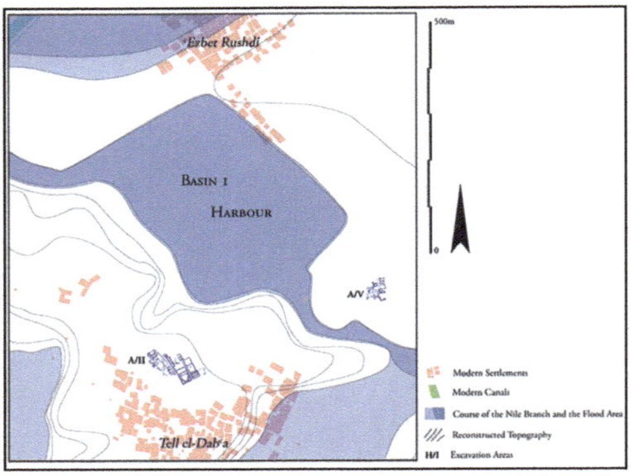

Figure 4 (left). The orientation of the eastern edge of the harbour basin is still preserved in a tangential track passing the modern village of 'EzbetRushdi (graphic work Nicola Math); Figure 5 (right). The harbour basin and the orientation of the Canaanite Temple precinct and adjoining enclosures (graphic work Nicola Math).

Important for the identification of Tell el-Dab'a/Avaris as the naval base of Peru-nefer, apparently installed by the Thutmosid kings at the already existing harbour of Avaris, are, besides its geographic position, the large harbour basin and the palatial precinct, the chronology and the stratigraphy of the site (see Figure 6). During the early Eighteenth Dynasty we find significant storage facilities and military camps.[28] Later, the New Kingdom palace was built. Finally a hiatus in occupation followed after the reign of Amenhotep II, lasting probably until

26 For the temple precinct see Bietak,2009c.

27 Forstner-Müller, 2014.

28 Bietak, 2010 c, p. 56, fig. 27; Bietak/Dorner/Jánosi, 2001, p. 59-74.

the reign of Amenhotep III or the Amarna period.[29] It may be significant that this hiatus in the occupation of the site finds parallels in the complete lack of references to Peru-nefer in inscriptions between Amenhotep II and the late Eighteenth Dynasty. It seems that Horemheb constructed an enormous defence system, perhaps even a fortress surrounding the harbour basin. Its northern wall was found by excavation and was followed over a length of several hundred meters by geophysical surveying. The wall cuts off an old access canal from the Pelusiac branch of the Nile issuing into the harbour basin. For this reason this canal has been considered non-existent,[30] even though it was identified by Josef Dorner's core drillings and its existence was verified by further core drillings by the research group of the University of Lyon 2. It seems to me that this was indeed an access canal as it provided a current from an active Nile arm to a more sluggish eastern branch. Without this canal, the basin would have been a sedimentation trap. It must have fallen into disuse during the time of the New Kingdom. Another access from the active Nile branch was found during the geophysical prospection work of the Austrian Archaeological Institute south of Ezzawin,[31] providing a fresh water current to the old, largely stagnant easternmost Nile branch.[32] This channel would, however, not provide a current for the harbour basin. It is also for this reason that one has to claim a special feeder canal for the harbour which must have silted up otherwise quickly under such circumstances.

Geomagnetic surveys suggest that the fortress of Horemheb was attached to an earlier fortress, the walls of which are clearly recognisable on the survey map (see figure 7). This older installation seems to have been in function until the construction of the fortress of Horemheb and in all likelihood dates prior to the Amarna Period or even to the reign of Amenhotep III. As a matter of fact, it would fit perfectly into this time when the high official Amenhotep, Son of Hapu, was charged with fortifying the mouths of the Nile branches to secure the Delta against incursions of pirates – the earliest of the Sea People.[33]

29 Ibid., p. 101-102.

30 FORSTNER-MÜLLER et al., 2010, p. 73-74, 84-85; FORSTNER-MÜLLER, 2014, and personal information.

31 Ibid.

32 For investigations about the current of the active and more stagnant water branches in the environment of Avaris see TRONCHÈRE, 2010.

33 Biography on a statue of Amenhotep, son of Hapu (Cairo Museum JE 38368 = CG 42127), VARILLE, 1968, p. 36, 41; OCKINGA, 1986, p. 33-34.

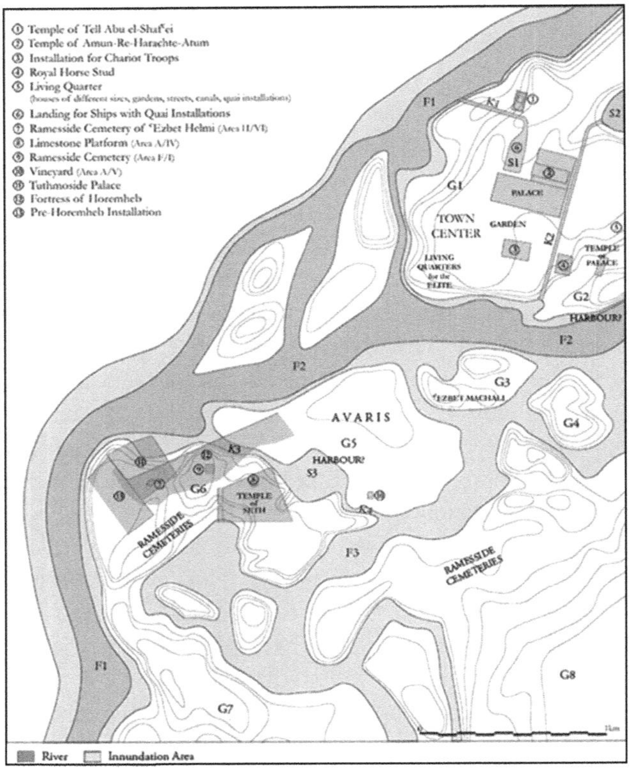

Figure 6. Reconstruction of the geography of Avaris-Peru-nefer and Pi-Ramesse (after BIETAK, *2010b, fig. 2.7).*

After Horemheb followed the establishment of Pi-Ramesse, the Delta residence of the Ramessides, which seems to have already been built to some extent by Sety I, with a palace at Qantir, 2 km north of Tell el-Dab'a/Avaris.[34] In order to keep control on the Near Eastern interests of Egypt it was necessary to relocate the political centre of the country to such a geographic position as to facilitate quick reaction to Near Eastern political developments and to have quick access to the Levant. The residence of pharaoh was moved to just 2 km north of the site of former Avaris and the site of Peru-nefer. From an eulogy on Pi-Ramesse

34 HAMZA, 1930, p. 64; BOREUX, 1932, p. 410; HAYES, 1937, p. 5-7, 17, 29-30; HABACHI, 1974; ID., 2001, p. 51-53, 69, 107, 123-126, 130-132, 141-143, 186, 210, 229-230, 254.

in Papyrus Anastasi III we learn that the new residence was *"The marshalling place of thy chariotry, the mustering place of thy army, the mooring place of thy ship's troops"*.[35] This suggests that the function of the site as a naval and military base persisted. As new excavations at Tell el-Dab'a showed that the big harbour basin was reduced in size by sedimentation in its northern part and that buildings invaded its space[36], it is possible that, at that time, the main harbour was moved somewhere else within this sprawling city which covered approx. 600 hectares. However, the memory of Avaris as the site of a harbour was still alive during the Twentieth Dynasty. We know from inscriptions on naosdoors – today in the Pushkin Museum in Moscow – that there was a temple of Amun at the harbour of Avaris then.[37]

During the late Twentieth Dynasty it seems that the lower reaches of the easternmost branch of the Delta were silted up.[38] The harbour and the residence moved to Tanis.[39]

Figure 7. The fortress wall of Horemheb (in black) abutting against an older fortress wall (in grey), (graphic work Nicola Math).

35 P. Anastasi III, 7.5–6; translation: Caminos, 1954, p.101.

36 See note 30.

37 Moscow I.1.a.4867; see Turayev, 1913, with pl. 13.

38 Bietak, 1975, 216.

39 Ibid.

3. Conclusion

Summing up, in some way the reasons for the identification of Peru-nefer with the site of Tell el-Dab'a is closely connected to the identification of Avaris with Tell el-Dab'a and environment, and of Pi-Ramesse with Qantir – a debate which was finally settled 40-50 years ago with the renewed excavations at Tell el-Dab'a and Qantir.[40] All three were harbour towns of considerable importance and paleogeographic investigations showed that the easternmost Nile branch was besides the westernmost river, the most important in the second millennium B.C. The specific reasons for settling also Peru-nefer at this place are as follows:

1. The presence of an enormous harbour basin and other harbour facilities[41] which, however, began to silt up in its northern part and was used there as settling and burial ground during the Hyksos Period, but nevertheless a harbour for hundreds of ships in Avaris is mentioned in the second Kamose stela[42]. That Avaris was a, if not *the* harbour of Pi-Ramesse – more than 200 years later – one could gather from naos door inscriptions originating from the Temple of Amun, "great of victories" in the Harbour of Avaris, dating to the Twentieth Dynasty.[43]
2. Paleogeographic studies showed that the easternmost Nile branch was the most important in the second millennium B.C. and the most convenient connection to the Near East.
3. The Canaanite cults in Avaris from the Second Intermediate Period[44] seem to have a continuum in Peru-nefer in the Eighteenth Dynasty[45] and in Pi-Ramesse during the Nineteenth and Twentieth Dynasty[46].
4. The presence of the site in reach of the sea during the period of drought.[47]

40 Bietak, 1975; Pusch et al., 1999.

41 Forstner-Müller et al. 2007; Ead., 2014.

42 See note 19.

43 See footnote 37.

44 Bietak, 2009c.

45 Stadelmann, 1967, p. 32–47, 99-110, 147-150; Colombert/Coulon, 2000, p. 217, Tazawa, 2009.

46 Daressy 1928-29, p. 326; Id., 1971, p. 446-73; Stadelmann, 1967, p. 148-150; Uphill, 1984, p. 200-202, 212, 233-234, 245 (Anta), 246 (Astarte),252 (Reshep), 252-3 (Seth); Tazawa, 2009; Coche-Zivie, 2011.

47 Bietak, 2010a; Id., 2010b.

5. The presence of one of the biggest palace precincts in Egypt which no doubt has royal dimensions and dates to the Thutmosid Period by scarabs of Thutmose I, Thutmose III and Amenhotep II and pottery.[48]
6. The appearance of the name of Peru-nefer in the texts naming Thutmose III[49] and Amenhotep II.[50] After an interval, the site is mentioned again in the late Eighteenth and the Nineteenth Dynasty.[51] The stratigraphy of the site parallels this information. The Thutmosid palatial compound was abandoned and the site was used again under Horemheb who constructed a big fortress and rebuilt the temple of Seth.[52] It seems even possible that he enlarged an older fortress which might go back to the Amarna Period.

Acknowledgements

For consultation I am obliged to Daniel Stanley (Smithsonian Institution, Washington), Jürgen Wunderlich (University of Frankfort) and Penelope Wilson (University of Durham). All mistakes, however, are mine. My thanks also go to the editors of this publication. For the ilustrations, if not specified especially, I am indebted to Nicola Math, Marian Negrete-Martinez and Silvia Prell.

Bibliography

Anders, Wolfgang/Wunderlich, Jürgen, Late Pleistocene and Holocene Evolution of the Eastern Nile Delta and Comparison with the Western Delta, in: Von der Nordsee bis zum Indischen Ozean, Ergebnisse der 8. Jahrestagung des Arbeitskreises "Geographie der Meere und Küsten", 1990, 13.-15. Juni Düsseldorf (Erdkundliches Wissen 105), ed. by Helmut Brückner/Ulrich Radtke, Stuttgart 1991, p. 121-130.

Badawi, Ahmed, Memphis als zweite Landeshauptstadt im Neuen Reich, Cairo.

48 See note 19.

49 The British Museum Papyrus BM 10056 has been dated by a new focused investigation by Roman Gundacker to the 51st year of Thutmose III. Publication planned for *Egypt and the Levant* 25 (2015).

50 See note 16 and 17.

51 Ibid.

52 Bietak, 1985, p. 267-278; Id., 1990, p. 12-14; Bietak/Dorner/Jánosi, 2001, p. 101-102.

BIETAK, MANFRED, Tell el-Dab'a II. Der Fundort im Rahmen einer archäologisch-geographischen Untersuchung über das ägyptische Ostdelta, Untersuchungen der Zweigstelle Kairo des Österreichischen Archäologischen Institutes vol. II, Vienna 1975.

ID., Ein altägyptische Weingarten in einem Tempelbezirk, (Tell el-Dabᶜa 1. März bis 10. Juni 1985), in: Anzeiger der Österreichischen Akademie d. Wissenschaften, Phil.-hist. Kl. 122 (1985), p. 267-278.

ID., Zur Herkunft des Seth von Avaris, in: Egypt and the Levant 1 (1990), p. 9-16.

ID., The Tuthmoside Stronghold Peru-nefer, in: Egyptian Archaeology 27/1 (2005a), p. 13-17.

ID., Neue Paläste aus der 18. Dynastie, in: Structure and Significance (Untersuchungen der Zweigstelle Kairo des Österreichischen Archäologischen Institutes vol. XXVII), ed. by PETER JÁNOSI, Vienna 2005b, p. 131-168.

ID., Peru-nefer; The Principal New Kingdom Naval Base, in: Egyptian Archaeology 34, (2009a), p.15-17.

ID., Peru-nefer: An Update, Egyptian Archaeology 35 (2009b), p. 16-17.

ID., Near Eastern Sanctuaries in the Eastern Nile Delta, in: Baal, Hors-Série, vol. VI, (2009c), p. 209-228.

ID., From where came the Hyksos and where did they go, in: The Second Intermediate Period (13th-17th Dynasties). Current Research, Future Prospects, (Proceedings of a Conference in the British Museum 2004) (Orientalia Lovanensia Analecta 192), ed. by MARCEL MARÉE, Leuven 2010a, p. 139-181.

ID., Minoan Presence in the Pharaonic Naval Base of Peru-nefer, in: Cretan Offerings: Studies in Honour of Peter Warren (British School at Athens Studies 18), ed. by OLGA KRZYSZOWSKA, London 2010b, p. 11-24.

ID., Houses, Palaces and Development of Social Structure in Avaris, 11-68, in: Cities and Urbanism, International Workshop in November 2006 at the Austrian Academy of Sciences Vienna (Untersuchungen der Zweigstelle Kairo des Österreichischen Archäologischen Institutes vol. XXXV), ed. by MANFRED BIETAK et al., Vienna 2010c, p. 139-181.

BIETAK, MANFRED/FORSTNER-MÜLLER, IRENE, Ausgrabungen Im Palastbezirk von Avaris, Vorbericht Tell el- Dab'a/'Ezbet Helmi Frühjahr 2003, in: Egypt and the Levant 13 (2003), p. 39-50.

ID., Ausgrabung eines Palastbezirkes der Tuthmosidenzeit bei 'Ezbet Helmi/Tell el-Dab'a, Vorbericht für Herbst 2004 und Frühjahr 2005, in: Egypt and the Levant 15 (2005), p. 65-100.

BIETAK, MANFRED, et al., Ausgrabungen in dem Palastbezirk von Avaris, Vorbericht Tell el-Dab'a/'Ezbet Helmi 1993-2000, in: Egypt and the Levant 11 (2001), p. 27-129.

BIETAK, MANFRED et al., Taureador Scenes in Tell el-Dab'a (Avaris) and Knossos (Untersuchungen der Zweigstelle Kairo des Österreichischen Archäologischen Institutes vol. XXVII), Vienna 2007.

BOREUX, CHARLES, Musée du Louvre, Departement des antiquités égyptiennes. Guide-catalogue sommaire II, Paris 1932.

BROX, NORBERT et al., Egeria Itinerarium Reisebericht, Fontes Christiani, Freiburg 1995.

CAMINOS, RICARDO A., Late Egyptian Miscellanies, London 1954.

COCHE-ZIVIE, CHRISTIANE, Foreign Deities in Egypt, in: UCLA Encyclopedia of Egyptology, ed. BY JACCO DIELEMAN/WILLEKE WENDRICH et al., Los Angeles 2011, http://digital2.library.ucla.edu/viewItem.do?ark=21198/zz0027fcpg.

COLLOMBERT, PHILIPPE/COULON, LAURENT, Les dieux contre la mer, le début du "papyrus d'Astarte" (pBN 202), in: Bulletin de l'Institut Français d'Archéologie Orientale 100 (2000), p. 193-242.

CLOT, ANTOINE BARTHÈLEMY, Aperçue général sur l'Égypte, Paris 1840.

COOPER, JOHN P., Nile Navigation: 'towing all day, punting for hours', in: Egyptian Archaeology 41 (2012), p. 25-27.

DARESSY, GEORGES, Les branches du Nil sous la XVIIIe dynastie, in: Bulletin de la Société Royale de Géographie d'Égypte 16 (1928-29), p. 225-254, p. 293-329.

DER MANUELIAN, PETER, Studies in the Reign of Amenophis II (Hildesheimer Ägyptologische Studien 26) Hildesheim 1987.

DORNER, JOSEF, Die Topographie von Piramesse, in: Egypt and the Levant 9 (1999), p. 77-83.

EDEL, ELMAR, Die Stelen Amenophis' II. aus Karnak und Memphis mit dem Bericht über die asiatischen Feldzüge des Königs, in: Zeitschrift des deutschen Palästina Vereins 69 (1977), p. 97-176.

FORSTNER-MÜLLER, IRENE, Providing a Map of Avaris, in: Egyptian Archaeology 34 (2009), p. 10-13.

EAD., Avaris, its Harbours and the Peru-nefer Problem, in: Egyptian Archaeology 45 (2014), p. 32-35.

FORSTNER-MÜLLER, IRENE et al., Geophysical Survey 2007 at Tell el-Dab'a, in: Egypt and the Levant 17 (2007), p. 97-106.

EAD., Preliminary Report on the Geophysical Survey at Tell el-Dab'a/Qantir in Spring 2009 and 2010, in: Jahreshefte des ÖsterreichischenArchäologischen Institutes in Wien 79 (2010), p. 67-85.

FRIHY, OMRAN E., Holocene sea level changes at the Nile Delta coastal zone of Egypt, in: Geo Journal 26/No. 3 (1992), p. 389-394.

GLANVILLE, STEPHEN, Records of a Royal Dockyard of the Time of Tuthmosis III. Papyrus British Museum 10056, in: Zeitschrift für ägyptische Sprache und Altertumskunde 66 (1931), p. 105-121.

ID., Records of a Royal Dockyard of the Time of Tuthmosis III. Papyrus British Museum 10056, in: Zeitschrift für ägyptische Sprache und Altertumskunde, 68 (1932), p. 7-41.

GOEDICKE, HANS, The 'Canaanite Illness', in: Studien zur Altägyptischen Kultur 11 (1984), p. 91-105.

HABACHI, LABIB, The Second Stela of Kamose and his Struggle against the Hyksos Ruler and his Capital (Abhandlungen des Deutschen Archäologischen Instituts Kairo, Ägyptologische Reihe 8), Glückstadt 1972.

ID., Sethos I[st] Devotion to Seth and Avaris, in: Zeitschrift für ägyptische Sprache und Altertumskunde 100 (1974), p. 95-102.

ID., Tell el-Dab'a I, Tell el-Dab'aandQantir, The Site and its Connection with Avaris and Piramesse (Untersuchungen der Zweigstelle Kairo des Österreichischen Archäologischen Institutes vol. II), ed. by EVA-MARIA ENGEL et al., Vienna 2001.

HAMZA, MAHMOUD, Excavations of the Department of Antiquities at Qantîr (Faqûs District). Season, May 21st – July 7th, 1928, in: Annales du Service des Antiquités de l'Égypte 30 (1930), p. 31–68.

HAYES, WILLIAM C., Glazed Tiles from a Palace of Ramesses II at Kantir, New York 1937.

HELCK, WOLFGANG, Der Einfluss der Militärführer in der 18. Ägyptischen Dynastie, Leipzig 1939.

ID., Die Beziehungen Ägyptens zu Vorderasien im 3. und 2. Jahrtausend v. Chr. (Ägyptologische Abhandlungen 5), 2nd ed., Wiesbaden 1971.

JEFFREYS, DAVID G., Perunefer: at Memphis or Avaris? in: Egyptian Archaeology 28 (2006), p. 36-37.

JEFFREYS, DAVID G./SMITH, HENRY S., Memphis and the Nile in the New Kingdom, in: Memphis et ses nécropoles au Nouvel Empire, ed. by ALAIN-PIERRE ZIVIE, Paris 1988, p. 55-66.

KAMISH, MARIAM, Foreigners at Memphis in the Middle of the 18th Dynasty, in: Wepwawet 1 (1985), p. 12-13.

EAD., Problems of Toponymy with Special Reference to Memphis and Prw-nfr, in: Wepwawet 2 (1986), p. 32-36.

LE PÈRE, JACQUES-MARIE, Mémoire sur la communication de la mer des Indes à la Méditerranée, par la mer Rouge et l'isthme de Soueys, in: Description de l'Égypte ou recueil des observations et des recherches qui ont été faites en Égypte pendant l'expédition de l'armée francaise 11, second edition, Paris 1822, p. 37-370.

OCKINGA, BOYO, Amenophis, Son of Hapu. A Biographical Sketch, in: The Rundle Foundation for Egyptian Archaeology Newsletter 18 (February 1986), p. 3-6.

PORTEN, BEZALEL/YARDENI, ADA, Textbook of Aramaic Documents from Ancient Egypt, Newly copied, edited and translated into Hebrew and English, Vol. 3, Winona Lake1993.

PUMPENMEIER, FRAUKE, Eine Gunstgabe von Seiten des Königs. Ein extrasepulcrales Schabtidepot Qen-Amuns in Abydos (Studien zur Archäologie und Geschichte Altägyptens 19), Heidelberg 1998.

RÖHRIG, CATHERINE H., The Eighteenth Dynasty Titles Royal Nurse (mn'.t-nswt), Royal Tutor (mn'-nswt) and Foster Brother/Sister of the Lord of the Two Lands, Diss Berkeley, Ann Arbor, Mich. 1990.

RÖWEKAMP, GEORG (ed.), Egeria itinerarium. Reisebericht, Fontes Christiani 20, Freiburg 1995.

SÄVE-SÖDERBERGH, TORGNY, The Navy of the Eighteenth Egyptian Dynasty, Uppsala/Leipzig 1946.

SHAFEI, A., Lake Mareotis, its past history and future development, in: Bulletin de l'Institut Fouad du Desert 1 (1952), p. 71-89.

SNAPE, STEPHEN R., Six Archaeological Sites in Sharqiya Province, Liverpool University Delta Survey No. 1, Liverpool 1986.

SPIEGELBERG, WILHELM, La ville de Prw-nfr dans le Delta, in: Revue de l'Égypte ancienne 1 (1927), p. 215-217.

STADELMANN, RAINER, Syrisch-palästinensische Gottheiten in Ägypten, Leiden 1967.

STANLEY, JEAN-DANIEL/TOSCANO, MARGUERITE A., Ancient Archaeological Sites Buried and Submerged along Egypt's Nile Delta Coast. Gauges of Holocene Delta Margin Subsidence, in: Journal of Coastal Research 25/1 (2009), p. 158-170.

STANLEY, DANIEL JEAN/ WARNE, ANDREW G., Nile Delta in its Destruction Phase, in: Journal of Coastal Research 14/3 (1998), p. 794-825.

STANLEY, DANIEL JEAN et al., Nile Delta, The Late Quarternary North-central Nile Delta from Manzala to Burullus Lagoons, Egypt, in: National Geographic Research & Exploration 8/1 (1992), p. 22-51.

TAZAWA, KEIKO, Syro-Palestinian Deities in New Kingdom Egypt. The Hermeneutics of Their Existence, (BAR I.S. 1965) Oxford 2009.

TRONCHÈRE, HERVÉ, Approche paléoenvironnementale de deux sites archéologiques dans le delta du Nil. Avaris et la branche Pélusiaque, Taposiris et le lac Mariout, Thèse présentée à l'Université Lyon 2 le 3 septembre 2010 en vue d'obtenir le grade de Docteur en Géographie, Lyon, Lyon 2010.

TRONCHÈRE, HERVÉ et al., Geoarchaeology of Avaris: First Results, in: Egypt and the Levant 18 (2008), p. 339-352.

TURAYEV, BORIS A., Dvertsy Naosa s molitvami boginye Tauert, N° 3914 Golenishchevskago sobraniya, in: Pamyatniki Muzeya izyashchnykh iskusstv imeni imperatora Aleksandra III v Moskve, fasc. III, Moscow 1913, p. 43-80.

UPHILL, ERIC, The Temples of Per Ramesses, Warminster 1984.

VAN DEN BRINK, EDWIN C.M., A Geo-physical Survey in the north-eastern Nile Delta, Egypt, The First Two Seasons, A Preliminary Report, in: Mitteilungen des Deutschen Archäologischen Instituts Abteilung Kairo 43 (1987), p. 7-31.

ID., Amsterdam University Survey Expedition to the Northeastern Nile Delta (1984-1986), in: The Archaeology of the Nile Delta. Problems and Priorities, ed. by EDWINVAN DEN BRINK, Amsterdam 1988, p. 65-110.

VARILLE, ALEXANDRE, Inscriptions concernant l'architecte Amenhotep, fils de Hapou (Bibliothèque d'Étude 44), Cairo 1968.

VRETSKA, Karl (ed.), Die Pilgerfahrt der Aetheria (Peregrinatio Aetheriae), Klosterneuburg 1958.

WILSON, PENELOPE/GRIGOROPOULOS, DIMITRIS, The West Delta Regional Survey, Beheira and Kafr el-Sheikh Province. (Excavation Memoir 86) London Egypt Exploration Society, London 2009.

WUNDERLICH, JÜRGEN/ANDERS, WOLFGANG, Late Pleistocene and Holocene Evolution of the Western Nile Delta and Implications for its Future Development, in: Von der Nordsee bis zum Indischen Ozean, Ergebnisse der 8. Jahrestagung des Arbeitskreises "Geographie der Meere und Küsten", 1990, 13.-15. Juni Düsseldorf (Erdkundliches Wissen 105), ed. by HELMUT BRÜCKNER/ULRICH RADTKE, Stuttgart 1991, p. 105-120.

YARDENI, ADA, Maritime Trade and Royal Accountancy in an Erased Customs Account from 475 B.C.E. on the Ahiqar Scroll from Elephantine, in: Bulletin of the American Schools of Oriental Research 293 (1994), p. 67-78.

ZIVIE, ALAIN-PIERRE (ed.), Aper-El et ses voisins: considérations sur les tombes rupestres de la XVIIIe dynastie à Saqqarah, in: Memphis et ses nécropoles au Nouvel Empire: nouvelles données, nouvelles questions. Actes du colloque international CNRS, Paris, 9 au 11 octobre 1986, ed. by ALAIN-PIERRE ZIVIE, Paris 1988, p. 103-112.

Development of the Memphite Floodplain

Landscape and Settlement Symbiosis in the Egyptian Capital Zone

JUDITH BUNBURY, ANA TAVARES, BENJAMIN PENNINGTON, PEDRO GONÇALVES

1. Introduction

The city of Memphis (see figure 1), although its exact location was lost for many years, was famed in antiquity as the capital of Egypt with an active scribal school. The large ruin field of Mit Rahina is part of Egypt's "Capital Zone" that extends from Abu Rawash and Giza in the north 80 km southwards to Maidum and the entrance to the Faiyum. The whole of this stretch is studded with pyramids and contains the locations of many necropoleis and settlements. In this paper we consider the fortunes of Memphis as part of this capital zone and examine how a number of environment factors have affected the area and may have affected the city as its fortunes waxed and waned. We also discuss the evolution of the Memphite floodplain and the evolving architecture of the Egyptian delta.

2. Memphis

The foundation of ancient Memphis was traditionally associated with a large intervention in the landscape – nothing less than diverting the Nile.[1] The city's subsequent development, over the millennia, was inextricably linked to its dynamic landscape. The memory of Memphis as the ancient capital of Egypt was

1 HERODOTUS II, 99, DIODORUS I, 505, see also JEFFREYS, 1985, p. 53–55.

preserved although its location remained uncertain[2] until the French Expedition identified the capital with the ruins of Mit Rahina.[3] Despite its importance only sporadic excavation has taken place at Memphis. Many of its dispersed monuments were accidental discoveries and much of the site remains under cultivation, urban sprawl and private ownership. Our current understanding of the city's development and the ancient environment, particularly the movement of the river, is due mainly to the work of the Survey of Memphis (SoM).[4] From its inception the project combined geomorphology and archaeology. This approach was continued in the Mit Rahina Field School (MRFS)[5] which recorded and excavated the Middle Kingdom settlement at Kom el-Fakhry. The results of the MRFS will fill a gap in our understanding of ancient Memphis and Egypt's "Capital Zone".

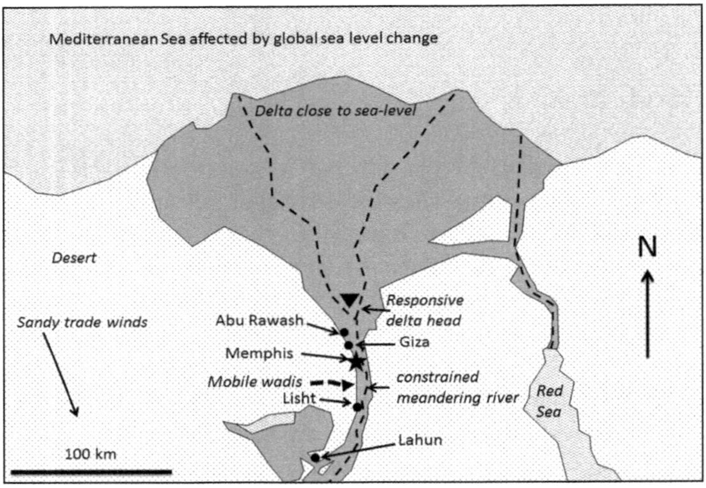

Figure 1. Location map of the "Capital Zone" showing the main locations including Lisht etc.

2 Memphis is aptly described by Jeffreys as "not a lost city, but a city temporarily misplaced", TAVARES/KAMEL, 2012, p. 5.

3 JEFFREYS, 2010, p. 63–66, p. 191–192.

4 JEFFREYS, 1985, GIDDY, 2012.

5 The Mit Rahina Field-School 2011 (MRFS) was a joint project of Ancient Egypt Research Associates (AERA), the American Research Center in Egypt and the Egypt Exploration Society (EES). The project is under the auspices of the Ministry of Antiquities.

2.1 Location and Landscape: Holocene Climate Change as a Driver of Change

From a geographical perspective, Memphis is located at the point where the desert cliffs of the Nile Valley broaden out and the delta starts to form, an area often described as the delta head. This location, at a landscape tipping point, means there is a complex interplay of geomorphological processes, mostly arising from climate changes that act upon the site. These include:

1. Sea-level change swamping the deltaic coast and the Nile hinterland
2. Migration of the delta-head in response to sea-level changes
3. Lateral migration of the river around the site
4. Vertical aggradation of the floodplain
5. Incursions of desert and wadi sand into the Nile valley in response to changing rainfall and erosion
6. Aeolian sandflux into the valley and the river around Memphis.

Much is known about the approximate geometry of each of these processes and the timescales in which each was active (see figure 2). To calibrate these processes and the way in which they combine from this region is given by detailed archaeological studies at key localities within the ancient capital zone.

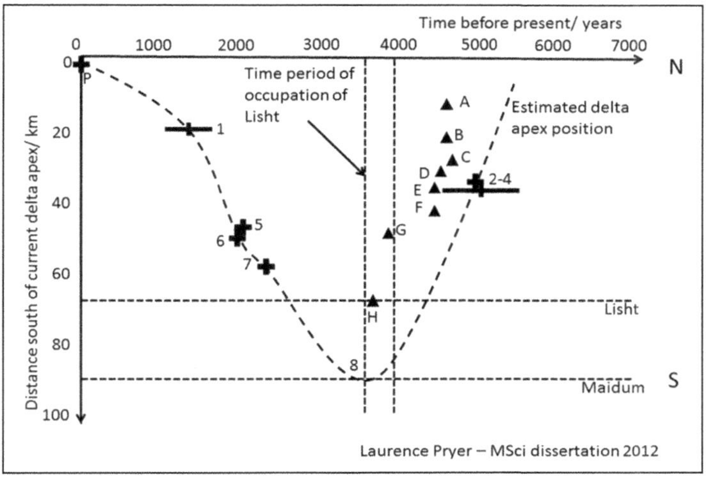

Figure 2. Delta head migration diagram after Pryer, 2012.

Combined studies of climate change proxies[6] suggest that the Holocene period has been characterised by a sharp rise in temperature from the last glacial maximum that peaked around 8,000 years ago during the wet phase in Egypt, designated the Saharan Neolithic. Following this peak there was a gradually oscillating decline in temperature until around 300 years ago when the post-industrial period caused temperatures to rise again. These global temperature changes generated two main effects on the Memphite area. The first was an increase in marshiness resulting from sea-level change in the Mediterranean coupled with the production of many river channels that formed an interconnecting distributary system.[7] The other effect of temperature change was humidification of the Saharan region as the equatorial belt widened and summer monsoon rains fell over a wider part of what is now the Sahara desert[8] with a subsequent decay of habitat as the rains retreated south. The equatorial monsoon in Ethiopia also affected the supply of water in the Blue Nile and hence the intensity and sediment content of the Nile flood.[9]

2.1.1 Sea-level change swamping the deltaic coast and the Nile hinterland

Global sea levels compiled from a number of sources[10] show a steep rise as ice-caps melted after the end of the last glacial maximum (see figure 3). This continued until around 6,000 years ago and was then followed by a period of very gentle rises in sea-level until present time. The coastal areas of deltas across the world were inundated,[11] as was the Egyptian Delta.[12] While marine incursion did not reach as far south as Memphis, at this time, fresh water in the Nile was retained in the valley making the area marshier and increasing the number of channels in the floodplain. Habitation seems to have been restricted to the Pleistocene sand "Gezirehs",[13] remaining from the previous high sand, and to the flanks of the Nile Valley[14] where the wadi mouths and low desert edge

6 Rohde, 2006.

7 Pennington et al., 2016.

8 Kropelin et al., 2008, Rodrigues et al., 2000, Stanley/Warne, 1993, Id., 1994, Id. 1998, Kuper/Kropelin, 2006.

9 Woodward et al., 2007.

10 e.g. Fairbanks, 1989.

11 Stanley/Warne, 1994.

12 Id., 1998.

13 Tristant, 2004.

14 Jeffreys/Tavares, 1994.

provided a refuge from the waters. After the marine incursion, new sediment started to rebuild the delta and the number of distributaries gradually fell and the network began to be more hierarchical dividing from a point in the capital zone area. Modelling of the landscape of the capital zone suggests that this distribution point, known as the "delta-head" migrated with time.[15]

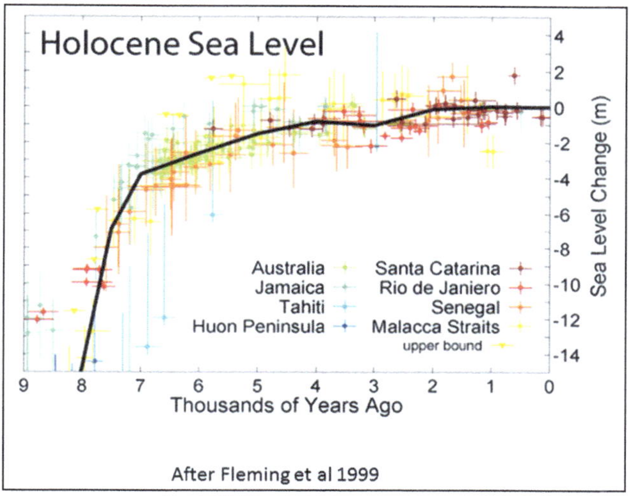

Figure 3. Holocene sea-level rise after the compilation of ROHDE, *2006.*

Intensive studies of the Rhine delta in the Netherlands, involving some 250,000 boreholes,[16] have shown that in an area of low gradient such as a river floodplain, a sea level increase of tens of metres can cause water to travel a hundred or more kilometres inland. The freshwater marshes that are created inland of the estuarine and coastal zone are a rich habitat and have a high nutrient availability. Thus at the same time that the Saharan region was interspersed with lakes and playa basins[17] there was also a rich habitat in the delta region. High levels of sediment accumulation in the area mean that much of this prehistoric inhabitation is

15 BUNBURY et al., in prep.

16 BERENDSEN, 2007, BERENDSEN/STOUTHAMER, 2001.

17 DRAKE, 2006, KUPER/KROPELIN, 2006.

cryptic but a few sites are known including Sais,[18] and several from the north-eastern delta that include Minshat Abu Omar.[19]

Our understanding of the protodynastic settlement is still in its infancy but it seems probable that settlement is reflected in the incidence of early cemeteries; interestingly these lie on both sides of the river (Saqqara, Giza, Helwan, Tura) whereas from the Old Kingdom (Third Dynasty) there is a marked preference for the western desert edge (pyramids and elite tombs).

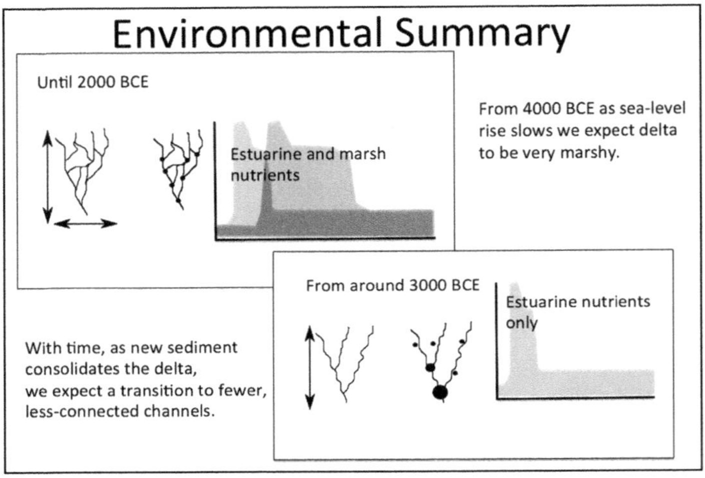

Figure 4. To show schematic river network and nutrient availability before (a) and after (b) c. 3000 B.C. from Pennington, et al., 2016.

In Egypt, the environment of diverse habitats with many interconnecting channels started to change around 4000–3000 B.C. with the marshes becoming marginalised towards the coast and a corresponding reduction in calories available. At the same time, the many anastomosing channels of the delta started to be replaced by more discrete meandering channels that divided from an upstream focus, the delta head. After this time there was a movement of settlements from the edges of the Nile Valley to the levees of the meandering Nile channel in the region of Memphis.[20] Borehole work by the Survey of

18 WILSON, 2006.

19 ROWLAND, 2012.

20 JEFFREYS/TAVARES, 1994.

Memphis[21] seems to suggest that there were two channels in the region of Memphis at this time, of which the western channel persisted until the Middle Kingdom.[22] However, constrained by the Nile Valley that narrowed to around 7 km (or less) at this point, the two channels were strategically close together and the "Capital Zone" starts to focus on Memphis.

2.1.2 Migration of the delta-head in response to sea-level changes

A distributary system is initiated as the base of a river channel reaches sea level when the channel divides into two smaller and shallower channels. These can continue to flow seawards until their bases reach sea-level and they, in turn, bifurcate. Thus the location of the delta head in the Nile is an inter-play between the amount of water in the river, which determines the size and depth of the channel, and sea-level. Factors causing migration of the delta-head inland include sea-level rise and increased water in the river, while factors that push the delta head sea-wards include aggradation of the floodplain and sea-level fall.

Observations of floodplain elevation suggest a rapid rise between the Old and the New Kingdom at Dahshur, which we expect to be reflected in a migration of the delta head seawards. Records of the location of the delta head from literature[23] can be combined with observations of Parcak[24] that the village of Lisht may have been located at the contemporary delta head, thus establishing a pattern of migration for this landscape feature during the development of Memphis (see figure 5). We seem to see two episodes during which the delta head was located at Memphis, broadly corresponding to the peaks of known activity in the area; the Old Kingdom necropolis of Saqqara and the New Kingdom expansion of Memphis.

21 Jeffreys, 1985.

22 Bunbury/Jeffreys, 2011.

23 Bunbury et al., in prep.

24 Parcak, 2011.

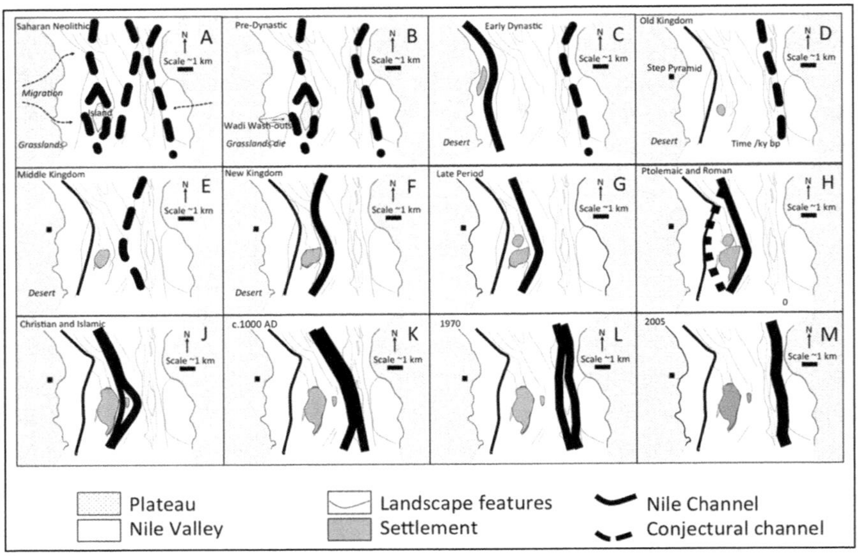

Figure 5. Time series to show the likely history of Nile migration across the Memphite Floodplain as inferred from boreholes of the Survey of Memphis.

2.1.3 Lateral migration of the river around the site

Lateral migration of the meandering Nile within the river floodplain was described by Butzer,[25] observed at Memphis[26] and studied further at Karnak[27] and in the Giza area.[28] Lateral migration of river bends, outwards and downwards across the floodplain, has a mean rate in Egypt of around 2 km/millennium, though lateral rates may reach up to 9 km/millennium in some areas and are frequently characterised by island production and capture.[29] The Survey of Memphis has bored around 150 cores amounting to some 2 km of sediment in a variety of locations across the mounds and in the surrounding floodplain. Facies analysis of these cores has suggested that in the Memphis area there has been broadly eastwards migration of the Nile across the floodplain during the past 6,000 years.

25 Butzer, 1976.

26 Jeffreys, 1985.

27 Bunbury et al., 2008.

28 Bunbury/Lutley, 2008.

29 Hillier et al., 2006.

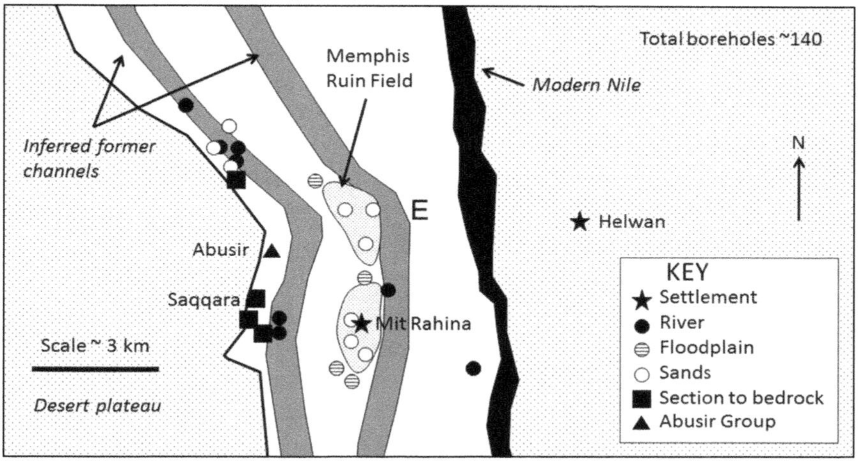

Figure 6. Summary map of the Survey of Memphis boreholes into the Memphite Floodplai showing erstwhile river positions (grey) and the current Nile location (black).

2.1.4 Vertical aggradation of the floodplain

The earliest work at Memphis, by Hekekyan in the 1850s,[30] was focussed on determining the rate of flood-plain silt accumulation in order to calculate the time since the recorded biblical flood. During his excavations Hekekyan made comprehensive notes and detailed observations of the sediments and monuments he encountered, making him the first geoarchaeologist of Egypt. Hekekyan was unable to determine the time since the flood but observations of sediment accumulation rates[31] give values around 1 m/millennium. This is a mean rate so there are areas such as river levees and settlement mounds where rates exceed this as well as areas like distal parts of the river plain where the sedimentation rate is lower. Comparison of the typical rate of vertical aggradation of the floodplain with the typical rates for lateral migration of the river channel, suggest that the latter is more rapid than the former (cf 1 m/millennium rise vs 2 km/millennium migration). None the less, results from the late Old Kingdom at Dahshur[32] suggest that floodplain rise was faster than that since the New Kingdom, which may be a product of an asymptotic approach to the base level of the river and a

30 JEFFREYS, 2010.

31 e.g. BORCHARD, 1907 and reviewed in BUNBURY et al., 2008.

32 ALEXANIAN et al., 2012.

large influx of sandy sediment to the Nile Valley during the Late Old Kingdom and Middle Kingdom.

2.1.5 Incursions of wadi sand into the Nile valley in response to rainfall and erosion

Studies of borehole cores drilled as part of the Cairo waste-water programme (AMBRIC) were examined by El-Senussi and Jones[33] and later Brandon.[34] The results revealed that the Pleistocene Nile canyon was filled by coarse sandy sediments that flowed out of the wadi mouths, that impinge upon the Nile Valley as rain fell directly in the Saharan region. Early settlements at Omari and Helwan were focussed on these paleo-fans. As the Holocene began and dark Nile silt accumulated above the sands the toes of the paleo-fans began to be covered. The presence of a water tank at Giza and of structures in the wadi at Dahshur[35] indicates that the wadis were stable during the early part of the Old Kingdom. However, around the end of the Fourth Dynasty the wadis seem to become unstable and as El-Senussi and Jones[36] and Dufton and Brandon noted,[37] successive tongues of sediment flowed out of the wadi mouths into the Nile valley. At around this time, settlement moved away from the wadi mouths and the terraces that flank the Nile Valley into the floodplain and occupied the levees of the Nile channel in somewhat extended "ribbon" developments.[38]

Studies of erosion rates in dry deserts[39] show that there is little erosion at high rates of rainfall since the rain sustains plentiful vegetation that stabilises the soil. When there is no rainfall, erosion is also low. However, at intermediate rainfall, around 200 mm/year erosion increases dramatically since the rain does not sustain sufficient vegetation to stabilise the sediment. We therefore infer that these sand tongues intruded into the Nile silts of the valley at the time when the climate was in transition between the wet early Holocene conditions and the drier conditions that were reached around 2000 B.C. The loss of vegetation from all but the refugia of the Saharan region focussed populations into the oases and the Nile Valley, as was shown by Kuper and Kropelin in their study of carbon

33 El-Senussi/Jones, 1997.

34 Branton, 2008.

35 Bebermeier, 2011, Alexanian et al., 2012, Ramisch, 2012.

36 El-Senussi/Jones, 1997.

37 Duffton/Branton, 2009.

38 Jeffrey/Tavares, 1994.

39 Goudie/Wilkinson, 1977, p. 88.

dates across the Saharan region.[40] As trade winds became established across the area,[41] sand dunes that had previously been locked in place by vegetation were released and started to move across the landscape, moving generally towards the south-east.

2.1.6 Aeolian sandflux into the valley and the river around Memphis.

Studies of the First Intermediate Period and the associated climate crisis have highlighted the influx of sand to the Nile Valley that occurred around this time.[42] However, from the results of Moeller[43] and other studies further south[44] it is clear that the north of Egypt was desiccated far earlier than the south of Egypt or the Sudan.[45] Neither can the transition be considered as instantaneous, with Kröpelin et al.[46] seeing a lag of some 2000 years between the death of the tropical vegetation in the area of Lake Chad and the ultimate establishment of trade winds across North Africa.

Large accumulations of wind-blown sand along the base of the escarpment at Saqqara seem to post-date the Early Dynastic occupation of the site but a much clearer picture of the amount and timing of sand arriving has been determined by Alexanian and colleagues in their excavations at Dahshur.[47] Here sand flux into the wadi below the valley temple of the bent pyramid began in the late Fourth Dynasty and peaked during the late Old Kingdom. Since the end of the Old Kingdom, sand has continued to accumulate but at a lesser rate. Sand accumulation along the base of the escarpment may have encouraged the early occupants of Memphis to move into the Nile Valley but there seems little evidence of aeolian sand deposited directly onto the site of Memphis from the borehole evidence and micro-morphological observations of Qin.[48] Qin's results suggest that sand deposited around Memphis at this time had been transported by river before its arrival at the site but there are traces of a former history as aeolian sand was still visible on the grain surfaces indicating that the Nile was

40 Kuper/Kropelin, 2006.

41 Kropelin et al., 2008.

42 Hassan, 2005.

43 Möeller, 2005.

44 Bunbury, 2010.

45 Woodward et al., 2001, Rodrigues et al., 2000.

46 Kropelin et al., 2008.

47 Bebermeier, 2011, Alexanian et al., 2012, Alexanian, et al., in press a and in press b., Ramisch, 2012.

48 Qin, 2009.

transporting sand that had recently been blown into the Nile valley. Additional islands and bars are likely to have formed in the river beds as the extra sand was flushed towards the sea.

While all five of these landscape factors affected the landscape of the capital zone a study of the archaeological material is required to discern which are the most important in the evolution of the city of Memphis.

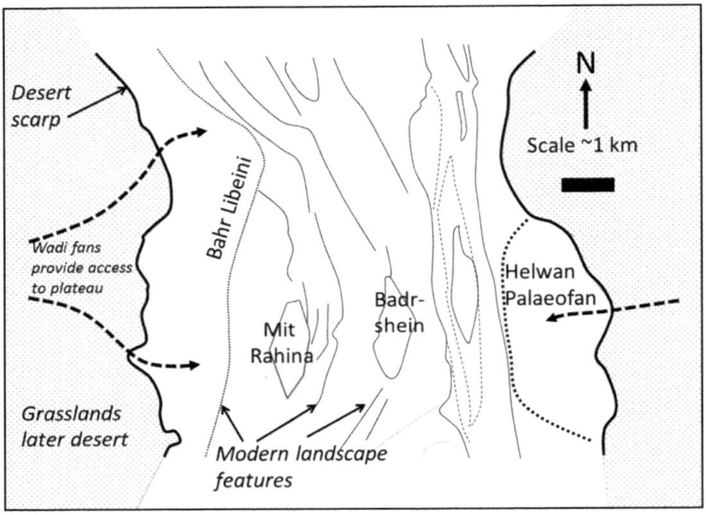

Figure 7. Diagram to show the effects of landscape processes at Memphis.

2.2 Dynamics of settlement at Memphis

A complex picture of the city of Memphis emerged from the work of the Survey of Memphis (SoM). The team provided a broad overview of the dynamics of settlement across the millennia,[49] as well as in-depth stratigraphic information on parts of the city.[50]

49 JEFFREYS, 1985, ID., 2010.

50 ASTON/JEFFREYS, 2007, JEFFREYS, 2006, GIDDY, 2012.

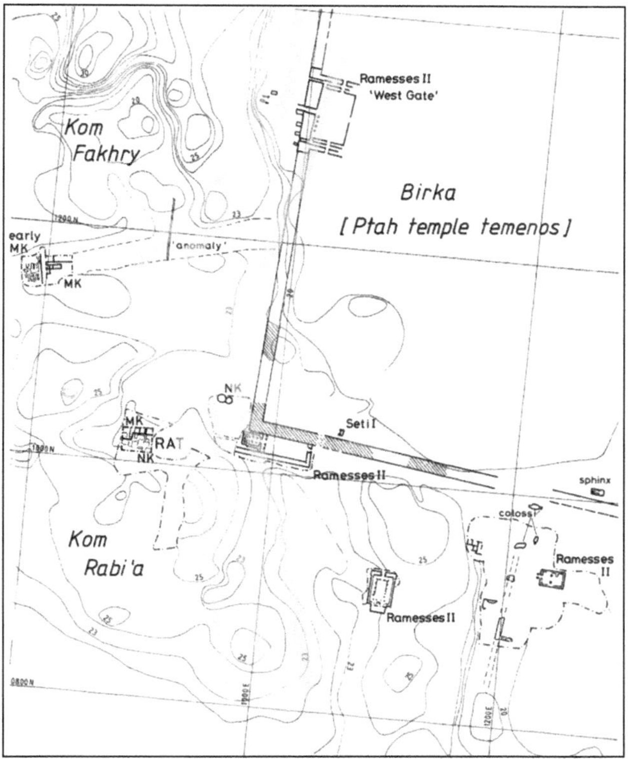

Figure 8. Map of the excavations and monuments in the Kom el-Fakhry and Kom Rabia area. After Giddy 2012, plate 1.

The most striking results concerned the movement of the Nile eastwards (see 2.1) above), and terracing of the settlement at Kom el-Fakhry and Kom Rabia.[51] The resulting topography preserved Middle Kingdom stratigraphy at a higher elevation than the New Kingdom remains to the east. For example the threshold of the West Gate of the Ramesside Ptah Temple is at elevation 18 m a.s.l., while the First Intermediate Period cemetery at Kom el-Fakhry just 100 m to the west is at elevation at 21 m a.s.l.. This 3 m difference in elevation between the two sites indicates an ancient slope with a gradient of 1: 30 or 1: 40. Jeffreys suggests an even greater gradient, up to 1: 10, in ancient times.[52] This slope suggests that the

51 JEFFREYS, 2010, p. 193–194, ID., 2008.

52 JEFFREYS, 2006, p. 1.

early town was on, or to the west of, Kom el-Fakhry.[53] It also created a window of opportunity to investigate early stratigraphy which elsewhere at Memphis is inaccessible as it lies under the water table.

The SoM excavations at Kom Rabia (site code RAT)[54] revealed a long sequence of occupation from the Third Intermediate Period[55] to the late Middle Kingdom.[56] Earlier occupation is attested but remains unexcavated as it lies below the water table. The Third intermediate Period levels consisted of a series of walls, an associated floor, a stone threshold, a pavement and a kiln. Earlier Ramesside architecture was still visible and possibly in use during the Third Intermediate Period.[57] The team also recorded two distinct phases of urbanism: an upper horizon of New Kingdom streets, houses, and a thick enclosure wall, representing part of an extra-mural priestly quarter,[58] and a late Middle Kingdom lower horizon of small rooms [houses?], streets and silos. The latter corresponded to an artisans' quarter, probably close to the Middle Kingdom waterfront, which was identified at the north-east corner of the excavation site.[59] The two urban layouts showed quite distinct alignments; the First Intermediate Period/late Middle Kingdom structures follow a north-south alignment, also found at Kom el-Fakhry, while the New Kingdom quarter respects the Ramesside Ptah temple enclosure aligned west-north-west to east-south-east.[60]

3. Mit Rahina and Kom el-Fakhry

As elsewhere at Memphis much of the ancient topography of Kom el-Fakhry is obscured by urban development or cultivated fields. Kom el-Fakhry lies immediately south of the modern village core of Mit Rahina.[61] The mound originally extended south into Kom Rabia but both mounds are now separated

53 Aston/Jeffreys, 2007, p. 1, Jeffreys, 1985, p. 6–10, p. 28–30, Kemp, 1976, p. 25–27, pl. I, Id., 1977, p. 192–195, fig. 7.

54 The current project uses the site codes attributed by Jeffreys, see Jeffreys, 1985, fig 7–8.

55 Aston/Jeffreys, 2007.

56 Giddy, 2012.

57 Aston/Jeffreys, 2007, p. 6–8, fig. 4–9.

58 Jeffreys, 2006.

59 Giddy, 2012, p. 4–7.

60 For a discussion of building alignments over time see Jeffreys, 1985, p. 65, fig. 15.

61 For sites and monuments in the southwest of the Ptah Temple see Ibid.

by the modern Saqqara-Bedrashein road.[62] Kom el-Fakhry was designated as "Tel el Moukalid" by Hekekyan during his pioneering geoarchaeological work in 1852–1854.[63] The mound is bounded on the east and west by lower ground: the Birka ("pool") on the east, corresponding to the New Kingdom Ptah temple enclosure;[64] and a cultivated plain "Hôd bahr al-qantara"[65] on the west, now almost entirely built up. Here remains of a limestone pavement, possibly an extension of the "Serapeum Way" into the Hellenistic town, were recorded.[66]

The Kom el-Fakhry mound has been reduced substantially since antiquity, in height and volume, due to digging for mudbrick and saltpeter. The ground rises from an elevation of 18 m a.s.l. (saltpeter pits) to 30 m a.s.l. under the Mit Rahina village. The accidental finds and principal excavations, up to 1981, around Mit Rahina are discussed in detail by Jeffreys,[67] and an overview of Middle Kingdom Memphis is provided by Giddy.[68] The cemetery at Kom el-Fahkry (site code FAC) was discovered accidentally during extension of the Saqqara-Bedrashein road and subsequently excavated, in 1954, by Abd el-Tawwabal-Hitta.[69] The adjacent settlement, dated to the Middle Kingdom, was excavated by Ashery in 1981,[70] while to the east, a Cairo University team under the direction of Gaballa Ali Gaballa[71] excavated large granary silos and an industrial area dated to the New Kingdom. The settlement and cemetery at Kom el-Fakhry represent the oldest *in situ* remains excavated to date at Memphis. Old Kingdom sherds were reported just south of Mit Rahina village.[72] Here the SoM recorded mudbrick walls and noted at least 12 m of intact settlement stratigraphy beneath the modern occupation.[73] This may well be the earliest accessible stratigraphy in Memphis, and merits further investigation.

62 Ibid., p. 28

63 Jeffreys, 2010, Id., 1985, p. 28–31, fig 7.

64 Extensively investigated see Id, 1985, p. 33–38.

65 Ibid., fig 4.

66 Ibid., p.47.

67 Ibid., p. 28–31, fig. 7.

68 Giddy, 2012, p. 4–7.

69 Al-Hitta, 1955, Dimick, 1959, p. 83, n. 18.

70 Smith et al., 1983, p. 35, Jeffreys 1985, p. 29, p. 68, fig. 20.

71 Gaballa, 1991.

72 Kemp, 1977, p. 194.

73 Jeffreys, 1985, p. 29–30, fig 21, 25.

3.1 Cemeteries in the Settlement

Two cemeteries have been excavated within the Memphite settlement. At Kom el-Fakhry tombs date to the First Intermediate Period/early Middle Kingdom (site code FAC), while those at Kom Rabia date to the Third Intermediate Period (site code RAC). A tantalising reference to burials was made by Burton: "The mounds appear to have been used as a Necropolis at some time, perhaps some later period ... And mummies and cases have been found in them particularly during the great rains that fell in 1824 when they were laid bare."[74] Jeffreys notes that this is unlikely to refer to either Kom el-Fakhry or Kom Rabia's cemeteries.[75]

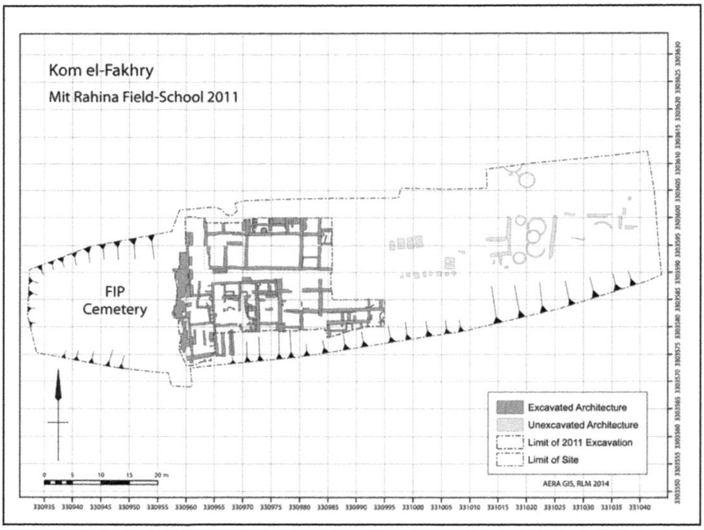

Figure 9. Settlement and cemetery at Kom el-Fakhry. (Prepared by Rebekah Mracle).

At Kom el-Fakhry tombs consisted of rectangular mudbrick chambers, aligned north south and topped by brick vaulting. The chambers were lined with large limestone slabs, which also formed a flat roof.[76] Some of the chambers were decorated. A square pit, also stone-lined, below the floor, served to deposit

74 Burton, Manuscripts British Museum, British Library MS 25618.84A.

75 Jeffreys, 2010, p. 73–74.

76 Id., 1985, p.29, p. 68, fig 19, Lilyquist, 1974, Willems, 1996.

funerary offerings or canopics. The tombs were built adjacent to each other with no apparent streets or access between groups, unlike tombs of the same period at Ehnasya el-Medinah which formed east-west streets.[77] The First Intermediate Period cemetery at Ehnasya also developed within the town. The area of cemetery exposed at Kom el-Fakhry shows a series of up to three tombs north to south, and at least six tombs adjacent to each other east to west. The finished faces of mudbrick walls indicate horizontal development, although eventually the tombs formed a single massive superstructure. There is also some vertical stratigraphy with earlier tombs emerging at a lower level but these may also be double chambered tombs. At Ehnasya individual funerary stela were set up, facing east, outside each tomb group. The Kom el-Fakhry tombs also seem to have had a communal frontage, on the east, where two false door stelae and 15 offering tables were excavated by Al-Hitta.[78] Here small mudbrick walls formed a chapel which held the funerary stelae of Impy-ankh, priestess of Hathor, and a man named Impy. Both stelae are currently in the Memphis open-air museum.

During the 2011 season we were able to clarify the stratigraphic relationship between the town and the cemetery, both at the north and south limits of the site. The central area of the site was previously excavated quite deeply, and therefore the stratigraphic information was removed. The earliest structures at the south are the southern and eastern boundary walls of cemetery. A series of rooms were built in the first half of the Twelfth Dynasty against the cemetery frontage. By the mid Twelfth Dynasty the earlier spaces were decommissioned, and overlaid with a sequence of make-up layers, floors, a bin and a hearth. Most of this building phase was removed by previous excavations, and survives only in small patches.

A second cemetery has been excavated within the Memphite settlement, at Kom Rabia.[79] Cist burials dated to the Twenty-First Dynasty were cut into the south enclosure wall of the Ptah temple. A group of stone roofed tombs, belonging to High Priests of Ptah, was built outside the south-west corner of the temple enclosure. Jeffreys points out that the deterioration of the temple enclosure wall may have freed land for funerary use.[80]

It is probable that the presence of these tombs within the settlement may correspond to a period of reduced urbanism when a lessening of urban pressure

---

77 Perez-Die, 2004.

78 Al-Hitta, 1955, Lilyquist, 1974, Daoud, 2005.

79 Anthes, 1959, see also Jeffreys, 1985, p. 22, p. 70–71, fig. 26–28. For tombs of the Twenty-second Dynasty see Anthes, 1959, p. 3–4, n. 1 and Jeffreys, 1985, p. 22.

80 Jeffreys, 1985, p. 70–71.

allowed for funerary structures to be built within the settlement. During the First Intermediate Period/Middle Kingdom there is a marked decline in the use of Saqqara as a burial ground. The Middle Kingdom is characterized by the development of important provincial cemeteries throughout Egypt. This, as well as the presumed move of the Twelfth Dynasty capital away from Memphis to Itiy-Tawy, contributed to the decline of Saqqara as a national cemetery in this period.[81] At Saqqara a few tombs dated to the First Intermediate Period/Middle Kingdom were built close to the pyramids of Teti, Unas and Merykare.[82] While to the south (South Saqqara to Mazghuna) necropoleis developed around the pyramids of the Twelfth and Thirteenth Dynasties.[83]

As Memphis contracted, at the end of the Old Kingdom, areas of high ground were freed for use as burial ground. River movement and the resulting change in landscape also played a part in this process. As part of the MRFS six auger corings were carried out in 2011 at Kom el-Fakhry. These filled a gap in the extensive Memphite geomorphological sample. Most of the augers reached depths of 11 m below the surface, about 10 m a.s.l., showing considerable depth of occupation as well as deposits indicative of river activity. An uneven sand of fluvial origin, recorded between 9 to 13 m a.s.l. may represent an island, or a bank of a palaeo-river channel, on which the cemetery and settlement were founded.[84] Finally, tomb owners in both the Kom el-Fakhry and the Kom Rabia cemeteries had a connection with the cult of Ptah and therefore proximity to the Ptah temple may have been desirable and/or permissible. It is possible that burials in the central Memphite area were reserved for the priesthood of the nearby temple.

3.2 Kom el-Fakhry Settlement

Part of the Kom el-Fakhry settlement was excavated by Ashery for the Antiquities Organisation in 1981.[85] The SoM reported two broad phases of architecture dated to the Middle Kingdom. Large rooms, some with fine limestone thresholds, were organised either side of an east-west street leading

81 Callender, 2000, Seidlmayer, 2000, Knoblauch 2008.

82 Daoud, 2005,

83 Lehner, 1997, p. 168–187. For further evidence of Twelth Dynasty Memphite monuments see Sourouzian, 1988, p. 229–254.

84 Gonçalves, 2012.

85 Jeffreys 1985, p. 29, p. 68, fig. 20.

to a courtyard with a basin installation.[86] The SoM team also noted that the eastern frontage of the cemetery, and its offering basins and chapels, became inaccessible as the settlement developed.[87] In 2011 the MRFS team recorded the architecture exposed by the previous missions, and excavated part the exposed settlement to gain an understanding of its character and date. The settlement sequence recorded runs from the first half of the Twelfth to the late Thirteenth Dynasty.[88]

The road works in 1954[89] and the excavations of 1981 left the site with a concave north-south profile. Thus later structures were exposed along the northern and southern limits of the excavation but have been removed in the central area of the site. These two areas are connected stratigraphically only through their relationship to earlier structures. The excavation reports of the MRFS 2011 are currently being prepared for publication and will refine the detail of the phases identified by the Survey of Memphis.[90]

Finally, New Kingdom remains are attested in the northern section by a series of walls and ash deposits. Kemp noted, on the north side of the Kom el-Fakhry cemetery, deposits sloping down markedly to the north, representing the destruction of both the cemetery and the settlement.[91] On the southern edge of the cemetery walls dated to the Late Period (Sixth century B.C.) were recorded just 0.5 m under the modern ground level.[92] Beneath these walls Eighteenth Dynasty sherds were noted, leading Jeffreys to conclude that the area had been levelled down in the first millennium B.C.[93]

4. Discussion

Memphis, at a geographically strategic point in Egypt, was subject to a range of landscape processes through time. In the early Holocene, the site was a marshy area at the head of the Egyptian delta. Recently swamped, as sea level rose, the rest of the delta was also marshy and had begun to recover when sea-level rises

86 JEFFFREYS, 1985, p. 29, p. 68, fig. 20 and GIDDY, 2012, p. 4.

87 GIDDY, 2012, p. 4.

88 TAVARES/KAMEL, 2012.

89 JEFFFREYS, 1985, p. 29.

90 ABD EL-AZIZ et al., 2011.

91 KEMP, 1977.

92 JEFFFREYS, 1985, p. 28–29.

93 JEFFFREYS, 1985, p. 29.

slowed around 6,000 years ago. The delta at this time is expected to consist of sand gezireh (or islands) surrounded by marshes, the source of abundant cattle and the opportunity for hunting and fishing. The Nile, fed by the equatorial monsoon and augmented by tributaries in Egypt and Sudan, was high and rich in sediment which, supplied to the floodplain, caused it to rise gradually with the many channels of the Nile focussing into a few channels with distinct levees suitable for habitation.

In the Early Dynastic and Early Old Kingdom settlement moved from the edges of the valley onto the Nile levees, where the earliest deposits at Memphis dating to the Old Kingdom are recovered. Declining rainfall during this period also attenuated the vegetation in the wadis that impinge upon the Nile Valley and the wadis became unstable, washing out into the Nile Valley on a number of occasions, making them less desirable for habitation and muting their topography. The stabilisation of the Nile channels was accompanied by a restructuring of the distributary system into a hierarchical network. The rising of the floodplain reducing marshiness coincides with the Old Kingdom and the earliest known sherds found by the SoM at Mit Rahina and in boreholes by the MRFS. Current excavations have not yet reached these early levels whose geography therefore remains speculative.

As aeolian sand flowed into the Nile valley at the end of the Old Kingdom, we expect rapid floodplain rise and other records from Dahshur[94] suggest that this was indeed the case. The effect at Memphis seems to have been to reduce the size of the settlement (by sediment on-lap) and by the First Intermediate Period at Kom el-Fakhry the settlement had been replaced by use of the site as a cemetery. Eastward Nile migration may also mean that mounds/islands further to the east were favoured for settlement and that the western mound was therefore given over to the dead. The slope of the land to the east coupled with the north-south elongation of the settlement are suggestive of a river levee and borehole analysis of the SoM archive suggests that a channel lay close by to the east.

During the mid-Twelfth Dynasty, the earlier cemetery was decommissioned and the settlement expanded into this area. Although by this period, texts and models of delta development suggest that the palace and the delta head were now further south, possibly at Itiy-Tawi (currently thought to be in the Lisht area),[95] Memphis re-emerged as a regional centre. The active national cemetery was no

94 Bebermeier, 2011, Alexanian et al., 2012, Alexanian, et al., in press a and in press b., Ramisch, 2012.

95 Parcak, BBC1 research project – Egypt's Lost Cities.

longer at Saqqara but the regional centre at Memphis seems to have retained some eminence[96] and grew to be a significant administrative centre by the mid Thirteenth Dynasty at Kom el-Fakhry. During this period the Nile continued to migrate eastwards and new land forming to the east no doubt accommodated some of the expansion of the city.

Delta-head migration northwards again from Lisht to Memphis was completed by the beginning of the New Kingdom and the burgeoning of activity around the Ptah temple, a little to the north-east of Kom el-Fakhry seems to have stimulated a revival, albeit one that was mostly removed during a period of first millennium B.C. levelling. It is likely that the sediment supply required to drive the delta head north also contributed to the consolidation of the delta reducing the area of marsh and increasing the agricultural potential. The construction of the Ptah Temple enclosure wall in the low ground of the Birka suggest that the Birka may have formed from an in-filled channel or bay that was by that time consolidated and dry.

By Roman times, the main waterfront (Hekekyan's Nilometer and Jeffreys' nymphaeum) was now around a kilometre away to the east of Kom el-Fakhry.[97] The river has continued to migrate eastwards since that time to its location near Helwan today. Some of the levelling of Kom el-Fakhry reported from the first millennium B.C. may be related to construction of new developments closer to the waterfront in the east or to the construction of the monumental mound of Kom Tuman for the palace of Apries to the north.

As the delta head continued to migrate northwards, Memphis was supplanted by Babylon and later Cairo. Eventually the ruin mounds were abandoned with the exception of Mit Rahina and satellite villages (Shimbab, Aziziya, etc).

5. Conclusion

The results of the borehole surveys of the Survey of Memphis when combined with observations of the processes of landscape change in Egypt and archaeological excavations suggest a time series of landscapes that have formed part of a dialogue between the city of Memphis and the landscape in which it is set. The pinning of archaeological excavation data to models of landscape change helps to provide a time-scale for the geological processes and a context for the development of the city.

---

96 SOUROUZIAN, 1988.

97 JEFFREYS, 2010.

Bibliography

ABD EL-AZIZ, ASHRAF. et al., Mit-Rahina (Memphis) Field School (MRFS) 2011. Data Structure Report (DSR) for the excavations at Kom el-Fakhry. Report on file Ancient Egypt Research Associates Archive, Boston 2011.

ADBEL HAMID, HUSSAM. Survey study of animal bones from Miet-Rahena, in: Travaux du Centre d'Archéologie Méditerranéenne de L'Académie Polonaise des Sciences, Warsaw 1990.

ALEXANIAN, NICOLE et al., The Pyramid Complexes and the Ancient Landscape of Dahshur/Egypt, in: Landscape Archaeology. Proceedings of the International Conference Held in Berlin, 6th – 8th June 201, ed. by WIEBKE BEBERMEIER et al., eTopoi, Journal for Ancient Studies, Special Volume 3 (2012), p. 131–133. http://journal.topoi.org/index.php/etopoi/article/viewFile/97/126, 28.04.2015.

IDEM., The Discovery of the Lower Causeway of the Bent Pyramid and the reconstruction of the Ancient Landscape at Daschur (Egypt), in: International Colloquium on Landscape Archaeology. Egypt and the Mediterranean World, Cairo, 19th–21st September 2010, Bibliothèque d'études (in press a).

IDEM., The Necropolis of Dahshur, (2002-2010) Excavation Report German Archaeological Institute/Free University of Berlin, Annales du Service des Antiquités de l'Égypte (in press b).

AL-HITTA, MOHAMED, Les grandes découvertes archéologiques de 1954, in: Revue du Caire 33, 175 (1955), p. 50-51.

ANTHES, RUDOLF, Mit Rahineh 1955, Philadelphia 1959.

ASTON, DAVID/JEFFREYS, DAVID. The Survey of Memphis III. The Third Intermediate Period Levels, London 2007.

BEBERMEIER, WIEBKE et al., Analysis of Past and Present Landscapes Surrounding the Necropolis of Daschur in: Die Erde 142, 3 (2011), p. 41-68.

BERENSDEN, HENK, History of geological mapping of the Holocene Rhine-Meuse delta, the Netherlands, in: Netherlands Journal of Geosciences, 86, 3 (2007), p. 165-177.

BERENDSEN, HENK/STOUTHAMER, ESTHER, Paleogeographic Development of the Rhine-Meuse Delta, The Netherlands, Assen 2001.

BORCHARDT, LUDWIG, Das Grabdenkmal des Königs Ne-user-re'. Ausgrabungen der Deutschen Orient-Gesellschaft in Abusir, 1902-1908; Band 1, Leipzig 1907.

BRANTON, THOMAS , The Development of the Memphite Floodplain from borehole data, Cambridge University Unpublished MSci dissertation 2008.

BUNBURY, JUDITH, The Development of the River Nile and the Egyptian Civilization. A Water Historical Perspective with Focus on the First Intermediate Period, in: A History of Water, Series II, vol 2. Rivers and Society. From Early Civilizations to Modern Times, ed. by TERJE TVEDT et al., London 2010, p 52-71.

BUNBURY, JUDITH et al., Stratigraphic landscape analysis. Charting the Holocene movements of the Nile at Karnak through ancient Egyptian time, in: Geoarchaeology 23, 3 (2008), p. 351-373.

BUNBURY, JUDITH/JEFFREYS, DAVID, Real and Literary Landscapes in Ancient Egypt, in: Cambridge Archaeological Journal 21 (2011), p. 65-76.

BUNBURY, JUDITH/LUTLEY, KATHERINE, The Nile on the move, in: Egyptian Archaeology 32 (2008), p. 3-5.

BUTZER, KARL, Early hydraulic civilization in Egypt: A study in cultural ecology, Chicago and London *1976*.

CALLENDER, GAYER, The Middle Kingdom Renaissance (c. 2055-1650 BC), in The Oxford History of Ancient Egypt, ed. IAN SHAW, Oxford 2000, p. 544.

DAOUD, KHALED, Corpus of inscriptions of the Herakleopolitan period from the Memphite necropolis. Translation, commentary, and analyses, Oxford 2005.

DIMICK, JOHN, Descriptive text for the Survey Map of Memphis, in: MitRahineh 1955, ed. by RUDOLF ANTHES Philadelphia 1959, p. 81-83.

DIODORUS SICULUS (Loeb trans. H Oldfather) Cambridge M.A. 1968, Vol I.

DRAKE, NICK/BRISTOW, CHARLIE, Shorelines in the Sahara. Geomorphological evidence for an enhanced monsoon from palaeolake Megachad, in: Holocene 16 (2006), p. 901-911.

DUFTON, DAVID/BRANTON, TOM, Climate Change in Early Egypt, in: Egyptian Archaeology 36 (2009), p. 2-3.

EL-SANUSSI ASHRAF/JONES, MICHAEL, A site of the Maadi Culture near the Giza pyramids, in: Mitteilungen des deutschen archäologischen Instituts, Abt. Kairo 53 (1997), p. 241–53.

FAIRBANKS, RICHARD, A 17,000 year glacio-eustatic sea level record. Influence of glacial melting rates on the Younger Dryas event and deep ocean circulation, in: Nature 342 (1989), p. 637-642.

GABALLA, ALI, Latest excavations in Memphis: progress report, in: Fragments of a shattered visage: the proceedings of an international symposium on Ramesses the Great, ed. by BLEIBERG, EDWARD/FREED, RITA, Memphis 1991, p. 25–27.

GIDDY, LISA, The Survey of Memphis VI. Kom Rabia. The late Middle Kingdom Settlement (Levels VI-VIII), London 2012.

GONÇALVES, PEDRO, Preliminary Report on Auger work. Mit Rahina Field-school 2011. Specialist Report February 2012. Report on file, Ancient Egypt Research Associates, Boston 2012.

GOUDIE, ANDREW/WILKINSON, JOHN, The Warm Desert Environment, Cambridge 1977, p. 88.

HASSAN, FEKRI, A River Runs Through Egypt, in: Geotimes April (2005), http://www.geotimes.org/apr05/feature_NileFloods.html, 28.04.2015.

HILLIER, JOHN et al., Monuments on a migrating Nile, in: Journal of Archaeological Science 34 (2006), p. 1011-1015.

HERODOTUS, Historia, Book II, Leiden 1988.

JEFFREYS, DAVID, The Survey of Memphis I. The archaeological report, London 1985.

ID., Archaeological implications of the moving Nile, in Egyptian Archaeology 32 (2008), p. 6-7.

ID. The Survey of Memphis V. KomRabia. The New Kingdom Settlement (Levels II-V), London 2006.

ID., The survey of Memphis VII. The Hekekyan Papers and other sources for the Survey of Memphis, London 2010.

JEFFREYS, DAVID/TAVARES, ANA, The Historic Landscape of Early Dynastic Memphis, in: Mittelteilungen des Deutschen Archäologischen Instituts, Abt. Kairo 50 (1994), p. 143-173.

JEFFREYS, DAVID/BUNBURY, JUDITH, Fieldwork, 2004-05. Memphis 2004, in: Journal of Egyptian Archaeology 91 (2005), p. 8-12.

KEMP, BARRY, A note on stratigraphy at Memphis, in: Journal of the American Research Center in Egypt 13 (1976), p. 25-28.

ID., The early development of towns in Egypt, in: Antiquity 51 (1977), p. 185-200.

KNOBLAUCH, CRISTIAN, The Memphite area in the late First Intermediate Period and the Middle Kingdom, in: Ancient Memphis: 'Enduring is the Perfection'. Proceedings of the international conference held at Macquarie University, Sydney on August 14-15, ed. by EVANS, LINDA, Sydney 2008, p. 267-278.

KROPELIN, STEFAN et al., Climate-Driven Ecosystem Succession in the Sahara. The Past 6000 Years, in: Science 320 (2008), p. 765-768.

KUPER, RUDOLPH/KROPELIN, STEFAN, Climate-Controlled Holocene Occupation in the Sahara: Motor of Africa's Evolution, in: Science 313 (2006), p. 803-807.

LEHNER, MARK, The complete pyramids, London 1997.

LILYQUIST, CHRISTINE, Early Middle Kingdom Tombs at Mitrahina, in: Journal of the American Research Center in Egypt 11 (1974), p. 27-30.

MOELLER, NADINE, The First Intermediate Period. A Time of Famine and Climate Change?, in: Agypten und Levante 15 (2005), p. 153-167.

PARCAK, SARAH, Egypt's Lost Cities, 2011, http://www.bbc.co.uk/programmes/b011pwms, 30.04.2011.

PENNINGTON, B.T. / BUNBURY, J.M. / HOVIUS, N., Emergence of Civilisation, Changes in Fluvio-Deltaic Styles and Nutrient Redistribution Forced by Holocene Sea-Level Rise, in: Geoarchaeology 31(3) (2016), p. 194-210.

PEREZ-DIE, MARIA, The ancient necropolis at Ehnasya el-Medina, in: Egyptian Archaeology 24 (2004), p. 21-24.

PRYER, LAURENCE, The Landscape of the Egyptian Middle Kingdom Capital Itj-Tawy, Unpublished MSci. dissertation, Department of Earth Sciences, University of Cambridge, Cambridge 2012.

QIN, YING, Landscape change in the Saqqara/Memphis area of Egypt from 3000 BC to the present. Unpublished MSci.dissertation, Earth Sciences, University of Cambridge, Cambridge 2009.

RAMISCH, ARNE et al., Fractals in topography. Application to geoarchaeological studies in the surroundings of the necropolis of Dahshur, Egypt, in: Quaternary International 266 (2012), p. 34-46.

RODRIGUES, DONALD et al., Seasonality in the early Holocene Climate of Northwest Sudan: interpretation of Etheriaelliptica shell isotopic data, in: Global and Planetary Change 26 (2000), p. 181-187.

ROHDE, ROBERT, Holocene Temperature Variations, in: Global Warming Art (2006), http://commons.wikimedia.org/wiki/File: Holocene_Temperature_Variations.png, 28.04.2015.

ROWLAND, JOANNE, Minshat Abu Omar, in: The Encyclopedia of Ancient History, 2012, http://onlinelibrary.wiley.com/doi/10.1002/9781444338386.wbeah15282/pdf, 28.04.2015.

SEIDLMAYER, STEPHAN, The First Intermediate Period (c. 2160-2055 BC), in: The Oxford History of Ancient Egypt, ed. by IAN SHAW, Oxford 2000, p. 544.

SMITH, HARRY et al., The Survey of Memphis 1981, in: Journal of Egyptian Archaeology 69 (1983), p. 30–42.

SOUROUZIAN, HOURIG, Standing royal colossi of the Middle Kingdom reused by Ramesses II, in: Mitteilungen des deutschen archäologischen Instituts, Abt. Kairo 44 (1988), p. 229-254.

STANLEY, DANIEL/WARNE, ANDREW, Recent Geological Evolution and Human Impact, in: Science 260 (1993), p. 628-634.

IDEM., Worldwide Initiation of Holocene Marine Deltas by Deceleration of Sea-Level Rise, in: Science 265 (1994), p. 228-231.

IDEM., Nile Delta in its Destructive Phase, in: Journal of Coastal Research 14 (1998), p. 794-825.

TAVARES, ANA/KAMEL, MOHSEN, Memphis. A City Unseen, in: AERAGRAM: Newsletter of Ancient Egypt Research Associates 13, 1 (2012), p. 2-7.

TRISTANT, YANN. L'Habitat Pre-Dynastique de la Vallee du Nil: Vivre sur le rives du Nil aux V et IV Millenaires, Oxford 2004.

WILLEMS, HARCO, A note on the date of the early Middle Kingdom cemetery at Ihnâsiya al-Medîna, in: Göttinger Miszellen 150 (1996), p. 99–109.

WILSON, PENELOPE, Survey of Sais (Sa El-Hagar) 1997-2002, London 2006.

WOODWARD, JAMIE, et al., The Holocene fluvial sedimentary record and alluvial geoarchaeology in the Nile Valley of Northern Sudan, in: River Basin Sediment Systems. Archives of Environmental Change, ed. by DARREL MADDY et al., Rotterdam 2001, p. 327-356.

IDEM., The Nile Evolution, Quarternary River Environments and Material Fluxes, in: Large Rivers, Geomorphology and Management, ed. by AVIJIT GUPTA, Chichester 2007, p. 261-291.

Karnak's Quaysides

Evolution of the Embankments from the Eighteenth Dynasty to the Graeco-Roman Period

MANSOUR BORAIK, LUC GABOLDE, ANGUS GRAHAM

1. Introduction

The results presented by Luc Gabolde and Angus Graham at the symposium held at Mainz in March 2013[1] have in part already been published or are in print.[2] The authors proposed to the editors – who were very kind to accept it – a re-orientation of their contribution to the proceedings focused on the recent results gained through archaeology, history, geoarchaeology and geophysical survey on the evolution of the Nile embankments/quaysides at Karnak from the Eighteenth Dynasty onwards. This has led to the inclusion of the recent and fruitful research carried out by Mansour Boraik on the western and southern parts of the site.[3]

The idea that quaysides, embankments and river banks around the temples of Karnak were located differently than they are today can be traced back to

1 The communications were entitled as follows: A. Graham, *"The Origins of Karnak – Geoarchaeological and Geophysical Survey Results"*; L. Gabolde, *"The Origins of Karnak – Archaeological, Astronomical, Textual and Theological Sources"*.

2 GABOLDE, *in press* and an overview which has been presented in GABOLDE, 2014, p. 13–35; BUNBURY et al., 2008; GRAHAM, 2010a; GRAHAM, 2010b; GRAHAM/BUNBURY, 2005.

3 BORAIK, 2013a; BORAIK, 2013b; BORAIK, 2010b; BORAIK et al., 2010.

the work of Legrain and Pillet.[4] They investigated short sections of buried embankments and proposed some hypotheses on former positions of the river.

The evidence of important movements of the Nile during the New Kingdom until the Roman period can be ascertained by a number of different kinds of sources which are reviewed and analysed in this paper enabling us to propose some new hypotheses as to the historical evolution of the embankments in front of the temple.

2. The textual and iconographical sources

Textual sources reveal that the Egyptians had a wide empirical knowledge of the dynamics of their river as their wisdom literature attests.

The *Teaching of Ani*, most likely composed in the Nineteenth Dynasty,[5] mentions the unpredictable drainage of a channel:

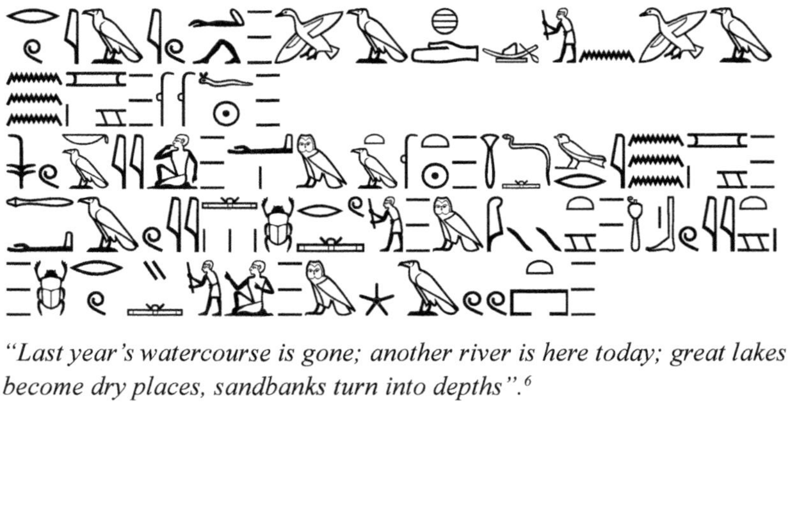

"Last year's watercourse is gone; another river is here today; great lakes become dry places, sandbanks turn into depths".[6]

4 LEGRAIN, 1906a; LEGRAIN, 1906b; PILLET, 1924, p. 84–86. See also CABROL, 2001, p. 427–430.

5 QUACK, 1994, p. 61–62; LICHTHEIM, 1976, p. 135, suggested Eighteenth Dynasty.

6 *Teaching of Ani*, LICHTHEIM, 1976, p. 142; QUACK, 1994, p. 110–111, 318–319.

The *Teaching of Merikara*, known to us from late Eighteenth Dynasty papyri was most probably composed in the early Middle Kingdom.[7] It recalls the natural creation of new channels and the inevitability of the phenomenon:

"like a channel (is) being replaced by (another) channel, there is no river that suffers to be constrained: it (the river) is the releaser of the water-branch (lit. 'arm') it had hidden himself inside".[8]

Several literary texts make clear allusions to islands located off Karnak and Luxor while others mention an island of Amun connected to the god, although the locations of these islands are less precise.

A Ramesside hymn, P. Turin 54031, evokes the string of islands stretched between Karnak and Luxor:

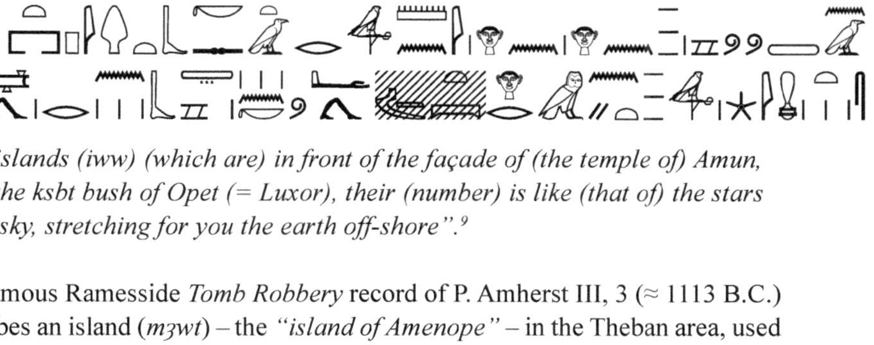

"The islands (iww) (which are) in front of the façade of (the temple of) Amun, up to the ksbt bush of Opet (= Luxor), their (number) is like (that of) the stars in the sky, stretching for you the earth off-shore".[9]

The famous Ramesside *Tomb Robbery* record of P. Amherst III, 3 (≈ 1113 B.C.) describes an island (*mȝwt*) – the *"island of Amenope"* – in the Theban area, used

7 QUACK, 1992, p. 114–136 favours an early Twelfth Dynasty date.

8 *Teaching of Merykara*, translation based on QUACK, 1992, p. 74–75, 193–194; GARDINER, 1914, p. 33–34 translated: *"As the mud-flat (?) is replaced by a flood. There is no river that suffers itself to be concealed; but it loosens the dam (?) by (?) which it lay hid".*

9 P.Turin 54031, V, 20, I, 8; cf. CONDON, 1978, p. 14, 22. The translation given here is L. Gabolde's.

as a shelter by the robbers, where they stopped in order to divide quietly the booty of their theft.[10]

The ostracon O. Uppsala 608, dated to 115 B.C., deals with the collection of taxes on the territory of eastern Thebes *"Taxes on the harvest and (on the) extra (quantity payable) by the farmers for the Island of Amun named 'the southern district'"*, while the late Ptolemaic O. Leipzig 2200 mentions oil mills and cultivations of oleaginous seeds in the Theban area located by reference to *"the Island of Aberhu"*, to the *"sand (the silting up ?) of the canal (?) Hel-mu"*, and the *"island of Amran"*.[11]

The *"islands of Amun"* connected with the cultivation of oleaginous seeds mentioned in papyri originating from Gebelein are not precisely located.[12] Whatever their location was, within or outside the territory of Thebes, their name, often elaborated as *"the Island of Amun ... "* shows that these new lands were generally attributed to divine properties and especially to the estate of Amun, as Yoyotte has already noted.[13]

Nine different *"islands of Amun"* are mentioned on the recto of P. Amiens and P. Baldwin (BM EA 10061) dated to the late Ramesside period (≈ 1150-1070 B.C.).[14] Whilst all of the islands appear to be located in the Ninth or Tenth nome of Upper Egypt, Janssen's commentary on the text raises the important point that the terms *iw* and *m3wt* are difficult to differentiate.[15] Furthermore, Gardiner[16] makes a very interesting point concerning the difficulty of deciding whether an expression such as *t3 m3wt* is part of the composite place name or whether it is simply a descriptive term. In the case of "the New land of Samĕ" (P. Wilbour (B 10, 5)), Gardiner[17] argues that there is no certainty that the "new

10 Peet, 1997, p. 49 (3,3), 61 (r° 1,6), 152 (10, 4–5): *"the Island (?) of Amenemope"* and p. 162 comment on *mȝwt.*

11 Kaplony-Heckel, 2010, p. 132, n° 55 and 142–143, n° 68.

12 Kaplony-Heckel, 2009, p. 577–579 notes that the designation *"islands of Amun"* does not make reference to deities of Gebelein but to Amun of Thebes.

13 Yoyotte, 2013.

14 Janssen, 2004, p. 4–5, 32; Gardiner, 1940, p. 95; Gardiner, 1941, p. 37–43.

15 Janssen, 2004, p. 19 n. 5, 33). See Graham, 2007, p. 299 for some initial comments on differentiation of the terms and also Eyre, 1994, p. 75–76; Schenkel, 1978, p. 60–65.

16 Gardiner, 1948, p. 29.

17 Ibid.

land" was actually new at the time of Ramesses V as it appears that this was an established local name. We can see today that toponyms on maps can also record a 'memory' of past geomorphology of the floodplain.[18]

The walls of the "Chapelle Rouge" of Hatshepsut (≈ 1450 B.C.) present descriptions of the embankments and of the procession roads used during the *Opet* and the *Valley festivals*.

The portable bark of Amun is installed onboard the great riverine bark of Amun departing for Deir al-Bahari for the *beautiful Valley festival*, at the embankment, described as a *tp-itrw* (*"above the river"*; *"overhanging the river"*, an expression which may define both the position of a boat, like the *User-hat*, as that of a quay):

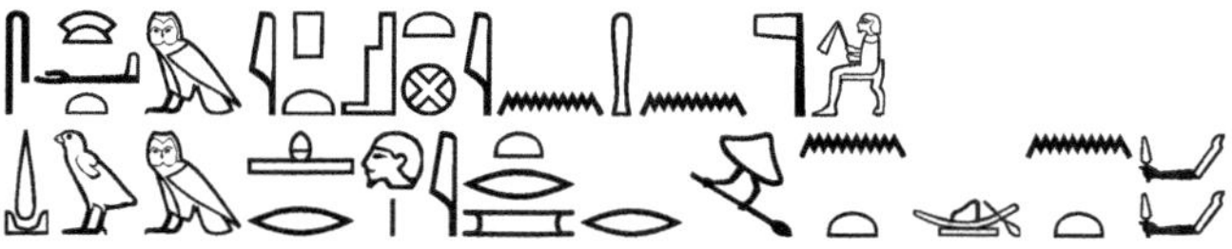

"Appearance in procession out of Karnak by the Majesty of the august god. Going in peace at the (place) above the river for the navigation of Deir al-Bahari".[19]

Cabrol argues that *tp-itrw* refers to the basin at the end of the canal, but that it does not have to have a "tribune / platform" associated with it.[20]

The landing place is also described in the scene of the arrival of the great riverine bark returning to Karnak from Luxor:

"Landing in peace at the tribune (?) of Karnak".[21]

However, the exact place of this tribune cannot be precisely determined from the texts and representations.[22]

18 BUNBURY et al., 2008, p. 356; GRAHAM, 2010a, p. 133; JEFFREYS, 1985; SMITH/JEFFREYS, 1986.

19 LACAU/CHEVRIER, 1977, p. 170, § 227; LARCHÉ/BURGOS, 2006, p. 98.

20 CABROL, 2001, p. 631-634.

21 LACAU/CHEVRIER, 1977, p. 185, § 263; LARCHÉ/BURGOS, 2006, p. 61.

22 LACAU/CHEVRIER, 1977, p. 186.

On the earthly way from Karnak to Luxor, the procession stops before the Mut temple in a bark station named

"the first station (named) 'the stairs / platform of Amun in front (?) of the house of the chest'".[23]

The remains of this chapel have been identified by Ricke in front of the temple of Mut and facing the *Kamutef-Heiligtum.*[24] Interestingly, a gate of Hatshepsut was found in the vicinity, inside the precinct of Mut[25] and it most probably opened directly on the processional road to Luxor.

A text of Montuemhat (≈ 660 B.C.) in his crypt at the Mut Temple seems to describe the building of an embankment in order to prevent the flood from threatening the temple of Khonsu: [26]

[22]

"(... I ordered an embankment to be built for the god Khon)su (?) in beautiful light sandstone in order to repel from (him) the flood of the River when it comes".[27]

Dated between the reign of Ptolemy V Epiphanes (reign 204-181 B.C.) and that of Ptolemy IX Soter II (116-110 B.C., 109-107 B.C. and 88-81 B.C.),[28] the archive of Hermias' trial has revealed that, west of the dromos of Khonsu and of the forecourt of the Opet Temple, a north-south road existed named the *"royal*

23 Ibid., p. 161, § 207; 163, § 214; LARCHÉ/BURGOS, 2006, p. 51–52.

24 RICKE, 1954, p. 42 and pl. 13d.

25 FAZZINI, 2001, p. 61–74.

26 LECLANT, 1961, p. 215 l.22, 219 and 227, n. bj.

27 The restitution « *[Khon]sou* » has been suggested by WRESZINSKI, 1910, pl. III, col. 393, [22], but, without a divine determinative, it remains questionable. One may also understand *"[... I decided to built an embankment, and I made] it in beautiful light sandstone ..."*

28 See BRUGSCH/REVILLOUT, 1880; PESTMAN, 1993, p. 385–409.

street". This street ran parallel to the eastern quay of a branch of the Nile, called the *"canal of the island of Amun"*. The quay thus faced an island called *"the island of Amun"* named *"Tamaut"*.[29]

A passage of the book of Nahum in the Old Testament (*Nahum* 3.8) deserves to be quoted here as it seems to provide us with a description of Thebes before 663 B.C., lying among branches of the Nile: *"Are you better than Thebes, situated on the Nile, with water around her? The river was her defense, the waters her wall"*. This passage led Egli to reconstruct canals around the city, with the main branch of the Nile flowing to the east but without any basis of geomorphological investigation (see figure 1).[30]

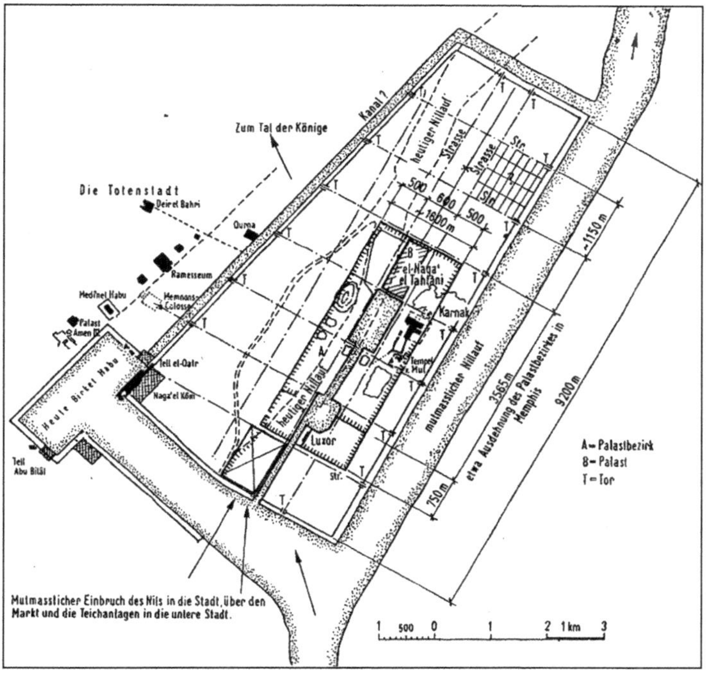

Figure 1. Reconstruction of the Nile by EGLI, 1959 based on Nahum 3.8.

29 On *tȝ-mȝwt, "New land"*, see YOYOTTE, 2013, p. 231–237; BONNEAU, 1971, p. 70 n. 311, 115 and 193; MEEKS, 1972, p. 56, n. 18; GASSE, 1988, p. 148 et n. 4; EYRE, 1994, p. 75–76; GRAHAM, 2007, p. 299.

30 EGLI, 1959, p. 40–43.

Three iconographical sources appear to attest to canals and lakes associated with Karnak. Issues of perspective (i.e. images showing *primae facie* plan and elevations) of Egyptian pictorial evidence must be borne in mind at all times when interpreting images.[31]

The Theban tomb of Neferhotep (TT 49 dating to the late Eighteenth Dynasty) shows a widely known representation of the layout of the river Nile connected to a basin by a canal in front of Karnak[32] (see figure 2). The pylon at the end of the processional way linking the tribune has been argued to be the third pylon constructed by Amenhotep III.[33] However, the precise position and even the existence of the processional way, the tribune and the basin with its canal linked to the Nile all remain to be determined. One of the most informative features of the picture is the presence of an island surrounded by navigable channels in front of the entrance of the canal linking the river to the tribune of the dromos.

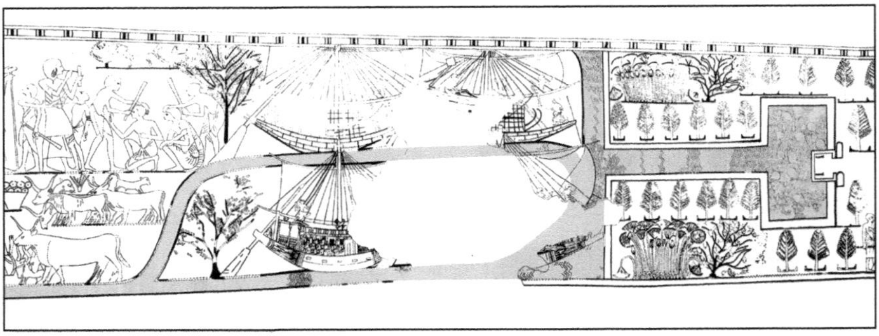

Figure 2. TT 49 (Neferhotep): basin and tribune of Karnak with an island in front of the canal (after Davies, waterways shaded in grey. NB not all the banks are extant in the image and are approximations).

The representation of Khâbekhenet's tomb, dated to the reign of Ramses II[34] (≈ 1250 B.C.), adds some information concerning the dromos in the area of the Mut temple and the environment of the *isheru* lake, though the specific point of

31 SCHLÜTER, 2009, p. 143–161.

32 TT 49. PORTER/MOSS, 1960, p. 93, [15–16]; DAVIES, 1933, I p. 28–32 and pl. XLI–XLII and II, pl. III; CABROL, 2001, p. 433–436 and CABROL/TRAUNECKER, 1993, p. 19–25.

33 CABROL/TRAUNECKER, 1993, p. 24 and n. 41; cf. LOEBEN, 1992.

34 TT 2 Khâbekhenet see CABROL, 1995.

view of the artist did not result in representing the Nile embankment itself (see figure 3).

a. Schéma exécuté d'après photos et complété après la visite de la tombe de Khâbekhenet.

Figure 3. Representation of the tomb of Khâbekhenet after C*ABROL*, *1995, pl. V a.*

A block formerly attributed to Piânkhy (but more likely belonging to the reign of Psamtek I) (≈ 660 B.C.) and discovered in the Mut Precinct[35] shows an interesting feature (see figure 4): a tree is represented between the prow of the riverine bark of Amun and the tribune. Benson and Gourlay suggested the tree represented "those growing on fertile soil beside the water".[36] In effect the block could represent an elevation view of a plan similar to that depicted in Neferhotep's tomb, with the trees set around the basin and canal. Alternatively, it could represent the presence of some kind of dry sandy bar between the tribune and the boat.

35 PORTER/MOSS, 1972, p. 257–258 (8); BENSON/GOURLAY, 1899, pl. XXII 5; FOUCART, 1924, pl. IX b and p. 118–119 (who wrongly sees a representation of Heliopolis); LECLANT, 1965, p. 114–115 [32, B]. On the precise date of the blocks, see PERDU, 2010 and PERDU, 2011, where a more convincing attribution of the blocks to the reign of Psamtek I is proposed.

36 BENSON/GOURLAY, 1899, p. 258.

Figure 4. Block from the temple of Mut showing a tree between the bark of Amun and the tribune (after BENSON/ GOURLAY, *1899, pl. XXIIb).*

In summary we can see that a number of Ramesside texts attest to islands in the vicinity of Karnak. This is no surprise given the propensity of island formation in the Nile.[37] The pictorial evidence provides crucial information on the tribunes and basin as formal access to the temple complex, but the location of the basin remains unresolved.

3. The archaeological sources

Complementing the epigraphical and iconographical sources, excavations in several areas of Karnak have brought to light various features connected with embankments and dated to various periods.

37 GRAHAM, 2010a.

3.1 The excavations of the Nineteenth and Twentieth centuries

Extensive excavations were carried out on the western dromos of Karnak and its tribune (key plan see figure 23), under the supervision of Lauffray.[38] The extensive reports of this scholar are not always clear and often remain inconclusive, but the subsequent comments of A. Cabrol have greatly improved our understanding of these excavations.[39] The dromos seems to have been re-organized after the reign of Pinedjem I (≈ 1060 B.C.) of the Twenty-first Dynasty, whereas the tribune is apparently later than the reign of Sheshonq I (943-923 B.C.) of the Twenty-second Dynasty. A phase of restoration seems to have occurred between the Twenty-third and the Twenty-fifth Dynasties.

Mudbrick wall structures within the sandstone tribune must date to the Twentieth Dynasty as they covered – and thus postdated – layers which seem to date to the period between the end of the Eighteenth Dynasty and the Twentieth Dynasty, based on the blue painted ceramics they contained (layers d, d' and d'').[40] It is not clear if these mudbrick structures represent part of the construction of the tribune or relate to an earlier structure or embankment. Considering the notion that the basin and canal depicted in the tomb of Neferhotep were on the central axis of the temple, we cannot yet rule out the possibility that these mudbrick walls could have existed on the east bank of an island lying directly in front of the Amun-Re complex.

Between 72.40 m and 73.50 m a.s.l., few occupation levels were identified, some covered with lime, with abundant ashes, (undated) potsherds and chips of sandstone, but no trace of any architectural structure. It seems likely that this stratigraphy represents the dumps from workshops and a temporary occupation of craftsmen from the beginning to the end of the Eighteenth Dynasty. An older layer was identified below, around the levels 72.20–72.40 m a.s.l., consisting of ashes and incorporating conical bread moulds. In the absence of drawings and photos, it is alas not possible to evaluate their date. However, its composition – the ashes and the bread moulds – evoke a temple's bakery, and its low level, compared to similar levels of the central part of the temple[41] or from the forecourt

38 Lauffray, 1971; Lauffray et al., 1975, p. 43–76.

39 Cabrol, 2001, p. 117–136, 581–590.

40 Caution is, however, necessary in interpreting Lauffray's dating as the whole assemblage of finds from each stratigraphic level has not been fully published.

41 Lanoë, 2007, p. 374–375, sp. pl. V–VII.

of Opet,[42] suggests that they could correspond to some kind of workshops from the Middle Kingdom or Second Intermediate Period. However, the possibility of a dump from a nearby bakery located at a higher level cannot be ruled out.

3.2 Tribune

In the 1890s, George Legrain excavated to a level ten courses of ashlars below the protruding pavement course of the tribune, a height of c. 71 m a.s.l.,[43] in order to record the Nile levels on the face of the tribune (see figure 5). The notion that the tribune functioned as a quay was pointedly refuted by Clarke[44] stating that these structures do not meet the criteria for being quays, as they do not accommodate the rise and fall of the Nile and the parapets around the platform make loading/unloading and embarkation/disembarkation very problematic.

Figure 5. Georges Legrain recording inscriptions on the western tribune (Image no. 40405_1 © CFEETK).

3.3 North Quayside Wall

Chevrier[45] found a wall linked to the north side of the platform extending northwards in line with the west face of the platform (key plan see figure 23).

42 Charloux, 2010, p. 202, fig. 7; Charloux, et al., 2012, p. 40–43.

43 Lauffray, 1971, p. 85.

44 Clarke, 1921, p. 70 n. 1.

45 Chevrier, 1947, p. 157-158.

He believed this wall was used as a quay with a set of stairs as well as notches or loops at two different heights in the stone that allowed barks to be tied up (see see figure 6).[46] Of the five loops, three are cut in the fourth and fifth courses of the wall and two on the eighth and ninth courses of the wall (see see figure 6-7). The upper part of the loophole appears to have wear grooves where the ropes were tied through the hole (see figure 7 close-up).

An important aspect of these notches not previously raised is that the lower ones extend further north along the wall than the higher ones (see figure 6). Whilst not precisely at the same angle as the steps to the north, they are almost certainly designed so that the bow or stern was tied up to the mooring loop to enable (dis)embarkation amidships by way of the steps. The loops would allow vessels to moor at varying heights of the annual river cycle. The fact that there are mooring loops at the same height along the wall might suggest that they were designed to either accommodate vessels of varying lengths or they may have served to enable rafting up to occur. They could also have been used for spring lines. Fenders would have been used between the quay wall and the hull.

Figure 6 (left). North quay wall at Karnak showing the steps and two mooring loops (after CHEVRIER, *1947, pl. 48); Figure 7 (right). North-west corner of the western tribune with a mooring stone in the fifth course of the north wall (after 39685 ©CFEETK).*

46 Ibid., p. 158, pl. 48.

The lower loops are at c. 71.45 m a.s.l. whilst the upper loops are at c. 72.85 m a.s.l.[47] This would allow mooring at water levels certainly lower than 69 m a.s.l., assuming the crew member used full stretch to pass the rope through the loop and with some freeboard. When the river was at its peak, a level marked by the many inscriptions on the facade of the tribune (see figure 8), the boat would have been able to tie up to a mooring post or stone mooring loop located on the top of the north wall, which is at c. 74.5 m a.s.l. The freeboard of the ship would have enabled the passengers and crew to step on and off the boat with relative ease onto the top of the wall when the inundation lay in the region of 73.5–74 m a.s.l. (see see figure 8). Below this they would have used the stairs. In year 6 of Taharqa's reign with the inundation reaching 74.4 m a.s.l. the passengers would have needed a gangplank to descend from the boat, but even this highest of all marked floods would have left the top of the wall largely dry except for the lapping over of wind and boat-generated waves. In brief, this quayside was very well designed to enable boats to tie up and passengers to embark and disembark via the staircase.

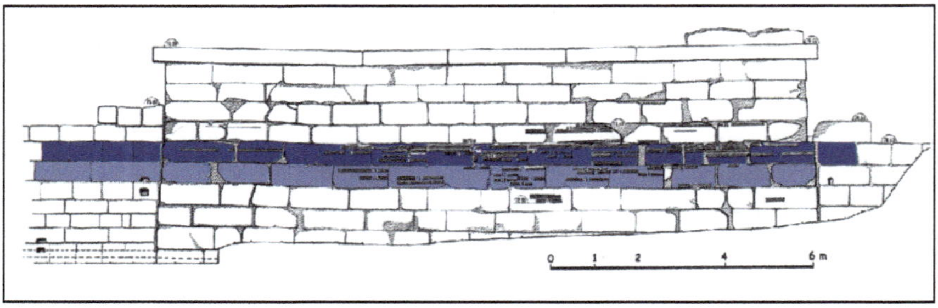

Figure 8. Facade of the western tribune at Karnak with inundation heights; the royal blue course of stones mark the flood heights between c. 73.5-74 m and the grey blue course of stones those between c. 73-73.5 m (after Lauffray, *1971, fig. 6bis).*

3.4 The recent excavations of the SCA (2007-2014)

Over seven years, the Supreme Council of Antiquities surveyed and excavated the area in front of the First Pylon of Karnak Temple. The excavations extended both north and south of the tribune to expose an embankment/quay wall at least 6.5 m high, running roughly perpendicular to the axis of the Amun-Re temple

47 Lauffray, 1971, fig. 6bis.

(key plan see figure 23).[48] The wall was built of roughly cut blocks of sandstone. The soundings have uncovered over 500 m of its length from north to south.

3.5 Quays north of the Tribune

The extension of excavations north of the tribune (key plan see figure 23) have revealed a second flight of steps integrated into the wall coming down from north to south and facing the steps earlier exposed by Chevrier. Both flights are 1 m wide and 20 steps were revealed in both descending 4.5 m.[49] Augering carried out by Matthieu Ghilardi (CNRS/CFEETK) suggested a platform might exist between the two sets of steps at ~ 69.2 m a.s.l.[50] If a platform does exist, given the freeboard of the large ceremonial boats the king and priests could embark and disembark with ease at levels between the highest inundation down to ~ 68.7 m a.s.l. A stepped gangplank could have assisted (dis)embarkation if the low Nile were even lower.

Different extensions of the embankment were added after the Kushite Dynasty with the reuse of blocks inscribed with the name of a divine votaress Amenirdis from the Twenty-fifth Dynasty being used in the construction. It remained in use until the end of the dynastic period when the Ptolemaic baths were constructed on top of the wall (see below).

3.6 Baths built after Ptolemy VIII and the Roman thermae

Excavations have revealed that the area to the west of the embankment was widely used during the Ptolemaic-Roman period (key plan see figure 23). Different phases of occupation were uncovered, including houses, workshops, and baths.[51] The occupation levels to the west of the embankment continued to a depth of only 1 m next to the wall and were superimposed on different layers of Nile silt. Occupation also extended north-south along the western facade of the First Pylon. It seems most likely that the Ptolemaic buildings were burnt in a large fire at the beginning of the Roman period. On the top level many

48 Boraik, 2010b; Boraik et al., 2013.

49 Boraik, 2010b, p. 72.

50 Augerings HA7 and HA8 were terminated as they hit upon sandstone fragments at 69.22 m and 69.37 m a.m.s.l. respectively (Boraik/Ghilardi, 2011, p. 3213, fig. 8). The 0.15 m difference in height could also suggest that rubble may have been encountered rather than a platform.

51 Boraik, 2010b, p. 73–75; Boraik, *et al.* 2013.

houses were found with their wooden roof beams burned and almost everything in situ, including pottery, grinding stones, terracotta figurines, oil lamps, tools of daily life, and coins. Among the important discoveries from the Ptolemaic-Roman period – houses, workshops, and industrial activity buildings – three bath complexes were unearthed. Two of them date to the Ptolemaic period and one to the Roman period. The three baths are located to the north of the tribune of Karnak Temple.

3.7 The Ptolemaic Public Bath

The Ptolemaic public bath was found 75 m to the north of the tribune ~ 1.50 m below the modern surface (see figure 9 and key plan see figure 23). Under topsoil at an elevation of approximately 77.20 m a.s.l., a very rich layer of pottery sherds and ceramic debris was encountered, probably the backfill from older excavations. It covered traces of masonry consisting of foundations of fired brick and reused blocks of the dynastic period. It is thought that they belong to a Roman or medieval domestic area. At a lower elevation, 76.64 m a.s.l., in the eastern part of the sounding, a large mudbrick wall was unearthed. It was apparently oriented north-south and associated with other brick masonry visible in the eastern and western baulks. Against one of these mudbrick construction blocks and at a level near the foundation at an elevation of 75.46 m a.s.l., a hoard of 316 coins was discovered. This enabled these mudbrick structures to be dated to the second half of the second century B.C. (Ptolemy VIII Evergetes II ≈ 145 to 117 B.C.). These features most likely sealed the destruction level of the bath.[52] The construction of this massive feature undoubtedly caused the systematic levelling and razing of the older bath, which was only preserved to a maximum height of approximately 0.70 m. The bath itself had been built directly over the uppermost course of the embankment.

The building of the Ptolemaic baths on top of and to the west of the quayside wall points to a westward migration of the river. Furthermore, the lowest point of the furnace serving the baths is measured at 72.46 m a.s.l.[53] Assuming it was constructed to stay dry all year round it provides a proxy for the flood levels of the Ptolemaic period suggesting that the highest floods of the time were somewhat lower than those recorded on the Western tribune during the Third Intermediate Period. This concurs with the large-scale interpretations of two

52 Boraik, 2010b, p. 73; Boraik/Faucher, 2010.

53 Boraik et al., 2013, p. 49 n. 11

studies, but contradicts Seidlmayer's conclusions.[54] The strontium isotope ratios from a single sediment core taken in the Manzala Lagoon as a proxy for Nile discharge suggest a fall in flood height during the Late Period.[55] Furthermore, a recent meta-analysis of radiocarbon and Optically Stimulated Luminescence (OSL) dated Holocene fluvial units in the Nile catchment suggests that a large scale hydroclimatic shift in the Nile catchment occurred in c.450 B.C. The study argues that after 450 B.C. water levels did not exceed those recorded before 500 B.C.[56]

Figure 9. The bath built after Ptolemy VIII.

The bath was built on a rectangular area, measuring 13 m by 18 m. The building is characterized by circular rooms (*tholoi*).

The floor of the southern room consists of two concentric decorative zones. The inner circle is 2.42 m in diameter and comprises mosaic components of small multi-coloured pebbles (white, red, black, and brown) set in red mortar. It is separated from the bath seats by a second zone which has a white band made of mosaic pieces of white stone flakes. The mosaic floor of the northern *tholos* is made of white stone flakes also set in red mortar and decorated with figures of dolphins and tilapia fish in sequence and a rosette in the centre (see figure 9). Three figures are made of small, coloured pebbles and outlined with lead strips

54 SEIDLMAYER, 2001, p. 72.

55 STANLEY et al., 2003, fig. 2.

56 MACKLIN et al., 2015.

that emphasize their shape. From this we can see that the bathing establishment at Karnak stands out for the luxury of its mosaics and wall paintings.

The bath was probably built between the end of the third century and the first decades of the second century B.C. and was therefore only in use for a very short period. It is to date, the southernmost thermal installation known in Egypt and apparently among the oldest archaeologically attested ones.[57]

3.8 The Great Roman Bath

As the excavation extended northwards from the private Ptolemaic bath at the nearby site called El-Hassasna (key plan see figure 23), a Roman bath came to light. These Roman *thermae* are made of fired brick with a floor of large sandstone blocks. It covers some 3000 m^2 with many well-preserved architectural features, such as the bathing pools. Most superstructure walls stand less than a metre high but some wall elements in the substructure are three metres tall. The *thermae* were remodelled and redecorated over what appears to be a long period of use. The Roman bath complex is partially built over late Ptolemaic buildings with the Ptolemaic settlement extending to the north. The *thermae* were probably built during the third century A.D. and were subsequently used for a long period of time.

3.9 Geophysics across North quayside wall

Three short Ground Penetrating Radar (GPR) profiles (G030, G031 and G032 respectively located c. 184 m, 190 m and 205 m northwards from the centre of the tribune) were carried out in February 2013 using a GSSI 200 Mhz antenna in the northern part of the excavated area of the Roman baths (see below). They were conducted in order to see to see if it was possible to identify the continuation of the quayside wall below the excavations (key plan see figure 23).

All of the profiles demonstrated considerable 'ringing' in the data, most probably caused by increased salt content in the soil derived from ground water to the west of Karnak.[58] In spite of this, a number of possible features were noted in the data. These include strong reflections in G030 approximately 14–16 m along the profile to a depth of 2.5–3.0 m, marking the possible location of the top of the quayside wall in the northern area of the MSA excavations of the Roman baths. This feature also appears to be represented in profile G031 as a

57 BOUCHAUD/REDON, in press.

58 CONYERS, 2004, p. 50; DOOLITTLE/COLLINS, p. 2004, 99.

faint reflection in the data some 18-20 m along the profile, and in profile G032 some 2–4 m along the profile, starting at a depth of 1.5–2.0 m below the surface of the ground. This work suggests the wall extends over 200 m northwards of the tribune.

An Electrical Resistivity Tomography (ERT) profile (P 14, see figure 23) across the line of embankment wall between the Ptolemaic and Roman baths excavations was carried out in February 2012 with the probes at 4 m spacing providing a vertical resolution of c. 2 m in the resistivity readings. The profile crosses the quayside wall at c. 68 m along the profile length and higher resistance readings are found down to about 12–13 m below the surface or 66–65 m a.s.l. This is a little lower than the level of the platform suggested by Ghilardi's augering between the facing staircases.

3.10 Ramps

Two of the three lateral ramps to the south of the tribune (key plan see figure 23) were investigated during Lauffray's research.[59] The central one was recognized as the work of Taharqa (690–664 B.C.) though the titulary of the king had been chiselled out by Psamtek. Lauffray thought that the ramp of Taharqa was older than the ramp located to the north,[60] but the recent research shows it was the contrary and that Taharqa's building post-dated the ramps which surround it. It is indeed now clear that, to the east, Taharqa's ramp had been inserted in the pre-existing sloping pavement while to the west it was founded on it. The Twenty-fifth Dynasty ramp therefore sloped down less steeply than its predecessors and thus ended originally more to the west, probably at the western limit of the surrounding quays, but this western end is now destroyed.

The recent SCA excavations in this area have found traces of an ancient waterway in the section of the excavation opposite the ramp of Taharqa. This waterway has an east-west orientation, but its base lies above the lower end of the ramp. It indicates that the Nile had shifted westward after the Twenty-fifth Dynasty (see figure 10). Different fragments of inscribed blocks and statues were uncovered and many bronze nails were found in the Nile sediments. The depiction already mentioned of the Karnak tribune with an obelisk, a sphinx, and a kneeling royal statue is engraved on a block found in the Mut Temple (now in the Cairo Museum) and is dated to the reign of Psamtek I (see

59 LAUFFRAY, 1971, p. 101–106; TRAUNECKER, 1972.

60 LAUFFRAY, 1971, p, 103–104.

figure 4).[61] The presence of a tree between the bark of Amun (i.e., the Nile river) and the tribune may suggest that the course of the Nile was moving westward at that time. However it is not the only possible interpretation of this tree which may also result of an artistic perspective of the background.

Mooring loops were found in the north and south walls of the main ramp for embarkation and disembarkation of the boats.[62] Two scenarios for mooring are possible, either perpendicular to the ramp or Mediterranean fashion tying up alongside the north or south wall bow first at the ramp. This latter seems more likely as the length of the boat would not be restricted. The slope over the upper 9 m of the ramp is slightly steeper than the 'central' ramp, which was measured at 7.5 cm/m. It then steepens to 25.5 cm/m.[63]

It should also be noted that the northern wall of the ramp has a fragment of a hymn and title of ritual that priests had to recite during a ceremony in which they filled theomorphic vases of the Theban triad with water.[64] The ramps may have also been used in a procession of a sacred vase of Amun to access the waters during a festival to mark the end of one year and the beginning of the new one.[65]

Figure 10. The ramps.

3.11 Ground Penetrating Radar (GPR) survey of the ramps

A Ground Penetrating Radar (GPR) area survey (20 x 20 m) was conducted using a GSSI 200 Mhz antenna at the bottom of the ramps in February 2013 in order to determine whether they extended further west below the excavated ground

61 Benson/Gourlay, 1899, p. 258, 378 and pl. XXII (5); Perdu, 2010; Perdu, 2011.

62 Boraik, 2010b, p. 70, 72.

63 Lauffray, 1971, p. 101, fig. 17.

64 Ibid., p. 101.

65 Traunecker, 1972, p. 234–235.

level. The survey was carried out using traverses spaced at 0.5 m intervals. The GPR data was processed in Reflex2DQuick and GPR Slice software. All profiles were processed to remove background noise, and a regain function was applied to strengthen the deeper responses to the radar signal. All data was then sliced and resampled to produce a series of timeslices through the site (key plan see figure 23).[66]

The GPR survey in area AG006[67] at the bottom of the exposed ramp adjacent to the tribune indicates several linear anomalies running from west to east on the line of the ramp (see figure 11). A single high amplitude linear anomaly [AG006_1] measuring 14 m in length seems to indicate a continuation of the exposed ramp revetment/foundations. Several linear and rectilinear anomalies to the north of this [AG006_2], [AG006_3] mark a parallel wall and possible structural elements [AG006_4] on the north side of the ramp. A fainter anomaly [AG006_5] marks a possible continuation of the exposed southern edge of Taharka's ramp, with other linear anomalies [AG006_6], [AG006_7] and [AG006_8] marking a possible earlier ramp revetment or small structures or platforms associated with the ramp. The results seem to indicate a continuation of the ramp under the alluvium to the west of the exposed structure, and possibly different phases of ramp. These anomalies seem to run up to c.7 m below the current ground level, i.e. 62 m a.s.l. The extension of the ramp may well have facilitated the loading and unloading of the barque of Amun-Re for the Theban festival processions. How far the ramp extended and to what height above sea level it descended is unknown.

66 The survey was carried out by Dominic Barker (Southampton University) and Sarah Jones (Museum of London Archaeology) and the data was processed by Kristian Strutt (Southampton University) as part of the collaboration between the Supreme Council of Antiquities research led by Mansour Boraik and the Theban Harbours and Waterscapes Survey led by Angus Graham. GRAHAM et al., 2013, p. 50–51; GRAHAM/STRUTT, 2013, p. 7.

67 GRAHAM/STRUTT, 2013, p. 6–7.

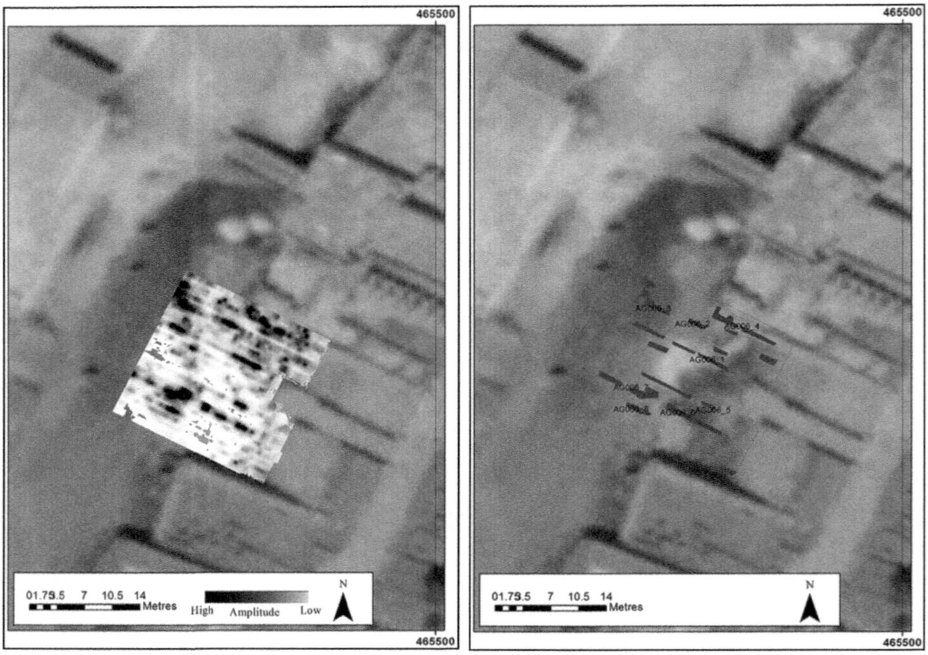

Figure 11. Structures revealed by Ground Penetrating Radar (GPR) in front of the ramps (K. D. Strutt).

3.12 The quay to the south of the tribune and its date

The excavations south of the ramps have revealed the quayside (*key plan* see figure 23) for a distance of around 110 m[68] (see figure 12). Though several phases of building can be identified, the main masonry work seems to date to the Kushito-Saite period (722-525 B.C.), based upon the parallels of the kind of masonry and the chisel marks on the ashlar blocks recorded at North Karnak.[69]

The two upper courses of the wall were constructed after the reigns of Akoris/Psamouthis, 393-380 B.C., with a sandstone lintel bearing their names being reused in the upper part of the wall.[70] Furthermore the stones in the two upper courses of the wall bear evidence of iron chisel marks suggesting a Ptolemaic-Roman date of construction. The uppermost course is characterised by headers

68 Boraik, 2010b, p. 67–69.

69 Gabolde/Rondot, 1993, p. 258–260.

70 Boraik, 2010b, p. 67.

and the second course by stretchers, whereas the wall built before this addition is composed of headers and stretchers within each course. The height of the wall prior to these two additional courses was measured at 74.40 m a.s.l., which is precisely the same height as the highest flood recorded on the face of the tribune dating to year 6 of Taharka's reign.[71] This is circumstantial evidence suggesting that this may not be a coincidence, but that the exceptional flood was used as a level for the height of the wall (see figure 13).

Two staircases leading from south to north were exposed during the excavations. Again mooring loops were revealed cut into the stone blocks of the wall positioned roughly parallel to the slope of the steps inset into the wall enabling boats to moor at the quay at a range of water levels throughout the annual cycle of the Nile.

Figure 12. The quay south of the tribune.

71 Legrain, 1896; von Beckerath, 1966; Broekman, 2002; Gozzoli, 2009.

Figure 13. The sondage on west facing side of the wall. The arrow head at the base of the second course indicates the height of the inundation (74.40 m) during year 6 of Taharka's reign. Note the second course is made up of stretchers and the upper course is a series of headers. (photo: A. Graham, February 2007).

3.13 The southern extension: the double-curved quay

Further south, an extension of the embankment in the shape of a double curved quay was discovered during the work undertaken next to Neferhotep's temple (key plan see figure 23).[72] As mentioned before textual sources may indicate that this wall was built by Montuemhet, Fourth prophet of Amun under the reign of Taharqa. To the west of the Ram Avenue in front of the Khonsu Temple, an extension of this embankment wall was discovered. This extension was built with small squared off blocks of sandstone and curved south toward the Avenue (see figure 14). This indicates that the Karnak complex was then on land partly protruding to the west and had been framed by this huge embankment intended to protect it from the flood erosion.

72 Neferhotep's temple has been dated to Ptolemy IV through a nearby naos, which however was not in situ. Barguet, 1962, p. 10–11; Porter/Moss, 1972, p. 224–225; Goyon/Traunecker, 1982, p. 300 and n. 3; Thiers, 1997, p. 264. For the reliefs of Ptolemy IV, see Lepsius, 1849, Bl. 15, b-c.

Figure 14. The double curved wall under Neferhotep's temple (view towards the south-west).

3.14 The quay of the island west of Nectanebo's dromos (Saïte ?)

Another part of a wall going under Nectanebo's sphinx-lined avenue was found (see figure 15 and key plan see figure 23). This embankment can be roughly dated to the Kushito-Saïte Period, based on the shape and workmanship of the blocks.[73] Interestingly, the slope of the masonry clearly shows that it represents the embankment on the eastern side of an Island.

Figure 15. The quay of the island west of Nectanebo's dromos (view towards the south).

3.15 Stratigraphical observations related to Nectanebo's dromos

The excavations have revealed the complete absence of older layers under Nectanebo's dromos. The excavators only encountered Nile mud from Nile

73 Gabolde/Rondot, 1993, p. 258–260.

silting.[74] The first conclusion to draw from this observation is that the road to Luxor in Hatshepsut's time was certainly not yet there, and most probably lay 100 m to the east. The second conclusion is that Nectanebo's dromos apparently occupied for its greatest part an abandoned channel of the Nile, a great opportunity which provided the king with a long strip of free land (2 km long) perfectly suitable for a new road.

3.16 The channel traces on Nectanebo's dromos

The filling up of the depression on which the monumental 2 km long dromos of Nectanebo I (380-362 B.C.) was installed, straight from the Mut temple complex to Luxor temple, was obviously not stabilised over its total length and, in some places, the pavement sunk down tens of centimetres. This resulted in the appearance of some sort of incidental ramps, subsequently incised by deep wheel marks (see figure 16).

Figure 16. The raising of the pavement on Nectanebo's dromos.

3.17 The processional road to Luxor Temple

In the precinct of Mut the discovery in 2000 by team of Richard Fazzini (Brooklyn Museum) of a gate of Hatshepsut opening to the west in the north-

74 Boraik, 2010a, p. 51 and personal observations.

west sector of the precinct (key plan see figure 23)[75] is of some interest to our discussion as it is likely to have opened on the processional road to Luxor, as has already been mentioned.

3.18 The processional road and tribune of Montu Temple

An extant tribune was built, certainly by the Twenty-sixth Dynasty, at the northern end of the Montu dromos (key plan see figure 23).[76] However, augering approximately midway between the tribune and the Montu enclosure has revealed that this area of North Karnak was still part of the Nile channel, certainly during the Middle Kingdom, and perhaps until the early New Kingdom. The ceramic fragments in the lower fining upwards sequences of Auger Site 2 (AS02, see figure 23) had been dated from the Middle Kingdom to the Second Intermediate Period with one sherd dated to the early New Kingdom.[77] However, a re-examination of the fragments in these lowest levels reveals that they are too rolled to be certain about their date. The fabrics are known from the Middle Kingdom to New Kingdom and probably later leaving the dating of the channel less precise.[78] It is not clear where the tribune (if there was one) of Amenhotep III was constructed. It remains a possibility that it was in the location of the extant tribune, but it could have been closer to the Temple of Montu.

3.19 The gate of Ptolemy XII Neos Dionysos

A gate of Ptolemy XII Neos Dionysos (≈ 81–59 and 56–52 B.C.) was discovered at the west end of the east-west dromos running westward from Mut temple, close to its intersection with what was the Chevrier/Legrain drain (see see figure 22 and key plan see figure 23).[79]

The continuous pavement of the road from Mut temple to that west end clearly shows that the previous island had then been joined to the right bank and

75 Fazzini, 2001, p. 61–74.

76 Pillet, 1924; Varille, 1943, p. 1 and pl. III [2–3]; Cabrol, 2001, p. 571–579.

77 Graham/Bunbury 2004; Graham/Bunbury 2005; Bunbury et al., 2008; Graham, 2010a, p. 134.

78 The ceramic fragments were re-examined by Marie Millet (Le Louvre) and Aurélia Masson (British Museum) in 2009 using the fabric collection from their 2001–2007 excavations at Karnak as a reference.

79 Boraik, 2010a, 46, Boraik, Thiers, 2015, 51-62.

that the new eastern embankment of the river had been shifted westward a little further towards the place where the gate was discovered.

4. The geoarchaeological data

4.1 Auger (AS16) in front of the tribune

In the 1890s Legrain excavated to a level 10 courses below the protruding course of the tribune (see see figure 5), reaching a height of c. 71 m a.s.l.,[80] in order to record the Nile levels on the face of the tribune. Chevrier's[81] excavations in the 1940s had reached the slightly greater depth of c. 70.50 m a.s.l.

Figure 17. Chevrier's 1940s excavation in front of the western tribune. The two mooring rope loops in course four and five are visible. The pump is located at c.70.50 m, five masonry courses below the stone loops (Photo no. 6112 ©CFEETK).

80 LAUFFRAY, 1971, p. 85.

81 CHEVRIER, 1947, pl. 48.

A hand auger (AS16),[82] carried out 1.54 m from the face of the tribune starting at 72.40 m a.s.l., reached a depth of 3.79 m down to 68.61 m a.s.l. when an obstruction terminated the work. Throughout the augering were found chronologically heterogenous ceramic fragments ranging from the Middle Kingdom to the Roman Period. All of the material was angular in terms of its wear pattern and not rounded by water transport. The only rounded sherds were cores from large coarse-ware vessels. The first 1.90 m was clearly Twentieth century backfill from excavations. Sample 09 with its bimodal light yellowish brown (2.5Y 6/4) desert sand and very dark greyish brown (10YR 3/2) mud would appear to mark the contact between the backfill and the unexcavated sediment at about 70.25 m a.s.l., which accords well with the excavations of Lauffray and Chevrier. Below this we have evidence of well-sorted silty sand/sandy silt water-lain deposits and well-sorted muddy sand; the ceramic fragments point to a *terminus post quem* date of the Roman Period.

4.2 Auger (AS39) behind the embankment

Auger Site 39 was undertaken at the base of a SCA/MSA excavation conducted under the direction of Boraik on the south side of the tribune, immediately behind the embankment wall.[83] The height at the top of the auger was 70.85 m a.s.l., terminating at 67.0 m a.s.l. due to pressure of the groundwater. AS39 aimed to investigate the process and chronology of the construction of the embankment wall.

The excavated sections revealed a Late Roman Period cut made behind the wall, which may have been an attempt to get to a foundation deposit at the corner of the tribune and the embankment wall. Such practices have been observed in several locations in the temple of Karnak. Unfortunately this hole may have destroyed any foundation trench that might have been present to the east of the embankment.

Primae facie it would seem that we have a sandbar or levee at the location of the tribune during the Middle Kingdom (see table 1). This finding has potentially considerable implications for understanding the position of the tribune and its possible date. However, with only a single auger on the east side of the embankment wall and a clear gap in the auger data between this location

82 AS16 was carried out in February 2004 by Morag Hunter (University of Cambridge) and Angus Graham (University College London) as part of the Karnak Land- and Waterscapes Survey directed by Angus Graham and Judith Bunbury (University of Cambridge). The ceramic fragments were studied by Sally-Ann Ashton (Fitzwilliam Museum, University of Cambridge).

83 GRAHAM et al., 2012, p. 41.

and the fifth pylon to the east and the court of the Opet Temple to the south it is difficult to clearly understand what this represents. Could it be a bar off the west side of the early island of Karnak? Could it represent the continuation of the bar deposits identified in the Opet Temple court augering? The ERT profile (P 3 see see figure 23) between the Khonsu temple and the 10th pylon court and hand auger along the profile revealed that a backwater or slough lay between the two areas during the First Intermediate Period.[84]

It also raises the question of when the tribune could have been constructed. The sandbar at the level it was would have been submerged for much of the annual cycle during the Middle Kingdom to early New Kingdom period.

Height m a.m.s.l.	Sedimentary description / interpretation		Chronological information
70.85	Backfill into the excavation		
70.50	Muddy, moderately sorted fine (sometimes very fine) sand with clasts of sandstone and concretions, sherds, quartz, and limestone. Small numbers of granite, dolerite, diorite, bone, flint, and shell fragments also occur		70.50 New Kingdom The ceramic fragments were mainly typical of the New Kingdom and not earlier than Thutmose I
69.50	Muddy medium and fine sands with an even greater variety of clast types.		68.86 Middle Kingdom – New Kingdom
68.30 67.00	A unit of less muddy, fine to very-fine sands that may represent levee or sandbar deposits		68.30 Middle Kingdom – early New Kingdom From identified fabrics, the bulk of the material is from the Middle Kingdom – early New Kingdom prior to the reign of Thutmose I. Fragments were less frequent and medium to medium-high rolled

Table 1: Sedimentary description/interpretation and chronology of Auger Site 39.

84 GRAHAM, 2010b.

4.3 Was the quayside part of a harbour basin?

The tribune and quayside wall exposed by Chevrier have previously been reconstructed as part of a harbour basin connected to the Nile by a canal.[85] Such a reconstruction has been influenced by the scene in the tomb of Neferhotep (TT49, see see figure 2). However, following two hand augers carried out in front of the wall – one to the south (AS31) and one to the north of the tribune (AS34)[86] – it can definitively be stated that the quayside wall was not part of a semi-enclosed basin but that it had been in direct contact with the Nile. If it had been in a semi-enclosed basin, one would have expected to find very fine sediment (silt and clay-sized particles) in the augers.

In AS31 fine sand was recorded throughout the auger cores. The sediments in the first 1.95 m (samples 01-17) down to 69.22 m a.s.l. are mostly fine sandy silt with some silt deposits. They are all well-sorted and deposited by the river. Below this a water-lain package of sands with medium sand at the base fining upwards to fine sand (samples 18-21) was recorded. These fining-upwards sands are typical of sandbar formations in the Nile. This lateral bar appears to have formed during the construction process. In samples 18–21 >20 % of the total core weight is fragments of sandstone, believed to be debris from dressing the stone blocks. An increased percentage of sandstone fragments at 70.05 m a.s.l. (sample 08) may represent a further dressing of the wall suggesting the wall was constructed over a number of years.

This interpretation is supported by the auger north of the tribune. The deposits encountered were all fining-upwards sand and silt packages with abundant yellow sandstone. The deposits in AS34 are generally finer than those recorded in AS31. This is consistent with the water being a little slower to the north of the tribune and can be understood by the curvature of the revetment wall. The tribune is the furthest westerly point of revetment wall and hence the water would have been slacker to the north of it.

85 For example Aufrère et al., 1997, p. 82–83, 86–87.

86 AS31 and AS34 were carried out as part of the collaboration of the SCA research excavations led by Mansour Boraik and the Karnak Land- and Waterscapes Survey led by A. Graham and J. Bunbury. AS31 was conducted at the base of the SCA sondage against the wall in the 'Madrasa site' excavations by Angus Graham, Romain Mensan (Cfeetk), Shima Montasser Abu El Haggag (SCA) and Rosemary LeBohec (Cfeetk) in February 2007. AS34 was carried out by Angus Graham, Judith Bunbury and Salah El-Masekh (SCA) in February 2008. Aurélia Masson and Marie Millet studied the ceramic fragments from both AS31 and AS34.

Dating the quay is more problematic. The study of the ceramic fragments in AS31 and AS34 has not led to definitive evidence of the date of construction of the wall. In AS31 there are in fact only 128 sherds > 4 mm in size in the whole core, which is a relatively low mean number of fragments per metre of depth. The sherds are heavily abraded, which is consistent with the coarse sediment matrix. Most of the material is dated to Middle Kingdom to Roman Period (MKRom), the abraded nature making tighter dating not possible. However, there is also material from the end of the Third Intermediate Period to Ptolemaic Period. There is no material specifically from the Middle Kingdom, New Kingdom, early Third Intermediate Period (Twenty-first – Twenty-second Dynasties), Roman or Late Roman Periods. The material dating from the end of the Third Intermediate Period (Twenty-fifth Dynasty) to the Ptolemaic Period suggests a terminus post quem for the construction of the wall. A late Third Intermediate Period/early Late Period date of construction (Twenty-fifth – Twenty-sixth Dynasty) would be contemporary with some of the later flood records on the western tribune.[87] An earlier construction date could not be ruled out as early Third Intermediate Period material may be present in the abraded MKRom corpus.

This conclusion is supported by the two deep water wells that were dug in 2009 with the support of the American Research Center in Egypt.[88] Remains of acacia wood were found at a depth of 17 m (z = 56 m a.s.l.). In the framework of a convention signed with the Institut Français d'Archéologie Orientale du Caire, ^{14}C testing was conducted, which indicated that these samples date to the New Kingdom (≈ 1494-1402 B.C. that encompasses the reign of Thutmose III). It is suggested that the river bed in the time of this king lay between 54.82 m and 58.82 m a.s.l.[89] Furthermore, it has been argued that this wood is from the remains of a wrecked boat, which sank down to the river bed. However, it needs to be stated that the dating of the wood does not date the time the boat sank as the boat may have had a considerable working life or the timber may have been reused in another younger vessel. Thus this finding does not conclusively prove that the river was in this location during the reign of Thutmose III.

87 Broekman, 2002; Legrain, 1896; von Beckerath, 1966.

88 Boraik et al., 2010; Boraik/Ghilardi, 2011; Boraik et al., 2012.

89 Gabolde, in press.

4.4 Conclusions on the western layout in the late New Kingdom, Kushite and Saite ages

To sum up, the recent excavations in front of Karnak show that, in the Late Period, the embankments did not form part of a basin, as it was probably the case earlier, following the depiction of the New Kingdom tomb of Neferhotep (TT49, ≈ 1300 B.C.). The Eighteenth Dynasty basin may well have been located in front of the Second Pylon in what is now the First Court. It seems possible that an area of low lying (marshy?) ground in this area was augmented and reshaped by the Egyptians to construct a basin and that a channel was dug to the Nile. This area of Karnak may have been the in-filled backwater observed in the augering and ERT profile (P3, see see figure 23) between the Khonsu temple and the court of the Tenth pylon[90] and also a previous auger (AS11) shows evidence of this backwater environment just west of the court of the ninth pylon.[91] Further geoarchaeological investigation is required in this area to clarify this hypothesis.

Returning to the massive wall running south and north of the extant western tribune, we suggest that the wall had two principal functions. Firstly, it was used as a quay to enable boarding and disembarkation of boats via the four staircases exposed thus far. The stairs are too narrow to safely load and unload the bark shrine of Amun on to his boat. However, the lateral ramps immediately south of the tribune provide a wide enough space to be able to load and unload the bark shrine for Amun-Re's journeys during the Opet Festival and Beautiful Festival of the Valley. This massive wall may well have also been constructed as a defensive measure in order to protect the temple complex against the erosive action of the river in flood. It would not have served to prevent water entering the temple precinct as water would still enter the temple by a rise in the groundwater level subject to the height of the inundation. It may have been constructed in response to an observed westward migration of the river and built as a barrier to prevent the river eroding the site of Karnak.

The extent of this massive wall reveals it to be a large-scale project. The aesthetic function of such a huge well-constructed wall should also been borne in mind. When the river was low it would have provided an impressive framing of the western extent of the site. Did it also serve as a clear boundary between the sacred space of Karnak and rest of the world?

The sandstone chippings found in the augers attest to more than one phase of construction and dressing of the stones in the wall. The findings of AS39 and the

90 Graham, 2010b.

91 Graham 2010a, p. 135–136, fig. 11.

recent excavations raise the possibility that the first phase of construction of this embankment began during the late New Kingdom. However, until the level of the base of the wall is known and further augering is carried out this proposition remains inconclusive. We propose that a subsequent phase of the wall was built by Taharqa in the Twenty-fifth Dynasty to the level of his year 6 inundation with his work on the area clearly attested by the central ramp. A further vertical extension of two further courses took place in the Ptolemaic Period.

5. Conclusion

The variety of data presented in this paper reveal numerous features in many geographical areas and belonging to different periods. We attempt here to sort them by date in order to propose a provisional and hypothetical view of the evolution of the temple's embankments, the relationship to the migration of the Nile, and the existence of channels and islands at the following epochs:

5.1 Hatshepsut-Thutmose III

A tribune was apparently located to the west of the "festival court" as can be postulated after the indications of the "Chapelle rouge".[92] The Nile or a canal must have flowed more or less parallel to the west walls of the court of the seventh and eighth pylons as was first suggested by Legrain[93] and more recently supported by the augering programme of Graham and Bunbury.[94] It is very likely that remains of the old eastern branch of the Nile had transformed into a backwater marshland to the north and the east. To the south, a slough lay between the main body of Karnak and a western bar upon which the Opet and Khonsu temples were subsequently founded.[95] The procession road to Luxor was following the axis of the seventh and eighth pylons, the dromos of Mut and then another dromos to Luxor located slightly west of the Hatshepsut door found at Mut temple.

The northern extent of the bar on the west side of Karnak depicted in figure 18 is conjectural. Furthermore, the extent of the backwater on the east side of Karnak is not yet fully determined. Further investigations are necessary to clarify both these features of the Karnak landscape during the mid Eighteenth Dynasty.

92 Lacau/Chevrier, 1977, p. 170, § 227; 185, § 263.

93 Legrain, 1906a, p. 112; Legrain, 1906b, p. 141.

94 Bunbury et al., 2008; Graham, 2010a; Graham, 2010b.

95 Graham, 2010b.

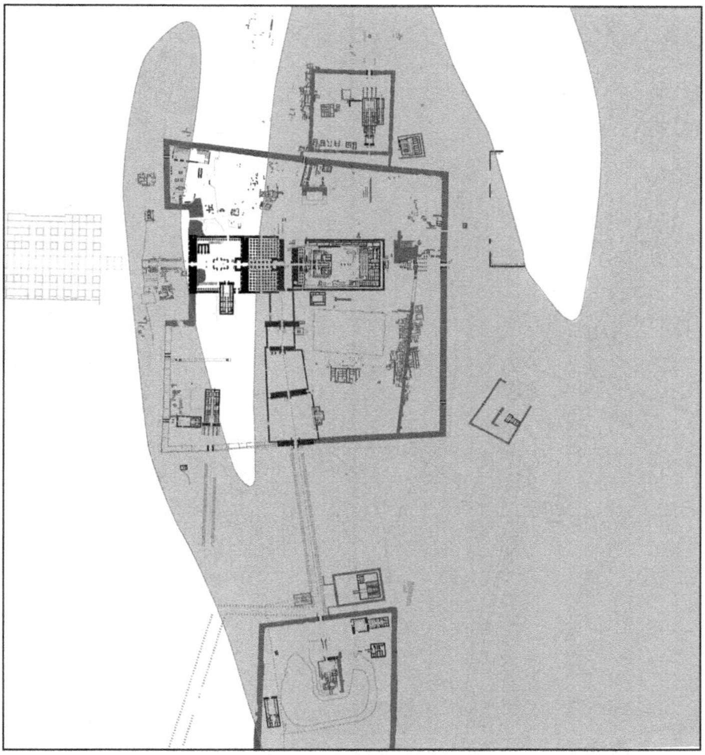

Figure 18. Hatshepsut-Thutmose III. Based on the cartographic fund of the architectural and topographical unit of the USR 3172 of the CNRS at Karnak (© CFEETK).

5.2 Tutankhamun to Ramses III

There is a scenario which presents the possibility of matching the representation of the Nile in Neferhotep's tomb with the geomorphological data: the former backwater of the Nile may have been transformed into a rectangular basin. Consequently a canal connecting the basin to the Nile might have been dug through the old bar. The location of the tribune slightly west of the second pylon and the location and dimensions of the basin as well as the position and geometry of the island to the west are hypothetical reconstructions.

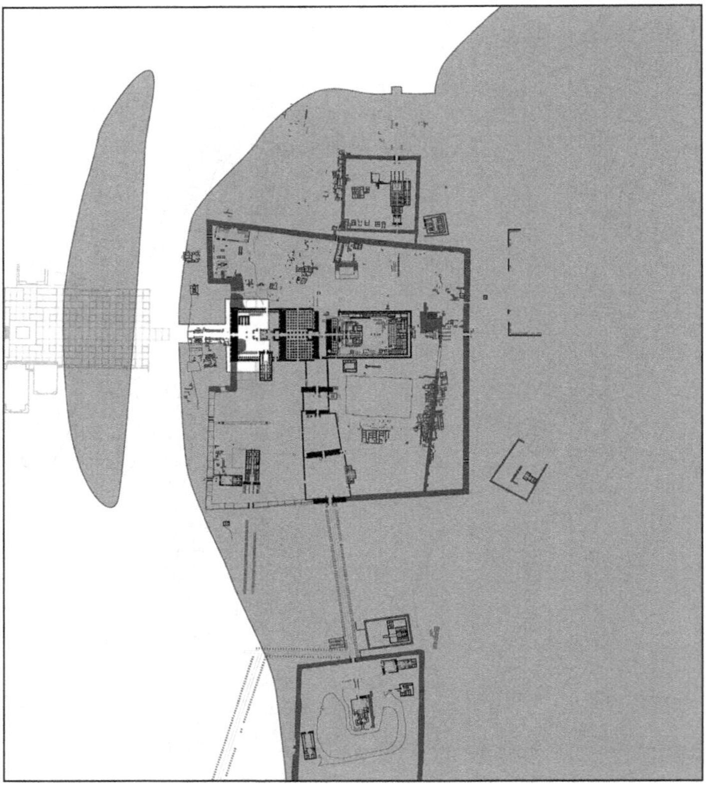

Figure 19. Tutankhamun. Based on the cartographic fund of the architectural and topographical unit of the USR 3172 of the CNRS at Karnak (© CFEETK).

5.3 Twenty-fifth to Twenty-sixth Dynasties

With the Twenty-fifth to Twenty-sixth Dynasties, the data available are more abundant as numerous traces of the embankments, tribunes, protective walls and quays have from place to place survived. The western tribune is still there, the quays north and south of it are preserved to a great extent, the double curved wall (possibly of Montuemhat) protecting Khonsu and Opet temples still shows remains of its lower courses. Last but not least, remains of the eastern quay of the island which lay west of Karnak are still preserved near Nectanebo's dromos. To the north, a stone masonry tribune was built, possibly by order of a king Psamtek, on the axis and along the dromos of the temple of Montu. The location and area of the basin fronting the tribune at North Karnak are conjectural and require further work to clarify this.

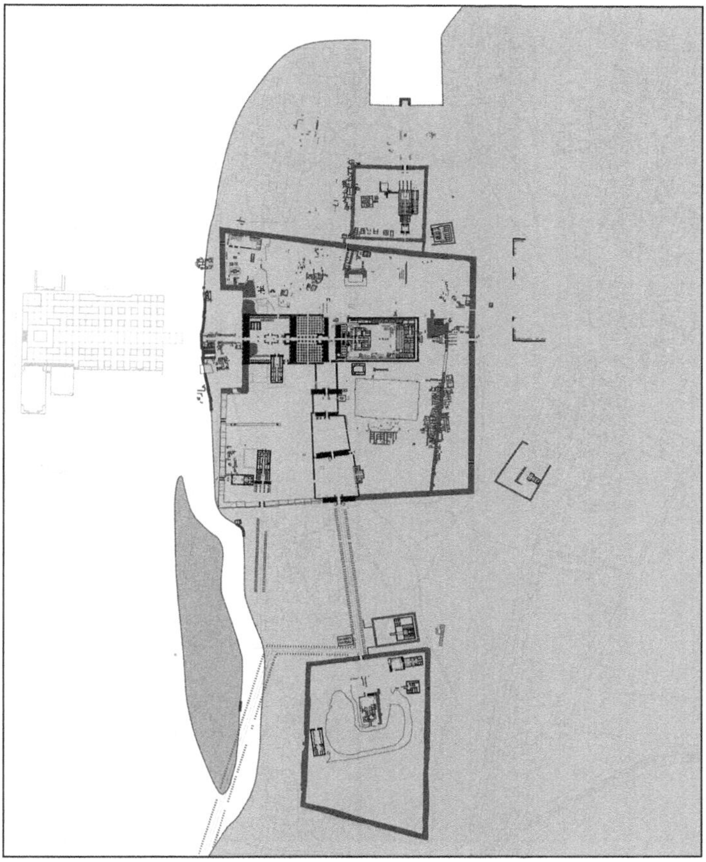

Figure 20. Twenty-fifth to Twenty-sixth Dynasties. Based on the cartographic fund of the architectural and topographical unit of the USR 3172 of the CNRS at Karnak (© CFEETK).

5.4 Nectanebo and early Ptolemies

The Saite channel appears to have been progressively filled up, providing new land and giving the opportunity to trace a monumental dromos straight to Luxor temple. At the same time, the westward Nile migration created a new channel (the "canal of Amun"), and a new island (dedicated to Amun and named *Tamaut*) as recorded in the archive of Hermias's trial. Again the location and geometry of the island to the west of Karnak remain conjectural.

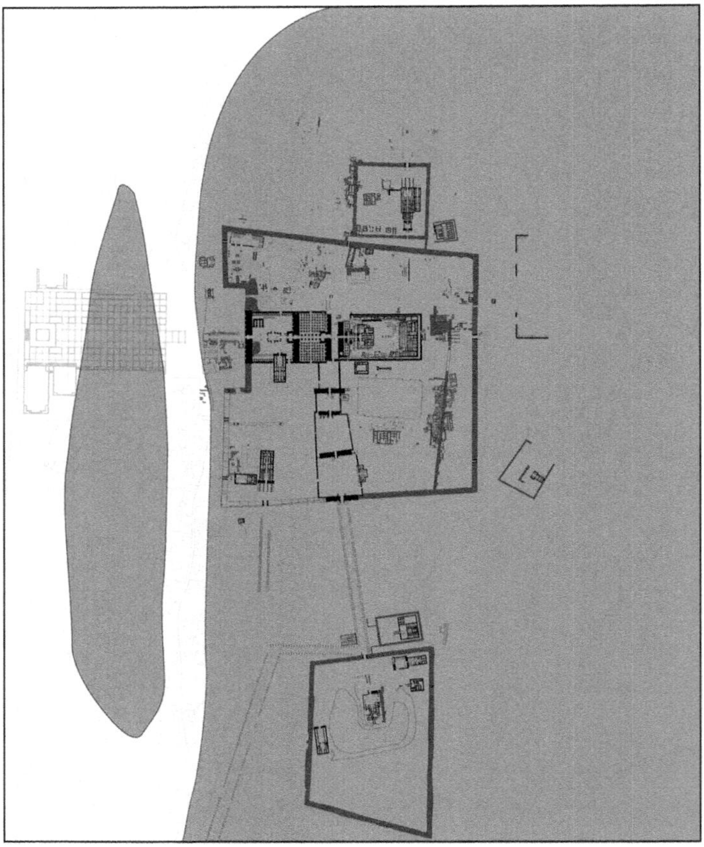

Figure 21. Nectanebo and early Ptolemies. Based on the cartographic fund of the architectural and topographical unit of the USR 3172 of the CNRS at Karnak (© CFEETK).

5.5 End of Ptolemaic and Roman ages

The canal of the Ptolemaic Period silted up and under Ptolemy VIII at the earliest, the old quay was partly dismantled and the baths were built over it and dumping over the wall took place in order fill up the former channel. The river bank had already migrated westwards. The position of the river bank and the island are conjectural.

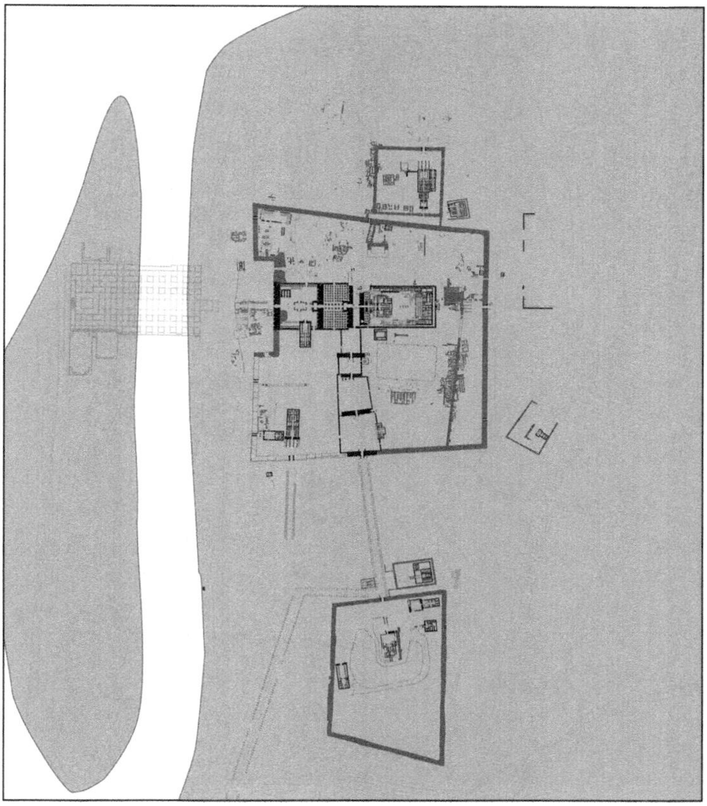

Figure 22. End of Ptolemaic and Roman ages. The gate of Ptolemy XII, which suggests the position of the embankment to the west end of Mut dromos, is indicated by a small black dot. Based on the cartographic fund of the architectural and topographical unit of the USR 3172 of the CNRS at Karnak (© CFEETK).

The five views of the evolution of the embankments of the temple of Karnak presented above are based upon current archaeological, geoarchaeological and textual interpretation, but remain conjectural. Further geoarchaeological survey is required to clarify the views presented.

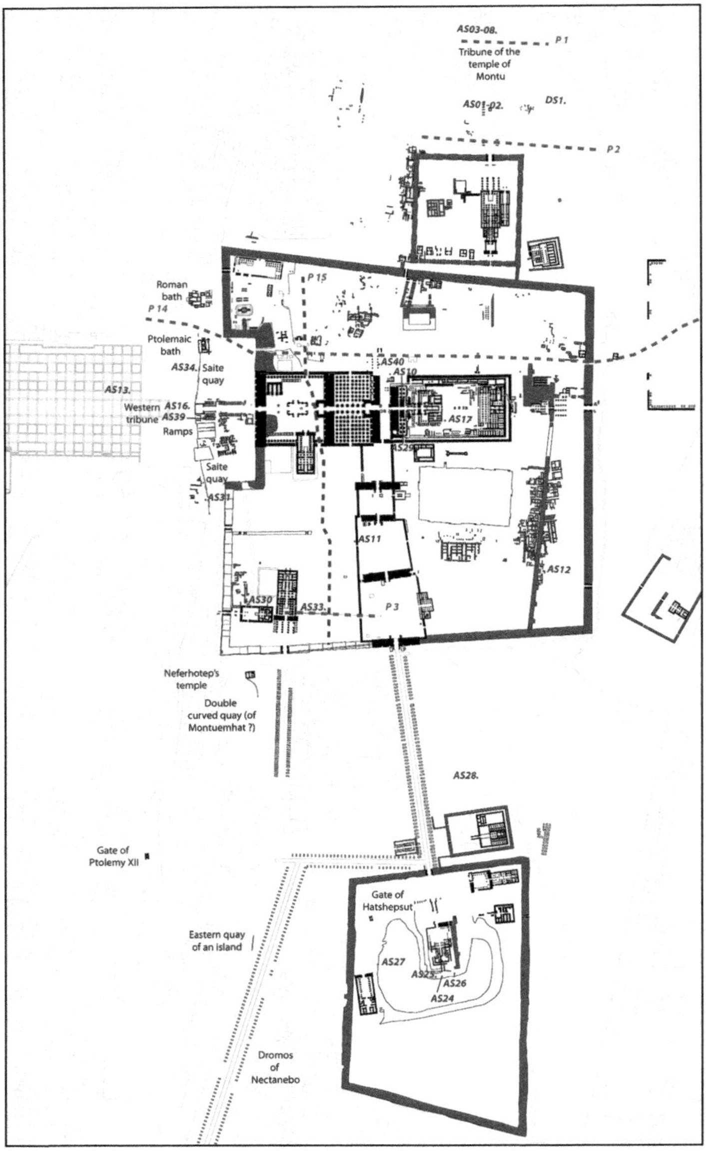

Figure 23. Key-plan of the temples of Karnak with indication of the archaeological units and of the augering sites. Based on the cartographic fund of the architectural and topographical unit of the USR 3172 of the CNRS at Karnak (© CFEETK).

Acknowledgements

Pierre Zignani, engineer of research at the CNRS and chief of the architectural and topographical unit of the USR 3172 of the CNRS at Karnak, very kindly provided us with a map of the layout of the area from Karnak to Luxor temples, from the cartographic fund of CFEETK. The SCA (Mansour Boraik and his successors on the site) offered all facilities to conduct the enquiries on the spot. The Franco Egyptian Centre, through the kindness of its director, Christophe Thiers, for the CNRS laboratory, provided us with the very valuable data of its rich archival documentation. Alban-Brice Pimpaud had the kindness to share with us information on the relative levelling in the surroundings of Karnak. Angus Graham would like to thank all those who have contributed to the augering and geophysical programme at Karnak (2002-09) as part of the Karnak Land- and Waterscapes Survey (KLaWS) especially his co-director Judith Bunbury, and since 2011 all the team of the Theban Harbours and Waterscapes Survey (THaWS) and all their colleagues and friends at Karnak who have generously collaborated with them since 2002.

Bibliography

AUFRÈRE, SYDNEY/GOLVIN, JEAN CLAUDE/GOYON, JEAN CLAUDE, L'Égypte restituée. Tome 1: Sites et temples de Haute Égypte. De l'apogée de la civilisation à l'époque gréco-romaine, Paris 1997.

BARGUET, PAUL, Le temple d'Amon-Rê à Karnak, essai d'exégèse (Recherches d'archéologie, de philologie et d'histoire 21), Cairo 1962.

BENSON, MARGARET/GOURLAY, JANET, The Temple of Mut in Asher, London 1899.

BONNEAU, DANIELLE, Le fisc et le Nil: incidences des irrégularités de la crue du Nil sur la fiscalité foncière dans l'Égypte grecque et romaine (Publications de l'Institut de droit romain de l'Université de Paris. Nouvelle séries 2), Paris 1971.

BORAIK, MANSOUR, Sphinxes Avenue Excavations. First Report, in: Les Cahiers de Karnak 13 (2010a), p. 45-64.

ID., Excavations of the Quay and the Embankment in front of Karnak Temples. Preliminary Report, in: Les Cahiers de Karnak 13 (2010b), p. 65-78.

ID., The Sphinx Avenue Excavations. Second Report, in: Les Cahiers de Karnak 14 (2013a), p. 13-32.

ID., A Roman Bath at Karnak Temples. A Preliminary Report, in: Les Cahiers de Karnak 14 (2013b), p. 33-46.

BORAIK, MANSOUR/FAUCHER, THOMAS, Le trésor des bains de Karnak, in: Les Cahiers de Karnak 13 (2010), p. 79-100.

BORAIK, MANSOUR/GHILARDI, MATTHIEU/BAKHIT, SAAD/HAFEZ, ABDEL/ALI, MOHAMED HATIM/EL-MASEKH, SALAH/MAHMOUD, ATTAIEB GARIB, Geomorphological investigations in the western part of the Karnak temple (quay and ancient harbour). First results derived from stratigraphical profiles and manual auger boreholes and perspectives of research, in: Les Cahiers de Karnak 13 (2010), p. 101-109.

BORAIK, MANSOUR/GHILARDI, MATTHIEU, Reconstructing the holocene depositional environments in the western part of Ancient Karnak temples complex (Egypt): a geoarchaeological approach, in: Journal of Archaeological Science 38 (2011), p. 3204-3216.

BORAIK, MANSOUR/GHILARDI, MATTHIEU/TRISTANT, YANN, Nile River evolution in Upper Egypt during the Holocene: palaeoenvironmental implications for the Pharaonic sites of Karnak and Coptos. Évolution du Nil en Haute Égypte au cours de l'Holocène: implications paléoenvironnementales sur les sites pharaoniques de Karnak et Coptos, in: Géomorphologie: relief, processus, environnement janvier-mars 1 (2012), p. 7-22.

BORAIK, MANSOUR/EL-MASEKH, SALEH/GUIMIER-SORBETS, ANNE-MARIE/REDON, BÉRANGÈRE, Ptolemaic Baths in front of Karnak Temples. Recent Discoveries (Season 2009-2010), in: Les Cahiers de Karnak 14 (2013), p. 47-77.

BORAIK, MANSOUR/ THIERS, CHRISTOPHE, Une chapelle consacrée à Khonsou sur le dromos entre le temple de Mout et le Nil?, in: Les Cahiers de Karnak 15 (2015), p. 51-62.

BOUCHAUD, CHARLÈNE/REDON, BÉRANGÈRE, (eds.), Current Research on Baths in Egypt: New Archaeological Discoveries, in: Actes de la journée d'étude, Cairo (in press).

BROEKMAN, GERARD P.F., The Nile Level Records of the Twenty-Second and Twenty-Third Dynasties in Karnak: A Reconsideration of their Chronological Order, in: Journal of Egyptian Archaeology 88 (2002), p. 163-178.

BRUGSCH, HEINRICH/REVILLOUT, EUGÈNE, Données géographiques et topographiques sur Thèbes, in: Revue Égyptologique 1 (1880), p. 172-180.

BUNBURY, JUDITH M./GRAHAM, ANGUS/HUNTER, MORAG A., Stratigraphic Landscape Analysis: Charting the Holocene Movements of the Nile at Karnak through Ancient Egyptian Time, in: Geoarchaeology 23,3 (2008), p. 351–373.

CABROL, AGNÈS, Une représentation de la tombe de Khâbekhenet et les dromos de Karnak-Sud : nouvelles hypothèses (les béliers du dromos du temple de Khonsou et l'intérieur de l'enceinte du temple de Mout), in: Les Cahiers de Karnak 10 (1995), p. 33-63.

ID., Les voies processionnelles de Thèbes, (Orientalia Lovaniensia Analecta 97), Leuven 2001.

CABROL, AGNÈS/TRAUNECKER, CLAUDE, Remarques au sujet d'une scène de la tombe de Neferhotep (TT 49): Les fonctions de Neferhotep, la représentation des abords ouest de Karnak et son contexte, in: Cahiers de Recherches de l'Institut de Papyrologie et d'Égyptologie de Lille 15 (1993), p. 19-30.

CHARLOUX, GUILLAUME, Rapport préliminaire sur la première campagne de fouilles du parvis du temple d'Opet à Karnak, in: Les Cahiers de Karnak 13 (2010), p. 195-226.

CHARLOUX, GUILLAUME/ANGEVIN, RAPHAËL/MARCHAND, SYLVIE/MONCHOT, HERVÉ/OBOUSIER, AGNÈS/ROBERSON, JOSHUA/VIRENQUE, HÉLÈNE, Le parvis du temple d'Opet à Karnak. Exploration archéologique (2006-2007), (Travaux du Centre franco-égyptien d'étude des temples de Karnak, Bibliothèque générale 41), Cairo 2012.

CHEVRIER, HENRI, Rapport sur les travaux de Karnak (1947-1948 [sic. 1946-1947]), in: Annales du Service des Antiquités de l'Égypte 47 (1947), 161-183.

CLARKE, SOMERS, El-Kâb and the Great Wall, in: Journal of Egyptian Archaeology 7 (1921), p. 54-79.

CONDON, VIRGINIA, Seven Royal Hymns of the Ramesside Period, (Münchner Ägyptologische Studien 37), Munich 1978.

CONYERS, LAWRENCE B., Ground-penetrating radar for archaeology, (Geophysical Methods for Archaeology 1), Walnut Creek, Calif. 2004.

DAVIES, NORMAN DE GARIS, The Tomb of Neferhotep at Thebes vol. I, New York 1933.

DAVIES, NORMAN DE GARIS, The Tomb of Neferhotep at Thebes vol. II, New York 1933.

DOOLITTLE, JAMES A./COLLINS, MARY E., Suitability of soils for GPR investigations, in Ground Penetrating Radar, 2nd Edition, Volume 1, ed. by DAVID J. DANIELS, London 2004, p. 97-108.

EGLI, ERNST, Geschichte des Städtebaues. Band 1. Die alte Welt, Zürich 1959.

EYRE, CHRISTOPHER, The water regime for orchards and plantations in Pharaonic Egypt, in: Journal of Egyptian Archaeology 80 (1994), p. 57-80.

FAZZINI, RICHARD, Some Aspects of the Precinct of the Goddess Mut in the New Kingdom, in: Leaving No Stones Unturned: essays on the ancient Near East and Egypt in Honor of Donald Hansen, ed. by ERICA EHRENBERG, Winona Lake 2001, p. 63-76.

FOUCART, GEORGES, Études thébaines. La belle fête de la Vallée, in: Bulletin de l'institut français d'archéologie orientale 24 (1924), p. 1-209.

GABOLDE, LUC, Les origines de Karnak et la genèse de la théologie d'Amon, in: Bulletin de la société française d'égyptologie 186-187 (2014), p. 13-35.

ID., La genèse d'un temple, la naissance d'un dieu, Cairo 2017 in press.

GABOLDE, LUC/RONDOT, VINCENT, Une catastrophe antique à Karnak-Nord, in: Bulletin de l'institut français d'archéologie orientale 93 (1993), p. 245-264.

GARDINER, ALAN HENDERSON, New Literary Works from Ancient Egypt, in: Journal of Egyptian Archaeology 1 (1914), p. 20-35.

ID., Ramesside Administrative Documents, London 1940.

ID., Ramesside Texts Dealing with the Transport of Corn, in: Journal of Egyptian Archaeology 27 (1941), p. 19-73.

ID., The Wilbour Papyrus. Volume II. Commentary. London 1948.

GASSE, ANNIE, Données nouvelles administratives et sacerdotales sur l'organisation du domaine d'Amon: XXe-XXIe dynasties, à la lumière des papyrus Prachov, Reinhardt et Grundbuch (avec édition princeps des papyrus Louvre AF 6345 et 6346-7), (Bibliothèque d'étude 104), Cairo 1988.

GOYON, JEAN CLAUDE/TRAUNECKER, CLAUDE, Une stèle tardive dédiée au dieu Neferhotep (CS X 1004), in: Les Cahiers de Karnak 7 (1982), p. 299-302.

GOZZOLI, ROBERTO B., Kawa V and Taharqo's *byꜣwt* : some aspects of Nubian royal ideology, in: Journal of Egyptian Archaeology 95 (2009), p. 235-248.

GRAHAM, ANGUS, Review: Grain Transport in the Ramesside Period. Papyrus Baldwin (BM EA 10061) and Papyrus Amiens. By Jac J. Janssen. Hieratic Papyri in the British Museum, 8. London, British Museum Press, 2004, in: Journal of Egyptian Archaeology 93 (2007), 298-300.

ID., Islands in the Nile. A Geoarchaeological Approach to Settlement Location in the Egyptian Nile Valley and the Case of Karnak, in: Cities and Urbanism in Ancient Egypt. Papers from a workshop in November 2006 at the Austrian Academy of Sciences, (Österreichische Akademie der Wissenschaften, Denkschriften der Gesamtakademie 40, Untersuchungen der Zweigstelle Kairo des Österreichischen Archäologischen Instituts, XXXV), ed. by. MANFRED BIETAK/ERNST CZERNY/IRENE FORSTNER-MÜLLER, Vienna, 2010a, p. 125-143.

ID., Ancient landscapes around the Opet temple, in: Egyptian Archaeology 36 (2010b), p. 25-28.

GRAHAM, ANGUS/BUNBURY, JUDITH, The ancient landscapes and waterscapes of Karnak, in: Egyptian Archaeology 27 (2005), p. 17-19.

GRAHAM, ANGUS/BUNBURY, JUDITH, Pottery from the Alluvial Environments at Karnak-North, in: Bulletin de Liaison du Groupe International d'Étude de la Céramique Égyptienne 22 (2004), p. 55-59.

GRAHAM, ANGUS/STRUTT, KRISTIAN D./HUNTER, MORAG A./JONES, SARAH/ MASSON, AURÉLIA/MILLET, MARIE/PENNINGTON, BENJAMIN T., Theban Harbours and Waterscapes Survey, 2012, in: Journal of Egyptian Archaeology 98 (2012), 27-42.

GRAHAM, ANGUS/STRUTT, KRISTIAN D., Ancient Theban temple and palace landscapes, in: Egyptian Archaeology 43 (2013), 5-7.

GRAHAM, ANGUS/STRUTT, KRISTIAN D./EMERY, VIRGINIA L./JONES, SARAH/ BARKER, DOMINIC B. Theban Harbours and Waterscapes Survey, 2013, in: Journal of Egyptian Archaeology 99 (2013), 35-52.

JANSSEN, JAC J., Grain Transport in the Ramesside Period. Papyrus Baldwin (BM EA 10061) and Papyrus Amiens (Hieratic Papyri in the British Museum 8), London 2004.

JEFFREYS, DAVID G., The Survey of Memphis I: The Archaeological Report (Egypt Exploration Society Occasional Publications 3). London 1985.

KAPLONY-HECKEL, URSULA, Land und Leute am Nil nach demotischen Inschriften, Papyri und Ostraka gesammelte Schriften, Volume 1 (Ägyptologische Abhandlungen 71), Wiesbaden 2009.

ID., Theben-Ost III, in: Zeitschrift für ägyptische Sprache und Altertumskunde 137 (2010), p. 127-144.

LACAU, PIERRE/CHEVRIER, HENRI, Une chapelle d'Hatshepsout à Karnak I, Cairo 1977.

LANOË, EMMANUEL, Fouilles à l'est du VIe pylône: l'avant-cour sud et le passage axial, in: Les Cahiers de Karnak 12 (2007), p. 373-390.

LARCHÉ, FRANÇOIS/ BURGOS, FRANCK, La chapelle Rouge: le sanctuaire de barque d'Hatshepsout. Volume 1: Fac-similés et photographies des scènes, Paris 2006.

LAUFFRAY, JEAN, Abords occidentaux du premier pylône de Karnak. Le dromos, la tribune et les aménagements portuaires, in: Kêmi 21 (= Les Cahiers de Karnak 4) (1971), p. 77-144.

LAUFFRAY, JEAN/SAUNERON, SERGE/TRAUNECKER, CLAUDE, La tribune du quai de Karnak et sa favissa. Compte rendu des fouilles menées en 1971-1972 (2e campagne), in: Les Cahiers de Karnak 5 (1975), p. 43-76.

LECLANT, JEAN, Montouemhat, quatrième prophète d'Amon, prince de la Ville (Bibliothèque d'étude 35), Cairo 1961.

ID., Recherches sur les monuments thébains de la 25e dynastie dite éthiopienne (Bibliothèque d'étude 36), Cairo 1965.

LEGRAIN, GEORGES, Textes gravés sur le quai de Karnak, in: Zeitschrift für ägyptische Sprache und Altertumskunde 34 (1896), 111-118.

ID., Fouilles et recherches à Karnak, in: Bulletin de l'Institut d'Égypte 4/6 (1906a), 109-127.

ID., Nouveaux renseignements sur les dernières découvertes faites à Karnak (15 novenbre 1904-25 juillet 1905), in: Recueil de travaux relatifs à la philologie et à l'archéologie égyptiennes et assyriennes 28 (1906b), p. 137-161.

LEPSIUS, KARL RICHARD, Denkmaeler aus Aegypten und Aethiopien. Band IX. Denkmaeler aus der Zeit der griechischen und roemischen Herrschaft, Berlin 1849.

LICHTHEIM, MIRIAM, Ancient Egyptian Literature. Vol.II. The New Kingdom, Berkeley 1976.

LOEBEN, CHRISTIAN E., Der Zugang zum Amuntempel von Karnak im Neuen Reich. Zum Verständnis einer zeitgenössischen Architekturdarstellung , in: The Intellectual Heritage of Egypt: studies presented to Lászlo Kákosy by friends and colleagues on the occasion of his 60th birthday (Studia Aegyptiaca 14), ed. by ULRICH LUFT, Budapest 1992, p. 393-401.

MACKLIN, MARK G. /TOONEN, WILLEM H. J./WOODWARD, JAMIE C./WILLIAMS, MARTIN A. J./ FLAUX, CLÉMENT/MARRINER, NICK/NICOLL, KATHLEEN/ VERSTRAETEN, GERT/SPENCER, NEAL/WELSBY, DEREK, A new model of river dynamics, hydroclimatic change and human settlement in the Nile Valley derived from meta-analysis of the Holocene fluvial archive, in: Quaternary Science Reviews 130 (2015), 109-123.

MEEKS, DIMITRI, Le grand texte des donations au temple d'Edfou (Bibliothèque d'étude 59), Cairo 1972.

PEET, THOMAS E., The great tomb-robberies of the twentieth Egyptian dynasty, New York 1997 (re-publication of the Oxford 1930 edition).

PERDU, OLIVIER., Le prétendu 'an V' mentionné sur les 'blocs de Piânkhi', in: Revue d'Égyptologie 61 (2010), p. 151-157.

ID., Les 'blocs de Piânkhi' après un siècle de discussions, in: La XXVIe dynastie: continuités et ruptures, Promenade saïte avec Jean Yoyotte (Actes du Colloque international organisé les 26 et 27 novembre 2004 à l'Université Charles-de-Gaulle – Lille 3), ed. by DIDIER DEVAUCHELLE, Lille 2011, p. 225-240.

PESTMAN, PIETER W., The Archives of the Theban Choachytes (second century B.C.): a survey of the Demotic and Greek papyri contained in the archive (Studia demotica 2), Leuven 1993.

PILLET, MAURICE, Rapport sur les travaux de Karnak (1923-1924), in: Annales du Service des Antiquités de l'Égypte 24 (1924), p. 53-88.

PORTER, BERTHA/MOSS, ROSALIND L. B., Topographical Bibliography of Ancient Egyptian Hieroglyphic Texts, Reliefs and Paintings I. The Theban Necropolis. Part 1. Private Tombs, Oxford 1960.

ID., Topographical Bibliography of Ancient Egyptian Hieroglyphic Texts, Reliefs and Paintings II. Theban Temples. 2nd edition, Oxford 1972.

QUACK, JOACHIM F., Studien zur Lehre für Merikare (Göttinger Orientforschungen 4, Ägypten 23), Wiesbaden 1992.

ID., Die Lehren des Ani. Ein neuägyptischer Weisheitstext in seinem kulturellen Umfeld (Orbis Biblicus et Orientalis 141), Freiburg 1994.

RICKE, HERBERT, Das Kamutef-Heiligtum Hatschepsuts und Thutmoses'III. in Karnak, (Beiträge zur ägyptischen Bauforschung und Altertumskunde 3,2), Cairo 1954.

SCHENKEL, WOLFGANG, Die Bewasserungsrevolution im alten Ägypten, Mainz 1978.

SCHLÜTER, ARNULF, Sakrale Architektur im Flachbild. Zum Realitätsbezug von Tempeldarstellungen (Ägypten und Altes Testament 78), Wiesbaden 2009.

SEIDLMAYER, STEPHAN J., Historische und moderne Nilstände: Untersuchungen zu den Pegelablesungen des Nils von der Frühzeit bis zur Gegenwart, Berlin 2001.

SMITH, HARRY S./JEFFREYS, DAVID G., A survey of Memphis, Egypt. Antiquity, 60/229 (1986), p. 88-95.

STANLEY, JEAN-DANIEL/KROM, MICHAEL D./CLIFF, ROBERT A./WOODWARD, JAMIE C., Nile flow failure at the end of the Old Kingdom, Egypt: Strontium isotopic and petrologic evidence, in: Geoarchaeology 18/3 (2003), p. 395-402.

THIERS, CHRISTOPHE, Un naos de Ptolémée II Philadelphe consacré à Sokar, in: Bulletin de l'institut français d'archéologie orientale 97 (1997), p. 253-268.

TRAUNECKER, CLAUDE, Les rites de l'eau à Karnak d'après les textes de la rampe de Taharqa, in: Bulletin de l'institut français d'archéologie orientale 72 (1972), p. 195-236.

VARILLE, ALEXANDRE, Karnak-Nord I (Fouilles de l'Institut français d'Archéologie orientale du Caire 19), Cairo 1943.

VON BECKERATH, JÜRGEN, The Nile Level Records at Karnak and their Importance for the History of the Libyan Dynasty, in: Journal of the American Research Center in Egypt 5 (1966), p. 43-55.

YOYOTTE, JEAN, À propos des "terrains neufs" et de Thmouis, in: Histoire, géographie et religion de l'Égypte ancienne. Opera Selecta (Orientalia Lovaniensia Analecta 224), Leuven 2013, p. 231-237 (re-edition of the same article previously published in two parts in Groupe Linguistique d'Études

Chamito-Sémitiques 8, 1961, p. 100-101 and Groupe Linguistique d'Études Chamito- Sémitiques 9, 1962, p. 5-9).
WRESZINSKI, WALTER, Die Inschriften des Monthemhet im Tempel der Mut, in: Orientalistische Literaturzeitung 13 (1910), p. 385-399.

Medamud and the Nile

Some Preliminary Reflections

FÉLIX RELATS MONTSERRAT

1. Introduction

The Nile is a structural element of the Egyptian landscape. It is widely assumed that communities originally settled on natural elevations in the floodplain in order to be shielded from the flood, and also that costly structures like temples should certainly be built well above flood level[1].

The link between temples and the Nile is visible in different levels: on the one hand, the friezes of Nile-gods ensure a symbolical connection[2], on the other, some constructions architecturally connect the river with the temples, for example processional roads lined with sphinxes, statues or enclosure walls[3]. When temples were situated near the river, their topographical nexus with the Nile was obvious. But what about temples located far from the river? The Theban floodplain offers several examples of temples situated at the edge of the desert plateau, and consequently, distant from the Nile. The royal mortuary temples

---

1 SEIDLMAYER, 2001, p. 82 has theorized on the relationship between assumptions on the expected flood height and building level. The location of the temples on elevations has a symbolic link with the image of the primeval mound *j3t*: SAUNERON/YOYOTTE, 1959, p. 35. For the case of Karnak, a succinct presentation is given by GOLVIN/GOYON, 1987, p. 28–31.

2 For a general presentation of the *soubassement* friezes: YOYOTTE, 1958, p. XI–XVI. More recently, LEITZ, 2004, p. 50–62.

3 The term *processional roads* is here used in the sense given to them by CABROL, 2001, p. 1–4. For a detail account of the pertinent structures, see *ibid.*, p. 330–418 (Sphinx), p. 424–7 (statues), p. 477–81 (enclosure walls).

of the New Kingdom can be considered paradigmatic cases. Most researchers assume the existence of channels between these temples and the river[4].

In this paper, I would offer some reflections on the case of one particular Theban temple, that of Medamud, raising the question of its relationship with the Nile.

2. Medamud, a temple far away from the Nile?

Medamud is known since the *Description de l'Égypte*, but the first archaeological studies started only in 1924, with the excavations of the French Egyptologist F. Bisson de la Roque, which lasted until 1932.[5] Cl. Robichon and A. Varille succeeded him, carrying out some campaigns between 1932 and 1939.[6] Their work offers a summary of the various phases of the temple's construction from the Middle Kingdom to the Roman era.[7]

The temple, which is still visible, is located 5 km north of Karnak, on the right bank of the river. It is also 4.9 km east of the present course of the Nile, as shown in the image below.

4 The monuments on the Theban west bank is detailed by PORTER/MOSS II, p. 340–538 and by CABROL, 2001, p. 653–6. GOLVIN, 1999, p. 40–41 and 64–71. The work of GRAHAM *et al.*, 2012 presents geomorphological studies on this matter.

5 All works in Medamud are published in the FIFAO collection of the Institut Français d'Archéologie Orientale: BISSON DE LA ROQUE, 1926; ibid, 1927; ibid, 1930; ibid, 1931; ibid, 1933; DRIOTON, 1926, ibid., 1927; BISSON DE LA ROQUE/CLÈRE, 1928; ibid., 1929; COTTEVIEILLE-GIRARDET, 1931; ibid., 1933; ibid, 1936.

6 ROBICHON/VARILLE, 1940 have never published all their work in Medamud, only providing a succinct overview of their research. In my PHD, I will publish all the documentation we have about their excavation, and I will especially reconsider the question of the so called *temple primitif.*

7 For a brief presentation, however to be corrected: REVEZ, 1999.

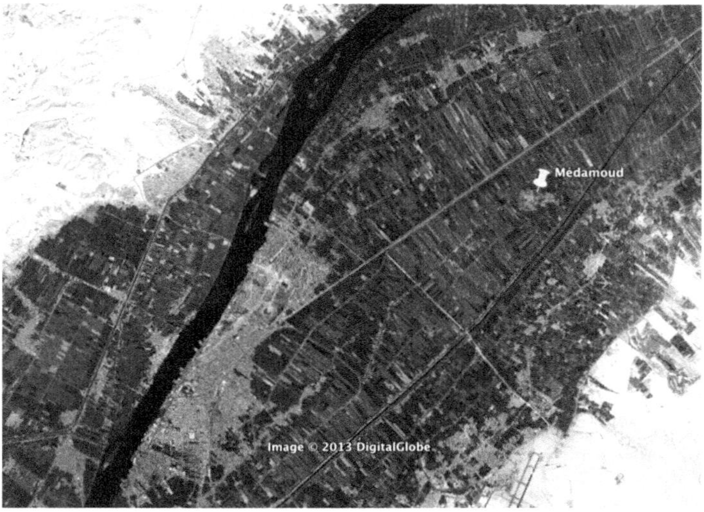

Figure 1. Satellite view of Luxor area, showing the actual location of Medamud and its distance from the river (© google-earth).

This view can be transposed in a diagram, highlighting the most important settlements in the region of Thebes,with the current location of the Nile.

Therefore, Medamud is not situated near the river, and at first glance the temple had been intentionally implanted far away from the Nile. However, the excavations have proved the existence of a relationship between the river and the temple. In 1932, F. Bisson de la Roque discovered a tribune[8] linked to the enclosure wall[9] with a *dromos*. This structure, to be discussed below, illustrates that an access to the Nile was planned during the construction of the latest phase of the Medamud temple.

8 Bisson de la Roque, 1933, p. 9–10. He at first analysed this structure as a quay, and called it "quai-tribune". However, Bisson de la Roque himself later recognised that this structure is too elevated to be considered as a quay. In fact the quay, never found, should be situated below the tribune. Cabrol, 2001, p. 565 has pointed out the terminological issue connected with the term tribune.

9 Bisson de la Roque, 1927, p. 127; Revez, 2004, p. 495. For the gate of Tiberius, a publication is prepared by an IFAO/Sorbonne mission supervised by Prof. D. Valbelle and the author.

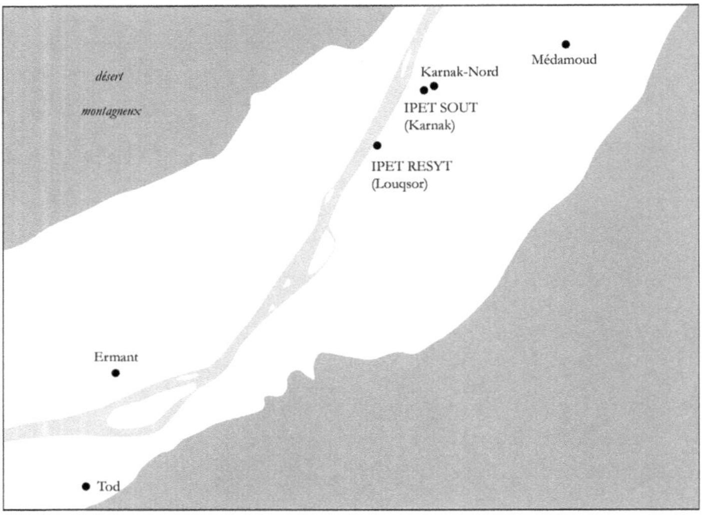

Figure 2. Theban temples with the current location of the Nile (© MASQUELIER-LOORIUS, 2013, carte 5).

3. The access to the Nile in the Ptolemaic temple: How was the tribune connected to the river?

The tribune[10], locatedwest of the *dromos*, is a rectangular, sandstone construction of 16.20 m x 13.40 m. It is an elevated building (with a floor level 1.70 m above that of the temple) and connected to the *dromos* by a ramp. In the excavator's opinion,[11] access to water was facilitated by two ramps beside the tribune.[12] These were adaptable to the water levels, being built with clay and mud brick.

The tribune of Medamud is similar to tribunes of other temples like Karnak, Karnak-North and Medinet Habu[13]. In all these examples, the temple is close

10 For an architectural description: BISSON DE LA ROQUE, 1933, p. 1–2 and 9–13.

11 BISSON DE LA ROQUE, 1933, p. 13: "Il est permis de supposer simplement des rives en terre pour le bassin d'accostage que la surélévation des eaux, par suite de l'apport de limon par l'inondation annuelle, obligeait de modifier sans cesse".

12 BISSON DE LA ROQUE, 1933, p. 110–111.

13 The most synthetic discussion of these three examples is offered by CABROL, 2001, p. 571–89 (Karnak/Karnak-North) and *ibid.* p. 611 (Médinet Habu). SCHENKEL, 1980, offers an overview of the known tribunes in Egyptian temples. The case of

enough to the Nile to postulate the existence of a waterway linking the landing stage to the Nile. In the case of Karnak, a direct access to the Nile is certain at least during the Kushite period. But what about Medamud?

3.1 The hypothesis of Bisson de la Roque: a canal linking Medamud and the Nile

To resolve the distance between the tribune and the river, the excavators have proposed the existence of an artificial channel, although they never found any trace of such a canal. Nor did they pronounce themselves on the date when it would have been dug.

This channel would have departed from Karnak-North and would have connected the two temples. According to Bisson de la Roque, "une avenue de sphinx conduisait de l'enceinte sacrée à un quai, où devait aboutir un canal unissant Karnak à Médamoud, pour les processions par voie d'eau".[14] É. Drioton advances a similar idea about the tribune, as being "élevée à cent cinquante mètres du temple. Cette tribune n'est point comme on l'a cru longtemps un quai d'embarquement. On assistait de là à l'arrivée et au départ du dieu sur le canal sacré".[15] Neither author underpins their assumptions by archaeological evidence, only stressing that this hypothetical canal comes from Karnak-North. Their hypothesis was probably rooted in the assumption that the temples of Medamud and Karnak-North were dedicated to Montou, composing what Drioton called the *Palladium* of Thebes.[16] According to their view, the temples were interconnected not only symbolically, but also physically through a channel. They were not, however, the first to discuss this hypothesis. The same suggestion had been made in 1924-1925 during the first excavations of the tribune and the *dromos*[17] of Karnak-North by M. Pillet.[18] According to him, the

Karnak, excavated by the CFEETK (Lauffray, 1971) and SCA (Boraik, 2010), is the most intensively studied example.

14 Bisson de la Roque, 1933, p. 1.

15 Drioton, 1932, p. 92. We can see in this quotation the same hesitation about the terms tribune and quay.

16 This expression is limited by Drioton, 1931 to the Third Intermediate Periode.

17 For the structures of Karnak-North; Porter/Moss, 1972, p. 1–2; Pillet, 1924; Id., 1925; Varille, 1943, p. 1; Gabolde, 1993, p. 248 and note 27; Cabrol, 2001, p. 571–579. For a global explanation on the excavations in Karnak-north: Cabrol, 2001, p. 9–12.

18 Pillet, 1924; Id., 1925.

tribune would provide access to a canal rather than to the river itself because it is not oriented towards the Nile (west) but to the North. That channel would have begun between Luxor and Karnak, branching off the right bank of the Nile, approaching the western quay of Karnak, then the quayof Karnak-Nord, finally reaching Medamud[19].

Although M. Pillet's hypothesis was not justified, it seems it was based on the same two considerations later brought up by F. Bisson de la Roque and É. Drioton: firstly, the religious bond between these temples would explain the existence of processions and, by extension, the canals linking them. Secondly, these authors were influenced by the orientation of contemporary channels supplying the Theban plain.

3.2 An anachronistic hypothesis?

The second of the two reasons just outlined was hence influenced by the contemporary hydrologic system. The image below (see fig. 3), reproducing part of the 1920 map produced by the Survey of Egypt (scale 1/100,000), presents the layout of the post-Mohamed Ali canals. As elsewhere in this volume,[20] these channels are oriented north/south, parallel to the river. Their digging, particularly in the case of the Theban plain, aimed to ensure the complete irrigation of the valley to the edge of the desert.

If this map is compared with the description given by the archaeologists cited, it is likely they have modelled the ancient landscape on the early twentieth century situation. At that time, a channel was crossing the eastern part of the Theban floodplain. They envisioned a similar layout for an ancient canal. Nevertheless the Egyptian channels have experienced notable changes in the early nineteenth century with the policy of Mohamed Ali who established a network of canals parallel to the Nile.The afore mentioned map shows that Medamud is located between two channels: one passing to the East of the modern village (*El*

19 This is what we understand in the confusing presentation PILLET, 1924, p. 86 gives in his excavation report: "On ne peut guère admettre, au contraire, que le fleuve fit à l'époque pharaonique, un coude tel qu'il vint baigner le quai nord et encore celui du temple de Médamoud, situé actuellement à plus de 3 kilomètres du Nil. Il semble beaucoup plus vraisemblable que les quais de Mantou *(sic)* et de Médamoud tout au moins, sinon le quai occidental lui-même ont été construits sur les rives d'un grand canal dérivé du Nil et prenant naissance entre Karnak et Louxor".

20 I refer to H. Willems' article in the present volume.

Bayadiya Canal[21]), and another to its west (*Ash-shi Saiyalet el el Gharbi*). These two channels both originate south of Luxor (at *El Bughâdi*), flowing parallel to the Nile and heading north to Khuzam. One of these channels (*Saiyalet el Ash-shi*) is even closer to Karnak-North and irrigates the plains between Medamud and the Nile.

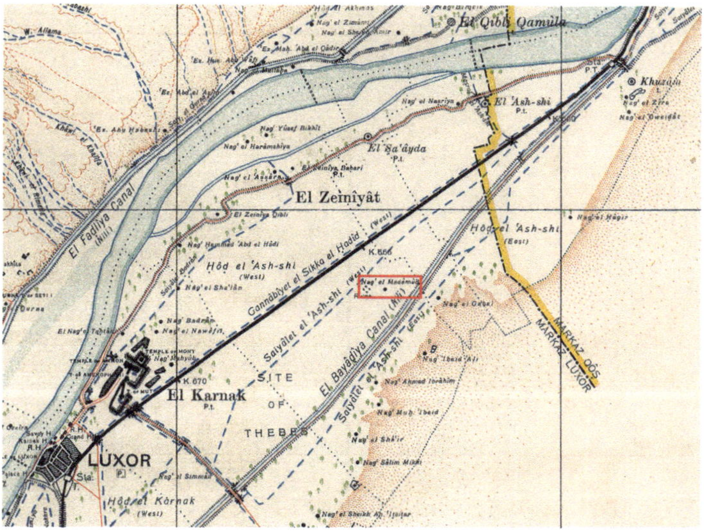

Figure 3. Survey of Egypt, 1920 (© Collège de France (cartes 18-32)).

G. Alleaume[22] was the first to point out the changes in the landscape due to Muhammad Ali's policy. This map can be compared with the one of the *Description de l'Égypte*, in which there is no channel parallel to the Nile[23]. Conversely, all channels are oriented East/West, crossing the plain. In the case of Medamud, a channel started in el Tahtâny, north of Karnak, and crossed the floodplain until reaching the northern part of the *Kom*.

21 I respect the spelling of the map.

22 ALLEAUME, 1992. The article of WILLEMS/CREYLMAN/DE LAET/VERSTRAETEN in this volume suggests that the plans of the *Description de l'Égypte* are not equally reliable for all parts of Egypt. Nevertheless, we are not looking for the course of the canals, but their overall direction. We also take into account the remarks of BALL, 1932, p. 130 for corrections to be made about the maps of the Description.

23 JOMARD/JACOTIN, 1818, Vol. II, pl. 1.

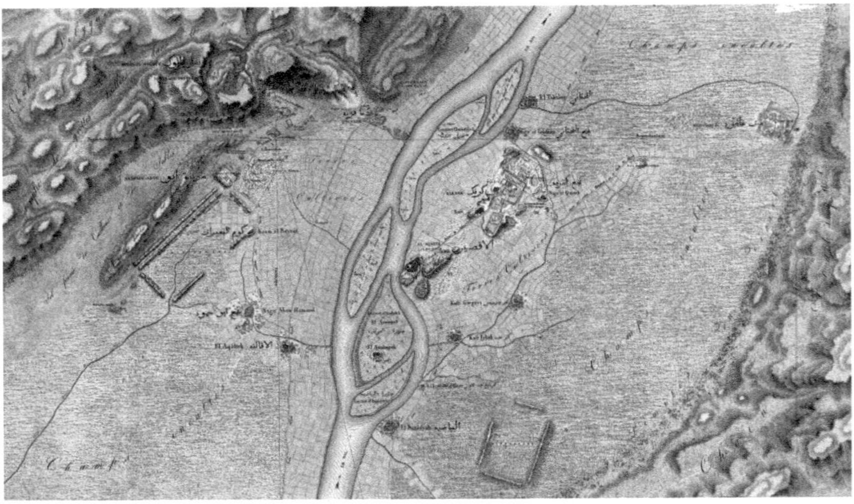

Figure 4. Map of the Theban region according to the Description de l'Égypte (see JOMARD/JACOTIN*, 1818, Vol. II, pl. 1).*

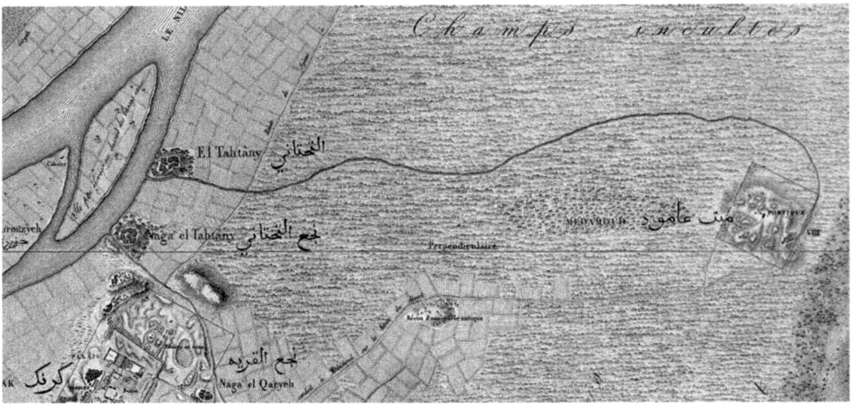

Figure 5. Detailed view of the canal linking Medamud with the Nile according to the Description de l'Égypte (see JOMARD/JACOTIN*, 1818, Vol. II, pl. 1, detailed view).*

The orientation of these channels is due to physical considerations, the profile of the valley being marked by an West/East dip from the river banks to the desert plateau.[24] The slope is formed by deposits of the Nile whose heaviest elements were accumulating on the banks. The channel of Medamud should have run only by the force of gravitation: the water flowed naturally from the highest part (the Nile) towards the inferior. Far from being crisscrossed by a network of canals, the Theban plain seems to be left uncultivated. This explains the testimonies of travellers describing the landscape before the creation of the modern hydrological system. R. Pococke, who traveled Egypt in the 1740s, is the first to describe the ruins of Medamud. He states he departed from Karnak, reaching Medamud over land. He depicts the landscape surrounding Medamud as a plain remaining poorly developed. He does not mention a structured system of canals,[25] referring rather to a series of small canals of irregular shape. In the eighteenth century, therefore, the Theban plain is irrigated by a network of canals similar to the image published by the *Description de l'Égypte*, very different from the modern network. Thus it appears that the hypothesis of a canal linking the Nile and Medamud via Karnak-North was largely influenced by the contemporary topographic network of the excavators.

4. The temple and the Nile: a localisation's problem

If such a canal never existed, two other explanations can be proposed for the presence of the tribune and quay at Medamud. Firstly, Medamud could have been linked to the Nile via an west-east canal similar to the one shown in the map of the *Description de l'Égypte*. But this hypothesis assumes the permanence of canals from antiquity to the early modern era without any supporting evidence. Also it attaches importance to human intervention in the construction of

24 For the double dip of the Nile valley see ALLEAUME, 1992, p. 303. For a cross-section of the Nile in Upper-Egypt, see the description given by SAID, 1993, p. 53–55, and fig. 1.23, p. 56 and BUTZER, 1976, p. 13–25.

25 POCOCKE, 1743, p. 96–97: "The plain to the east [of Karnak] naturally runs into coarse grass, much like a rush, great part of it lying waste; and where it is sown, the ground is laid in broad low hillocks, around which there are small irregular canals, the corn not being sown at the top of these hillocks, but only near the canals, in order, I suppose, that it may be more easily watered; for men raise the water out of the Nile into a small canal, which conveys it to all parts".

hydrological projects[26] and fails to explain why Medamud has been implanted at such a distance from the Nile.

In response to this objections, one might consider that a possible migration of the riverbed might explain the localization of the temple at such great distance from the modern Nile.

4.1 Medamud, a temple close to the river when it was founded?

The earliest structures found at Medamud date back to the Eleventh Dynasty. They are nowadays known as the *temple primitif* of Medamud.[27] Even though they have been only partially published,[28] and though the excavators have caused terminological confusion,[29] a new study of the ceramic material has established with certainty that the remains date from the reign of Montouhotep II.[30] I cannot

26 PILLET, 1924, p. 86, in his hypothesis of a canal connecting Karnak and Medamud, grants importance to the workers. For him, the land obtained from digging the canal was essential in the construction of the temple's boundary wall: "On peut croire, d'ailleurs, que la masse de terre nécessaire à l'élévation des murs d'enceinte des temples a été fournie par le creusement des canaux et, en particulier, d'un fossé large et profond, renforçant l'enceinte elle-même". Nowadays, the construction of boundary walls is explained differently: I refer to studies on the site of Karnak; GOLVIN et al., 1990, p. 926–927; GOLVIN, 1995, p. 33 and 41.

27 ROBICHON/VARILLE, 1940, p. 1–2.

28 In my PhD, I shall publish all the documentation of the excavations of Robichon and Varille kept in different archives. For a preliminary explanation: RELATS MONTSERRAT, 2014.

29 Admittedly the excavators recognised the ceramic as typical from the Eleventh Dynasty: ROBICHON/VARILLE, 1940, p. 11. Howener, at the same time they use the word *primitif* which confuses the epoch they were talking about. This expression "*primitif*" is used to compare the structures with the predynastic sanctuaries (SAINTE-FARE GARNOT, 1944, p. 68) and the old kingdom's temples (ROBICHON/VARILLE, 1940, p. 19). All researchers who have worked on Medamud have kept this terminology thereafter. For a further discussion see RELATS MONTSERRAT, *forthcoming*.

30 In her PhD, Z. Barahona Mendieta is currently studying all the ceramics of Medamud. She attributes the ceramics to the second part of the Eleventh Dynasty.

discuss here the question of whether these remains originate from a temple or from another kind of building, but will only take an interest in its location.[31]

Considering the issues discussed above, it is clearly vital to establish where the river bed was located at the time the temple was built, in order to know if there was a great distance between them. Indeed, comparison of the satellite image of the Theban plain (see fig. 1) with the map of the *Description de l'Égypte* (see fig. 4) proves that the trajectory of the Nile has evolved considerably over the last two centuries. In the time interval that elapsed between the moments these two pictures were made, the distance separating the Nile from the temple has increased. Moreover, the course of the river also changed. Thus, the two islands at the mouth of the *El Tahtâny* canal in front of Medamud have now disappeared. Such islands[32] are an obvious signal of the continuous changes in the course of the Nile. It is clearly conceivable that Medamud could have lain much closer to the Nile when it was founded than it does today.

4.2 The Nile's arms and the evolution of the river's flow

According to the work of Butzer, the Nile migrated from West to East.[33] But for the Theban part of the valley, this movement is opposite, from East to West, as A. Graham has shown.[34] Thanks to the geomorphological analysis of sediment cores, the westward movement of the river is estimated by this author at 2/3 km every millennium.[35] This movement is due to the piling up of solid elements on the banks, carried by the water; these deposits can build up islands. These latter modify the flow of the river, opening then a new river branch. Over time, islands can be united with banks, changing both their own profile and that of the river. This phenomenon can be called *défluviation.*[36] This dynamics causes a lateral

31 The existence of archaeological remains cannot be questioned from now (Relats Montserrat, 2014) given the photo archive. It is however clear that the interpretation given by Robichon/Varille, 1940 (a primitive osireion) has to be modified.

32 Bunbury/Graham/Hunter, 2008, p. 356–357.

33 Butzer, 1976, p. 33-36 and Jeffreys, 1985, p. 48–51.

34 Hillier/Bunbury/Graham, 2007, p. 1012. L. Gabolde, in his presentation at Mainz, argued similarly; see also Gabolde, 2014, p. 14–15. I thank him for the discussions he has granted me during the conference at Mainz, and during his lectures or stays in Paris.

35 Hillier/Bunbury/Graham, 2007, p. 1013 and 1015.

36 Levy/Lussault, 2008, p. 390.

displacement of the river, islands being gradually added to the banks, moving them westward[37].

If we apply this pattern north of Thebes, Medamud was – at least initially – much closer to the Nile or to one of its branches. This assumption was already made by A. Graham, being based on the depressions of the valley in the neighbourhood of Luxor. He postulates that Medamud, during the First Intermediate Period, was irrigated by the Nile.[38] He points out that "More ancient motions (> ~ 2000 years B.C) may be preserved in the landscape [...]. Interpreting the topographic lineations NE of Luxor [...] implies a large-scale NW migration. The main indicators of this are the two prominent lineations passing either side of Medamoud, and the current position of the river. So established during the First Intermediate Period (~2150 B.C.) Medamud may have originally had a riverside situation, and Karnak may have been founded on an island or spit".[39] It is likely that Medamud was near the river at the time of its first constructions even if the excavations of Robichon and Varille did not discover a structure linked with the Nile. Our knowledge of the *temple primitif* is

37 For a graphic representation of this phenomenon applied to the Theban bank I refer to figure 3 of the article of Angus Graham in Hillier/Bunbury/Graham, 2007 summarising the migration of the Nile over time. However, I do not figure out the reason that lead the author to qualify Medamud as "new land", that term being reserved for riverside structures. Other works of Graham provide useful supplements to understand this movement: Bunbury/Graham, 2005; Graham, 2010. A recent paper offers useful diagrams to explain the evolution of the Theban plain: Charloux/Mensam, 2012, p. 47–49.

38 Similar results were proposed by M. Ghilardi in a map rendering surface elevation in Upper Egypt, where a line is drawn between the south of Luxor and Medamud, indicating the presence of a possibly ancient Nile branch. But this author was only interested in Karnak in his paper (Boraik/Ghilardi, et al., 2010, fig. 1, p. 109).

39 Hillier/Bunbury/Graham, 2007, p. 1113. A. Graham follows the summary of the history given by Sambin, 2001, and his results recognize the changes in the river's bed. However, for the chronology, the article of Sambin follows the theories of Robichon and Varille dating the *temple primitif* at the beginning of the First Intermediate Period (at least before the Eleventh Dynasty – Sambin, 2001, p. 351). A. Graham propose *circa* 2150 B.C. which corresponds more or less with the beginning of the Eleventh Dynasty (Krauss, et al., 2006, p.482). His study has theredore to be modified.

limited to the structures inside the polygonal enclosure wall but other structures of this period seem to exist westwards.[40]

If this hypothesis is followed, Medamoud was founded near the Nile or one of its branches, and the course of the river gradually moved westward since. The aggregation of islands to the Nile banks and the emergence of new lands (*mȝwt*) distanced the former bank from the course of the river. This resulted in a gradual separation of the Nile from the temple.[41] Even under Sesostris III, who built a new temple,[42] the Nile should have been still closer to Medamud.[43] The western orientation of the temple[44] convinces us of a possible nexus with the Nile even if it has not been discovered yet.

However, it is difficult to establish precisely if Medamud was near the river rather than to one of its branches[45] and if its course can be pieced together again.

40 Sainte-Fare Garnot, 1944, p. 72–73.

41 This is certainly what happened between the nineteenth century and today: according to the *Description de l'Égypte,* islands were located in front of the village of El Tahtâny. They united the old river banks, pushing its course westward – which increased the distance between the river and temple.

42 The last archaeological study of the temple in the middle Kingdom is a short report of Robichon/Varille, 1939. The studies of Eder, 2002, p. 81–131 and Nivet-Sambin, 2008 claim to renew the subject, but their hypothesis are superficial because the lack of field work. We agree with Nivet-Sambin in her analysis of the middle kingdom's temple despite the lack of analysis of the architecture of the temple itself. For a new view: Relats Montserrat, *forthcoming.*

43 At the time of Sesostris III the main course of the Nile had to be more to the east. For some researchers, it passed west of the Middle Kingdom's temple of Karnak but one old arm was still east of the temple: Gabolde, 1998, § 201–205; Gabolde, 2013a, p. 6–8; Gabolde, 2013b; Gabolde, forthcoming; and his contribution at the symposium in Mainz). Graham is more circumspect about the idea of imagining Karnak as an island (Bunbury/Graham/Hunter, 2008, p. 364–365 and fig. 8). Nevertheless, he confirms that the Nile reached the *cour du Moyen Empire* inside the temple; and consequently that it flowed more to the west. Considering that, it is possible that the river was still closer of Medamud in the time of Sesostris III.

44 Nivet-Sambin, 2008; Relats Montserrat, *forthcoming.*

45 One should remember that geomagnetic measurements of the Theban plain allow to postulate that the Nile had discharged further east, almost on the edge of the Arabian desert's plateau (Hillier/Bunbury/Graham, 2007, fig. 3; Boraik/Ghilardi, et al., 2010, *fig.* 1, p. 109).

Considering that Karnak was, at the Eleventh Dynasty, surrounded by the Nile[46] it is possible that a Nile's arm, originated south of Karnak, circumvented the temple by the East and then reached Medamud. Nevertheless the investigations around the Thoutmosis I's treasure in Karnak encourage us to think that this Nile arm flowed to the north of Karnak[47] and not near Medamud.This hypothesis also raises the question of the migration of the river during the Middle Kingdom. No studies seem to confirm the continuous existence of a Nile arm to the east of Karnak during the Middle Kingdom because this temple was united with the East Bank[48] from this point forward. Therefore this branch could not reach Medamud under Sesostris III.[49] An alternative hypothesis would be that it began near Karnak-North, reaching the northern quay of the temple before heading for Medamud. This postulate is similar to the ones proposed by Bisson de la Roque and Robichon and Varille, but its implications are completely different: far from being an artificial canal created for processional reasons, the connection between Karnak-North and Medamud is a legacy of the geomorphological evolution of the river. However, this final statement must remain a hypothesis until a real geomorphological prospection is realised between Karnak-north and Medamud.[50]

46 The existence of an island at the end of the Old Kingdom has been suggested by Gabolde, 2013b, p. 24, fig. b and Gabolde, *forthcoming*. Graham, 2010, p. 136 considers Karnak as an island during the First Intermediate Period.

47 Bunbury/Graham/Hunter, 2008, fig. 7

48 Bunbury/Graham/Hunter, 2008, p. 364–365 and fig. 8

49 A canal seem to replace the former river branch at least until the Ramesside era (Cabrol, 2001, p. 430 et 579–581).

50 The only archaeological information has been collected by Pillet, 1925, p. 4 during the excavation of the quay at Karnak-North. He refutes the possibility of a branch of the Nile because of the absence of *sandbars* which are the sign of fluctuations of the Nile. Thus he says "les coupes de terrain faites entre le fleuve et les constructions de Karnak [...] n'ont jamais montré de ces bancs de sable témoins d'un cours ancien du fleuve. On semble donc être en droit de penser que le quai occidental de Karnak était, de même que ceux du nord et de Médamoud, établi sur un canal dérivé du Nil et non sur le fleuve lui-même". The indications given by the excavator should be verified: he didn't mention the shingles that are a signs of the aggradation of the Nile. However, given the research by A. Graham at Karnak-North, their absence in the excavations of M. Pillet seems highly suspect. For the thoughts of A. Graham on Karnak-north: Bunbury/Graham, 2005, p. 17–19.

4.3 A reinterpretation of the archaeological remains?

Must the existence of a canal therefore be ruled out completely? It is possible that a canal was dug later in order to counterbalance the movement of the river. The temple was possibly founded near a Nile branch, but its gradual movement may have required such an adjustment. Only on-site surveys may help answer this question without a doubt. Nevertheless, we can start analysing the results of Bisson de la Roque's excavations despite the imprecision of the archaeological information. He offers the following description of a survey carried out in 1932 to the west of the tribune:

> "À l'ouest de cette construction [the tribune] une coupe de terrain nous donne: une couche de limon noir au niveau -0m70, une couche de briques cuites brisées se terminant au niveau -0m80, une couche de briques crues brisées se terminant au niveau -1m20, une couche de terre argileuse jaune se terminant au niveau -1m50, une couche de limon noir se terminant au niveau -2 mètres sur un lit de terre argileuse jaune. Les deux couches de limon aux niveaux -0m70 et 2 mètres peuvent nous donner des lits de canaux postérieurs à l'abandon du dromos. Elles permettent de suppose à l'ouest de la tribune l'existence d'un canal."[51]

Two ideas can be retained: the depth of the survey and the mention of two layers of silt. First, Bisson de la Roque reached 3,16m below the level of the Ptolemaic temple (chosen as "level 0" in Medamud's excavations) but he assured having never dug deep enough to find the ancient flood levels.[52] However, the water level in antiquity must not have been so deep. It is difficult to estimate the date of the construction of the tribune. If we consider the constructions' program of the dromos and the Tiberius' gate, the tribune was probably constructed under Augustus.[53] We still need to compare the level of the Nile (in absolute height) with data from the excavations of Bisson de la Roque (in relative height). The

51 Bisson de la Roque, 1933, p. 9

52 "Ces quais [...] se trouvent sans doute à des niveaux, que les fouilles sans drainage ni pompe n'ont pas atteints" (Bisson de la Roque, 1933, p. 9).

53 There are no archaeological remains which would allow to date the tribune. One could accept Bisson de La Roque's postulate, dating the tribune to the same date as the enclosure wall (Revez, 2004, p. 495; Bisson de la Roque, 1933, p. 7–8). The construction of the so called *Porte de Tibère* dates from the reign of Augustus (Klotz, 2012, p. 242 – I refer also to the current publication of D. Valbelle and her team which will discuss the constructions of Augustus and Tiberius). Recently

Level 0 of the Bisson's excavations corresponds to the ground level of the Ptolemaic temple, which can be estimated to be 75,90 m.[54] This means that Bisson de la Roque came down to 72,74 m.[55] L. Gabolde and Cl. Traunecker have estimated the average level of the flood in the Middle Kingdom at Karnak to be 72,80 m[56] a.s.l. Following the same method, the average level of the flood at the end of the reign of Augustus can be estimated to be 74,69 m in Karnak.[57]

This method was questioned by Seidlmayer arguing the roman level should have been similar to those of the Third Intermediate Period (74,22 m).[58] If we adapt these measures to Medamud, the Nile should have flowed between 74,42 m and 73,95 m.[59] In both cases, Bisson seems to have reached the water levels of the time of the Tribune's construction. This also means that the water probably reached the Tribune.

More to the east, Bisson de la Roque describes "des couches de limon noir"[60] identified as Nile deposits transported by a canal during the Coptic period and he puts forward a hypothesis about the level of the ancient waters. In an unpublished report today kept in the Louvre archives, he dates these two silt layers to the fourth and fifth centuries AD based on the presence of a fired-brick construction.

DEVAUCHELLE, 2015 examined the demotic graffiti graved on the tribune dated it between the II BC and the I AC.

54 I thank E. Laroze for this information.

55 Ground level of the temple – Level Survey = 75,90 m – 3,16 m = 72,74 m.

56 For details I refer to the explanation given by GABOLDE, 1998, p. 196, n. 72: "Je me fonde sur les données rassemblées par Claude Traunecker (Karnak IV (=Kémi 21) 1971, p. 186–187) avec un maximum moyen atteint vers le 10 septembre grégorien à 76.50 m (moyennes entre 1873 et 1902) auxquelles je retranche l'exhaussement de la vallée évalué à 3.30 m (0.96 mm par an *cf.* GOYON/GOLVIN, 1987, p. 123); entre 1960 avant J.C. et 1900 après. J.C.".

57 The flood level under the reign of Augustus should be 1,82 m lower than the level maximum calculated by Traunecker for 1900 AD. This calculation includes the raising of the valley from the time of Augustus. I take the raising of 0,96mm annually established by J. Cl Goyon between the end of the reign of Augustus (14 AD) and 1900 AD.

58 SEIDLMAYER, 2001, p. 63–73. For the roman level see Ibid., p. 67.

59 SEIDLMAYER 2001, p. 67–68 (difference between Luxor and Karnak). A difference of 27 cm can be estimated between Karnak and Medamoud (6,75 cm/km). This measure is however approximated because the Nile's course is not established.

60 BISSON DE LA ROQUE, 1933, p. 9

Comparing mesures of the flood in Karnak and Medamud.		
Karnak's levels	Average level of the flood in the Middle Kingdom according to L. Gabolde.	72,80 m
	Average level of the flood evaluated under Augustus according to L. Gabolde	74,69 m
	Average level of the flood evaluated under Augustus according to Seidlmayer	74,33 m
Medamud's levels	Floor level of the Ptolemaic temple	75,90 m
	Level reached excavations west of the podium by Bisson de la Roque.	72,74 m
	Average level of the flood evaluated under Augustus according to L. Gabolde	74,42 m
	Average level of the flood evaluated under Augustus according to Seidlmayer	73,95 m

Table 1: Comparing measures of the flood in Karnak and Medamud

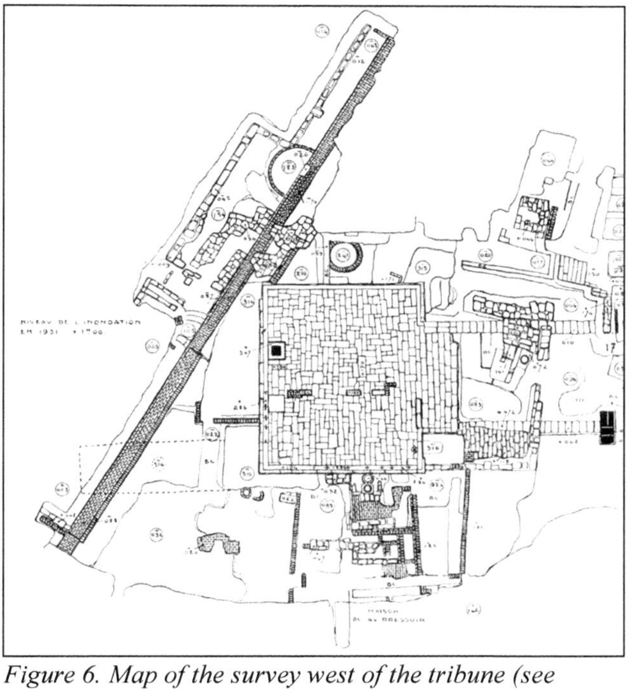

Figure 6. Map of the survey west of the tribune (see BISSON DE LA ROQUE, 1933, pl. I)

"Ce canal me paraît s'être déplacé obliquement en s'écartant vers l'ouest et en s'élevant au cours des siècles. Je crois reconnaître un parapet en briques cuites au niveau 2m25 où une monnaie de Constantin me fait supposer un nouvel aménagement du canal vers la fin du IVème s. Ce parapet donne l'apparence d'un trottoir. Il nous donnerait la ligne et le niveau du dernier canal ayant desservi la tribune. Un parapet en pierres provenant du temple, situé plus à l'ouest, à 1m 63 nous donnerait la ligne du canal vers la fin du Vèmes, à l'époque du village copte dont nous avons des restes de dallage dans ces parages. Ce dernier parapet formait l'arasement en grès placé sur la carte de l'expédition de Bonaparte. Le canal actuel passe à une petite distance à l'ouest".[61]

One photo is joined to the report illustrating the hypothesis of the excavator.

Figure 7. Photograph of the excavations of Bisson de la Roque, showing the emplacement of the silt layers and proposing a date (Louvre Archives, 2013 DAE1/9: 1931). The annotations are Bisson de la Roque's handwriting.

This picture shows what the excavator thought of the migration of the river. The fourth century canal lies east of the one of the fifth century. However, the silt levels seem too high to be interpreted as Coptic deposits of the Nile. These layers are at 75,20 m and 73,90 m a.s.l. (-0,70m and -2m respectively below the *level 0* according to the measurements taken by Bisson de la Roque). If these

61 Louvre Archives. 2013 DAE1/9: 1931 (report of F. Bisson de la Roque sent to P. Jouguet and Ch. Boreux). I thank V. Rondot, the head of the Egyptian department of the *Musée du Louvre*, for the permission to publish the records of the Medamud excavations kept in their archives. I also thank E. David, without the help of whom my work in the Louvre wouldnot have been possible.

measurements are compared with the estimated Roman flood level (74,42/73,95 m a.s.l.), one notes that these measurements are closer to levels of flood deposits under Augustus. Thus, the levels described by Bisson and attributed to the Coptic period could be in reality a deposit of an occasional Roman flood. Bisson de la Roque dates the silt layers only on the presence of a Roman coin and the fired-brick structure. But this structure might have been built later. Only an analysis of the silt could date the layers certainly. However Bisson gives no description of the contents of this silt (for example, is it really a silt level or did it carry rollers also?). It is impossible to offer a definitive answer to this problem until new excavations provide more information.

We diagrammed in the image below flood levels from the plane given by the excavators of the tribune.

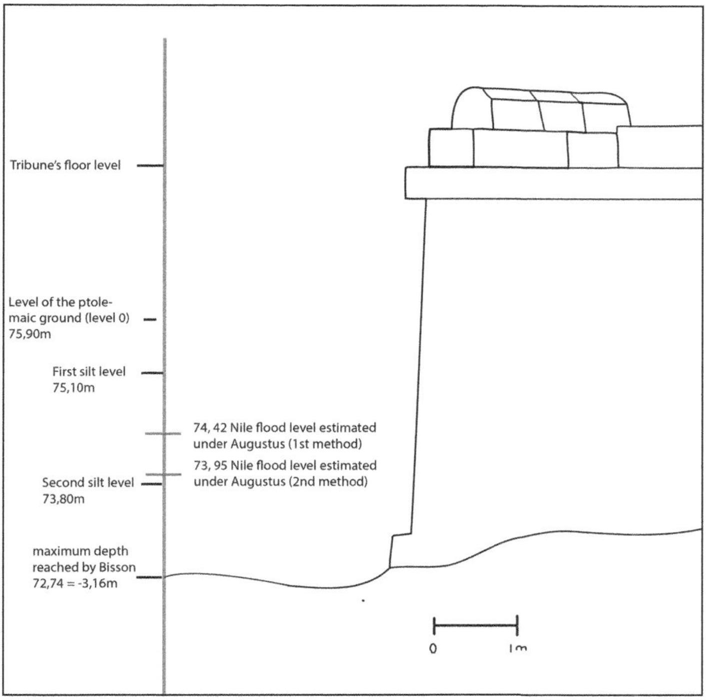

Figure 8. Assessment of the level's flow in Medamud according to the description of Bisson de la Roque and the estimation of the Nile's flood level in Karnak. Schema made by the author from a plan of the tribune by Cl. Robichon (see Bisson de la Roque, 1933, fig. 9).

To sum up, the datas given by Bisson de la Roque have to be modified. Surveys should be made to verify the stratigraphy, and specify the geomorphological stratigraphy to verify what level the water reached near the tribune. However, the silt levels can certainly correspond to Roman levels and not to Coptic ones as presumed. These surveys would also allow to find whether it was a channel or one of the Nile's arm that reached Medamud's tribune in Roman times.

Figure 9. Aerial view of Medamud when the River Nile was in flood (see PIERRE, 1978, p. 59). I thank P. Grandet for this reference. Note that in the original book, this photograph was reversed. I preferred to return the photograph to match the reality.

5. Conclusion

Medamud has been chosen as an example to understand the relationship between the Nile and a temple. Unlike what the excavators thought, migrations of the river bed may help to ascertain whether the foundation of Medamud was originally closer to the main branch of the Nile or to one of its side branches. The idea of a temple distant from the Nile should in that case be abandoned. Thereafter, throughout the history of Medamud, the link with the river would be preserved: the tribune of the Roman temple was used as an access to a waterway.

The existence and usefulness of the waterway can be proved with the blocs of the gate of Tiberius, carried by water[62].

In the future, new archaeological investigations will clarify the stratigraphy of the quay in order to determine if the tribune had a direct access to the Nile or one of its arms. However, as the colloquium in Mainz has shown, only a global framework of the Theban valley can give a clear answer to all these problems.

Addenda

Since I wrote this paper, the mission Médamoud-Porte de Tibère (IFAO/Sorbonne) has expanded its aims with a new Medamud mission which purpose is to restart the archaeological activities on site.[63] In September 2015 a geomagnetic survey of Medamud has been carried out through a partnership between the Sorbonne and the laboratory UPMC (University Paris VI Jussieu –UMR 7619 METIS)[64]. Its aims were to survey the archaeological areas never excavated by Bisson de la Roque and to reexamine some formerly excavated areas in order to verify his assumptions. Although the results are not completed yet (and go beyond this paper), we can advance that a zone in front of the tribune has been surveyed and seems to prove that the water reached the tribune, as we supposed, through a canal coming in a straight line from the west.

62 As Brand, 2000, p. 191 has proposed, some of these blocks may come from the temple of Sethy I in Qurna. The most likely explanation for their transportation is the use of the Sethy I temple's channel in order to bring these blocks to the Nile and then to Medamud. Otherwise the builders could use other monuments in ruins nearer to Medamud. It is precisely the waterway which facilitated the transport of these blocks. Note however, that unlike P. Brand, I do not believe that all blocks are from the temple of Qurna..

63 Directed by D. Valbelle and myself, under the auspices of the IFAO, the Sorbonne university and the Labex RESMED.

64 The survey was directed by J. Thiesson, F. Réjiba and R. Guérin. Chr. Sanchez has performed the survey.

Acknowledgements

This article could not have been possible without the help of Annie Forgeau and Chloé Ragazzoli. Jonathan, thank you for your patience.

Bibliography

ALLEAUME, GHISLAINE, Les systèmes hydrauliques de l'Egypte pré-moderne, essai d'histoire du paysage, in: Itinéraires d'Egypte. Mélanges offerts au père Maurice Marin (Bibliothèque d'études 107), ed. by CHRISTIAN DÉCOBERT, Le Caire 1992, p. 301-322.

BALL, JOHN, The "Description de l'Egypte" and the course of the Nile between Isna and Girga, in: Bulletin de l'Institut égyptien 14 (1932), p. 127-139.

BARGUET, PAUL, Le temple d'Amon-Rê à Karnak. Essai d'exégèse (Recherches d'archéologie, de philologie et d'histoire 21), 3rd ed., Cairo 2008.

BISSON DE LA ROQUE, FERNAND, Rapport sur les fouilles de Médamoud (années 1924-1925) (Fouilles de l'Institut français d'archéologie orientale III[1]), Cairo 1926.

ID., Rapport sur les fouilles de Médamoud (1926) (Fouilles de l'Institut français d'archéologie orientale IV[1]), Cairo 1927.

ID., Rapport sur les fouilles de Médamoud (année 1929) (Fouilles de l'Institut français d'archéologie orientale VII[1]), Cairo 1930.

ID., Rapport sur les fouilles de Médamoud (1930) (Fouilles de l'Institut français d'archéologie orientale VIII[1]), Cairo 1931.

ID., Rapport sur les fouilles de Médamoud (1931 et 1932) (Fouilles de l'Institut français d'archéologie orientale IX[3]), Cairo 1933.

BISSON DE LA ROQUE, FERNAND/CLÈRE, JACQUES-JEAN, Rapport sur les fouilles de Médamoud (1927) (Fouilles de l'Institut français d'archéologie orientale V[1]), Cairo 1928.

ID., Rapport sur les fouilles de Médamoud (1928) (Fouilles de l'Institut français d'archéologie orientale VI[1]), Cairo 1929.

BORAIK, MANSOUR, Excavations of the Qays and the Embankment in front of Karnak Temples. Preliminary Report, in: Karnak XIII (2010), p. 65-78.

BORAIK, MANSOUR/GHILARDI, MATHIEU, et al., Geomorphological Investigations in the western Part of the Karnak Temple (Quay and ancient Harbour). First Results, in: Karnak XIII (2010), p. 101-109.

BRAND, PETER, The Monuments of Sethi I, Epigraphic, Historical and Art Historical Analysis (Probleme der Ägyptologie XVI), Leiden 2000.

BUNBURY, JUDITH/GRAHAM, ANGUS, The ancient Landscapes and waterscapes of Karnak, in: Egyptian Archaeology 27 (2005), p. 17-19.

BUNBURY, JUDITH/GRAHAM, ANGUS/HUNTER, MORAG A., Stratigraphic Landscape Analysis. Charting the Holocene Movements of the Nile at Karnak through Ancient Egyptian Time, in: Geoarchaeology 23-3 (2008), p. 351-373.

BUTZER, KARL, Early Hydraulic Civilization, A Study in Cultural Ecology (prehistoric archaeology and geology), Chicago 1976.

CABROL, AGNES, Les voies processionnelles de Thèbes (Orientalia lovaniensia analecta 97), Leuven 2001.

CHARLOUX, GUILLAUME/MERSAM, ROMAIN, Karnak avant la XVIIIe dynastie, contribution à l'étude des vestiges en brique crue des premiers temples d'Amon-Rê (Études d'égyptologie 11), Paris 2012.

COTTEVIEILLE-GIRARDET, REMI, Rapport sur les fouilles de Médamoud (1930). La verrerie – les graffitis (Fouilles de l'Institut français d'archéologie orientale VIII[2]), Cairo 1931.

ID., Rapport sur les fouilles de Médamoud (1931), Les monuments du Moyen Empire (Fouilles de l'Institut français d'archéologie orientale IX[1]), Cairo 1933.

ID., Rapport sur les fouilles de Médamoud (1932), Les Reliefs d'Aménophis IV Akhenaton (Fouilles de l'Institut français d'archéologie orientale XIII), Cairo 1936.

DRIOTON, ÉTIENNE, Rapport sur les fouilles de Médamoud, Les inscriptions (Fouilles de l'Institut français d'archéologie orientale III[2]) Cairo 1926.

ID., Rapport sur les fouilles de Médamoud, Les inscriptions (Fouilles de l'Institut français d'archéologie orientale IV[2]), Cairo 1927.

ID., Les quatre Montous de Médamoud, palladium de Thèbes, in: Chronique d'Égypte XII (1931), p. 259-270

ID., Les fouilles à Médamoud, in: Chronique d'Égypte VII (1932), p. 92-93.

EDER, CHRISTIAN, Die Barkenkapelle des Königs Sobekhotep III in Elkab. Beiträge zur Bautätigkeit der 13. und 17. Dynastie an den Göttertempeln Ägyptens (Elkab VII), Tournhout 2002.

GABOLDE, LUC, Le "grand château d'Amon" de Sésostris Ier à Karnak (Mémoires de l'Académie des inscriptions et belles-lettres XVII), Paris 1998.

ID., L'Implantation du temple. Contingences religieuses et Contraintes géomorphologiques, in: Egypte, Afrique & Orient 68 (2013), p. 3-12.

ID., Les origines de Karnak et la genèse de la théologie d'Amon, in: Bulletin de la Société française d'égyptologie 186-187 (2013), p. 13-35.

ID., Karnak, Amon-Rê. La genèse d'un temple la naissance d'un dieu, (forthcomming).

GABOLDE, LUC/RONDOT, VINCENT, Une catastrophe antique dans le temple de Montou à Karnak-Nord, Bulletin de l'institut français d'archéologie orientale 93 (1993), p. 245-264.

GOLVIN, JEAN-CLAUDE, Enceintes et portes monumentales des temples de Thèbes à l'époque ptolémaïque et romaine, in: Hundred-Gated Thebes, Acts of a colloquium on Thebes and the theban area in the graeco-roman period, ed. by SVEN P. VLEEMING, Leiden 1995, p. 33 and 41.

ID., Voyage en Egypte ancienne, Paris 1999.

GOLVIN, JEAN-CLAUDE/GOYON, JEAN-CLAUDE, Les bâtisseurs de Karnak, Paris 1987.

GOLVIN, JEAN-CLAUDE, et al., Essai d'explication des murs "assises courbes". À propos de l'étude de l'enceinte du grand temple d'Amon-Rê à Karnak, in: Comptes rendus de l'Académie des inscriptions et belles-lettres 104 (1990), p. 905-945.

GRAHAM, ANGUS, Islands in the Nile. A Georarcheological Approach, in: Cities and urbanism in Ancient Egypt (Denkschriften der österreichischen Akademie der Wissenschaften Wien 40), ed. by MANFRED BIETAK, Vienna 2010, p. 125-140.

GRAHAM, ANGUS, et al., Reconstructing Landscapes and Waterscapes in Thebes, Egypt, in: Journal of Archaeological Science 39 (2012), p. 135-142.

HILLIER, JOHN/BUNBURY, JUDITH/GRAHAM, ANGUS, Monuments on a Migrating Nile, in: Journal of Archaeological Science 34, 7 (2007), p. 1011-15.

JEFFREYS, DAVID, Survey of Memphis I, London 1985.

JOMARD, EDME-FRANÇOIS/JACOTIN, PIERRE, Carte topographique de l'Egypte et de plusieurs parties des pays limitrophes levée pendant l'expédition de l'armée française par des ingénieurs-géographes, les officiers du génie militaire et les ingénieurs des ponts et chaussées, assujettie aux observations des astronomes, Paris 1818.

KLOTZ, DAVID, Cesar in the City of Amun (Monographies Reine Élisabeth 15), Bruxelles 2012.

LAUFFRAY, JEAN, Abords occidentaux du premier pylône de Karnak, le dromos, la tribune et les aménagements portuaires, in: Kêmi XXI (1971), p. 79-101.

LEITZ, CHRISTIAN, Quellentexte zur Ägyptischen Religion I. Die Tempelinschriften der griechisch-römischen Zeit (Einführungen und Quellentexte zur Ägyptologie 2), Münster 2004.

LÉVY, JACQUES/LUSSAULT, MAURICE, Dictionnaire de la géographie, Paris 2008.

MASQUELIER-LOORIUS, JULIE, Séthi Ier et le début de la XIXe dynastie, Paris 2013.

NIVET-SAMBIN, CHANTAL, À Médamoud, le temple du M.E. était orienté vers le Nil, in: Hommages à Jean-Claude Goyon (Bibliothèque d'études 143), ed. by LUC GABOLDE, Cairo 2008, p. 313-329.

PIERRE, BERNARD, Le Nil, des sources au delta, Paris, 1978.

PILLET, MAURICE, Rapport sur les travaux de Karnak (1923-1924), in: Annales du Service des Antiquités de l'Égypte 24 (1924), p. 51-88.

ID., Rapport sur les travaux de Karnak (1924-1925), in: Annales du Service des Antiquités de l'Égypte 25 (1925), p. 1-24.

POCOCKE, RICHARD, A description of the East, London 1743.

PORTER, BERTHA/MOSS ROSALIND, Topographical Bibliography of Ancient Egyptian Hieroglyphic Texts, Reliefs and Paintings, Vol. V, Upper Egypt sites, 2nd ed, Oxford 1962.

ID., Topographical Bibliography of Ancient Egyptian Hieroglyphic Texts, Reliefs and Paintings, Vol. II, Theban Temples, 2nd ed., Oxfrod 1972.

POSTEL, LILIAN, Le paysage monumental de la vallée du Nil sous le règne de Sésostris III, in: Sésostris III, pharaon de légende, Catalogue d'exposition, Lille, Palais des Beaux-Arts, 9 octobre 2014-25 janvier 2015, ed. by FLEUR MORFOISSE/GUILLEMETTE ADREU-LANOE, Lille 2014, p. 114-132.

RELATS MONTSERRAT, FELIX, La redécouverte du temple primitif de Médamoud, in: Göttinger Miszellen 240 (2014), p. 123-124.

REVEZ, JEAN, Medamoud, in: Encyclopedia of the Archeology of ancient Egypt, ed. by KATHRYN A. BARD, London 1999, p. 475-481.

ID., Une stèle commémorant la construction par l'empereur Auguste du mur d'enceinte du temple de Montou-Rê à Médamoud, in: Bulletin de l'institut français d'archéologie orientale 104 (2004), p. 495-510.

ROBICHON, CLÉMENT/VARILLE, ALEXANDRE, Médamoud, les fouilles du musée du Louvre, 1938, in: Chronique d'Égypte XIV, 27 (1939), p. 82-87.

ID., Description sommaire du temple primitif de Médamoud (Recherches d'archéologie, de philologie et d'histoire 11), Cairo 1940.

SAID, RUSHDI, The River Nile, Geology, Hydrology and Utilization, Oxford 1993.

SAINTE-FARE GARNOT, JEAN, Le temple primitif de Médamoud (Haute-Égypte), in: Comptes rendus de l'Académie des inscriptions et belles-lettres 88, 1 (1944), p. 65-74.

SAMBIN, CHANTAL, Medamud, in: The Oxford Encyclopedia of Ancient Egypt, ed. by DONALD REDFORD, Oxford 2001, p. 351-352.

SAUNERON, SERGE/YOYOTTE, JEAN, La naissance du monde selon l'Egypte ancienne, in: La naissance du Monde, SourcOr I (1959), p. 17-92.

SCHENKEL, WOLFGAN, Kai-Anlage, in: Lexikon der Ägyptologie III (1980), p. 293-295.

SEIDLMAYER, STEPHAN JOHANNES, Historische und moderne Nilstände. Untersuchungen zu den Pegelablesungen des Nils von der Frühzeit bis in die Gegenwart, Berlin 2001.

THIERS, CHRISTOPHE, Armant (Hermonthis), in: The Encyclopedia of Ancient History, ed. by SUSANNE BAGNAL, Oxford 2013, p. 720-722.

VARILLE, ALEXANDRE, Karnak Nord I (Fouilles de l'Institut français d'archéologie orientale XIX), Cairo 1943.

YOYOTTE, JEAN, Remarques sur les processions de génies au temple d'Opet, in: CONSTANT DE WIT, Les inscriptions du temple d'Opet à Karnak (Bibliotheca aegyptiaca XI), Bruxelles 1958, p. XI-XVI.

The Nile in the Fayum

Strategies of Dominating and Using the Water Resources of the River in the Oasis in the Middle Kingdom and the Graeco-Roman Period

CORNELIA RÖMER

1. Introduction

The Fayum is usually called an oasis, but unlike the *real* oases in the western desert, which are irrigated by the underground Nubian aquifer, the Fayum receives its water supply exclusively from the Bahr Yusuf, a side branch of the Nile which leaves the river in the area of Assiut and enters the Fayum at the *Lahun Gap*.[1] Thus, the Fayum was (and is) subject to the same conditions as the floodplains in the Nile valley.

The Fayum is a depression west of the Nile valley measuring roughly 70 x 60 km. At its lowest point, which lies in the north, it reaches to 53 m below sea level, while the Bahr Yusuf enters the depression in the south at approx. 25 m above sea level. These two points mark the extremes of a sloping, oval-shaped landscape, covered for millennia by a lake and a swamp with changing tides according to flood or drought seasons; only the plateau approx. 23 m of height in the south-east of the depression where the Bahr Yusuf first enters emerged from the swamp; on this plateau, the only settlement built in the centre of the Fayum prior to the Middle Kingdom was located: Shedet, from the Ptolemaic period onwards called Arsinoe or Krokodilopolis, renamed at the end of the second century B.C. Ptolemais Euergetis, and is now known as Medinet el-Fayum, the

1 The Lahun Gap lies west of the modern town of Beni Suef at 29°13'07.94" N, 30°57'21.27" E.

capital of the province. As long as no measures were taken to regulate the water stream coming in with the Bahr Yusuf, the *Land of the Lake* (= Fayum) was to remain largely covered by the huge swamp, home of the crocodile god Sobek.

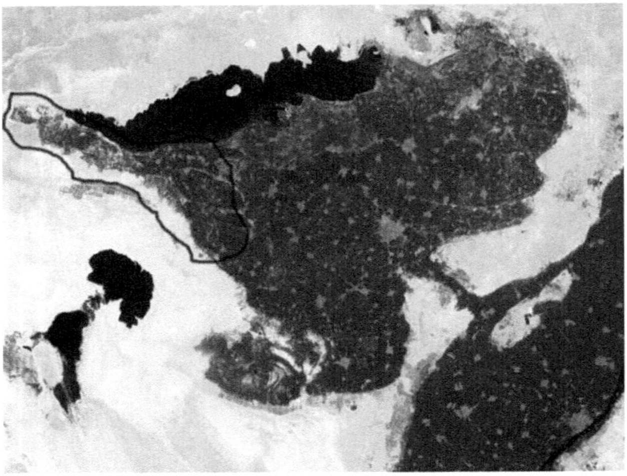

Figure 1. The Fayum Oasis; in the north-west, the area of the archaeological survey of the Fayum Survey Project is encircled. (Google Earth).

The problem here was and is that the Fayum, unlike the Nile valley, does not have an exit at the other end of the depression as does the Nile to the Mediterranean.[2] All water coming into the Fayum collects in the lake in the north, today the Birket Qarun, of which the deepest point lies at -53 m. Whoever wanted to make this sloping depression fertile and useful had to control the water coming into it, and by doing so diminish the swamp and the lake. The task was therefore, on the one hand, to make sure not to allow too much water into here in order to reduce the level of the lake and, on the other, to provide enough water for the fields which had emerged from the shrinking lake and the surrounding marshes.

The development of this swampy depression into one of the most fertile areas of Egypt with extensive fields bordering a shrunken lake to the north, from prehistoric to modern times can only be outlined briefly here.[3] There were two

2 For a supposed underground exit of the lake see below.

3 For a very short overview with impressive maps see HASSAN/TASSIE, 2006, p. 37–40; more elaborate and focussing on the prehistoric period: HASSAN, 1986, p. 483–501.

periods, in which the Fayum received special attention from the rulers of Egypt who saw the potential of this landscape.

The first highlight of the development of the Fayum dates to the period of the Middle Kingdom, the second to the early Ptolemaic period. At both times strategies were found to reduce the expansion of the lake and to enlarge the area of arable land between the plateau and the lake.[4]

2. The First Golden Age of the Fayum: the Middle Kingdom

The Pharaohs of the Twelfth Dynasty, in particular Amenemhet III (who was still venerated as a god in the Ptolemaic and Roman periods) managed to stabilise the level of the lake at between 17 and 20 m above sea level, while the first

4 I do not share the new view on the irrigation problems of the Fayum, as proposed by HAUG, 2013. Haug claims that the Fayum did not receive a perennial water supply from the Bahr Yusuf, but was irrigated only during flooding times, whereas the Bahr Yusuf would dry out in winter. This led, so Haug, to the unstable situation at the fringes of the Fayum until the nineteenth century when the Ibrahamiya Canal was built along the Bahr Yusuf in the Nile valley. While there is no doubt that the centre of the Fayum around Arsinoe has always been more fertile (and still is today), there is also no doubt that the fringes were highly developed agricultural land for many centuries during the Ptolemaic and Roman periods. Haug's view is mostly based on the observation of a British traveller who visited the Fayum in 1830: ST. JOHN, 1845; he seems to describe a barren Bahr Yusuf in the winter; however, both citations, introduced by Haug, fail to convince: on p. 182 St. John describes the scene at an unknown river bed, where a local man seems to have indicated a wrong name for the canal the travel group was crossing (a little later, we hear of rich water streams; where would that water come from, if not from the Bahr Yusuf ?); on p. 192 we hear of the northern continuation of the Bahr Yusuf at Maidum, not about the Bahr Yusuf in the Fayum. Throughout his travelogue, St. John hails the fertility of the Fayum. Also Haug's claim that "the irrigation system was unable, on the whole, to efficiently transport Nile mud to the margins in significant quantities", is not correct. During the drillings in the canals and excavations at Philoteris at the far end of a feeder canal (see below) we found rich silt sediments both in the canals and the adjacent fields (forthcoming in my "The Fayum Survey Project. The Themistou Meris"). Mud bricks used here had (at least in part) the same magnetic value as bricks in the Nile valley.

Ptolemies took more radical steps to make the lake shrink down to approx. 45 m below sea level which more or less equals the measure of today.

In his study *Contribution to the Geography of Egypt* of 1939, John Ball gave what – at that time – was considered a thoroughly investigated picture of the development of changing levels, evaporation and salinization of the lake from prehistory to his own days.[5] Even though the main results of that study are still valid, many details have been reconsidered since then.[6]

Ball understood that Amenemhet's new water management in the Fayum had stabilised the lake to a level of between 17 and 20 m above sea level,[7] after it had been on a lower level for at least some time after the Old Kingdom.[8] The figure of 17-20 m is corroborated by the observation of location heights of known Middle Kingdom sites in the Fayum (from north to west, then south, and north again)[9]: Qasr el-Sagha (31 m a.s.l.), Soknopaiou Nesos (c. 20 m), Quta (20 m), Medinet Madi (26 m), cemetery at Tebtynis (24 m), Abgig (18 m) and Biahmu (18 m). When Herodotus saw the two monuments of Amenemhet III at Biahmu and described them as standing in the middle of the lake (II 149)[10], he was obviously visiting there when the lake was at its peak after the flooding.

5 Ball, 1939, p. 178–289; similar conclusions are reached by Shafei Bey, 1940, p. 283–327 and map in front of article. As the title says, Shafei's article is mostly based on An Nabulsi who was the governor of the Fayum under the Aiyubite Sultan el Salih Negm el Din in 1245-1246; An Nabulsi often seems to be too negative (longing for his Syrian homeland), and not very clear in his descriptions; however, in Shafei's article, it is doubtful whether misunderstandings were due to Shafei, or to Nabulsi himself. It is clearly impossible that the two ravines in the Fayum had not existed in Nabulsi's time (p. 286 and 301; see further below). On An Nabulsi see also Ball, p. 219–225. An English translation of An Nabulsi (approx. 50 % of the whole book) can be found on the web page of Queen Mary College, University of London, Rural Society of Medieval Islam (translation by Y. Rapoport). The description of the dam at the Bahr Yusuf in Chapter 1 is clearly a misunderstanding of what was going on at Lahun.

6 For the history of research on this subject after Ball see Hassan, 1986, p. 483–485.

7 On the activities of the Pharaohs of the Twelth Dynasty in the Fayum see Ball, 1939, p. 199–210; Hassan/Tassie, 2006, show the lake at +10 m and add: "although it fluctuated by +10 m from this level".

8 See Hassan, 1986, p. 491.

9 Measures are from Google Earth; cf. measures given in Davoli, 1998.

10 ἐν γὰρ μέϲῃ τῇ λίμνῃ μάλιϲτά κῃ ἑϲτᾶϲι δύο πυραμίδεϲ.

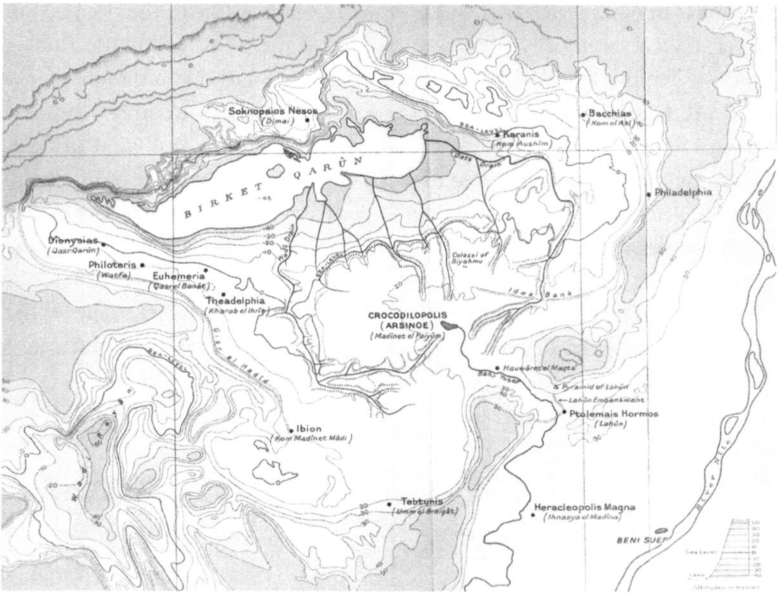

Figure 2. Map after BALL, *1939; p. 216.*

At that time (c. 425 B.C.) there may also have been the chance to observe that some of the water would flow back from the depression of the Fayum towards the Nile valley. Herodotus claims to have seen such a flow back (II 149)[11]. The high lake would have allowed that, according to Ball,[12] since the sediments in the Bahr Yusuf at the entrance to the Fayum at El-Lahun had not yet piled up to the height at which we find that area today (at 25 m). However, it seems possible that the extremely slow flow of some of the canals induced the beholders to assume that the water was flowing towards the Bahr Yusuf, not towards the lake.[13]

11 Τὸ δὲ ὕδωρ τὸ ἐν τῇ λίμνῃ αὐθιγενὲς μὲν οὐκ ἔστι (ἄνυδρος γὰρ δὴ δεινῶς ἐστι ἡ ταύτῃ), ἐκ τοῦ Νείλου δὲ κατὰ διώρυχα ἐσῆκται, καὶ ἓξ μὲν μῆνας ἔσω ῥέει ἐς τὴν λίμνην, ἓξ δὲ μῆνας ἔξω ἐς τὸν Νεῖλον αὖτις. "The water in the lake does not originate there (for the area is exceedingly arid), but is brought into it by a channel from the Nile; six months it flows into the lake, and six months back into the Nile".

12 BALL, 1939, p. 204.

13 Today, it is sometimes difficult to understand the direction of the water flow; for the minimal speed of the canals see also SHAFEI BEY, 1940, p. 290; he observed a water level drop of only 5 cm over 21 km in the Bahr Wardan.

Be that as it may, the idea that water was to flow back from the depression into the river makes sense only if we assume that at the point where the Bahr Yusuf entered the Fayum at El-Lahun, coming from the south, where there was also a channel allowing the water to escape toward the north; or in other words, a continuation of the Bahr Yusuf towards the north, if one did not want the water to proceed uncontrolled in that area. Such a continuation would finally re-join the river Nile. It is therefore strange that Ball does not indicate any sort of controlled waterway towards the north from the Bahr Yusuf on his otherwise splendid map (Figure 2). However, where else would the water go coming back out of the Fayum (certainly neither flowing upstream in the Bahr Yusuf, nor to the east towards the river Nile, because the path is blocked by a range of elevations)?

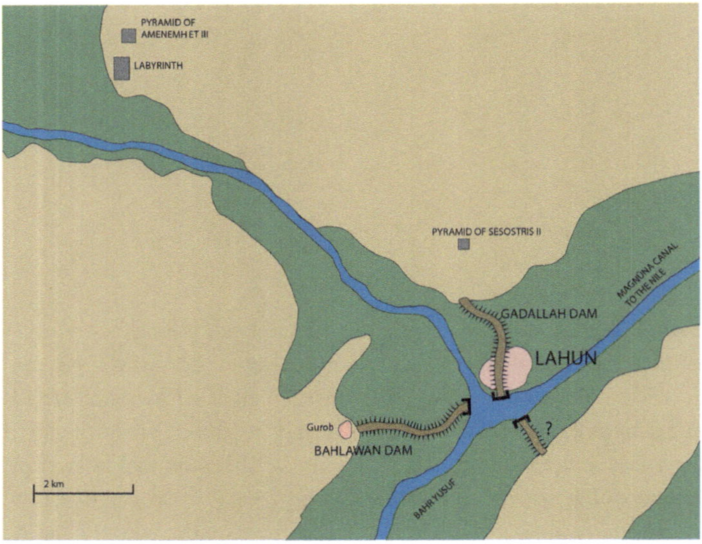

Figure 3. Map after S*hafei* B*ey*, *1940; drawing by I. Klose.*

The entrance to the Fayum at the *Lahun Gap*, where the Bahr Yusuf coming from the Nile valley cuts through a range of elevations of approx. 80 m in height,[14] is marked by two pharaonic settlements, Gurob in the south – predominantly settled in the New Kingdom, but perhaps also earlier –, and Lahun – a Middle Kingdom cemetery and pyramid – in the north. Between the two, remains of two dams are still visible which closed the gap in a curving, bell shaped, line

14 The first breakthrough of the Bahr Yusuf at this point is supposed to have taken place in the mid-Pleistocene.

towards the east (see figure 3).[15] One dam spanned the gap between the Bahr Yusuf and Gurob in the south (the Bahlawan Dam), the other ran parallel to the Bahr Yusuf between the village of El-Lahun and the height of the Lahun pyramid (the Gadallah Dam). After recent drillings in these dams, the Middle Kingdom date of the Gadallah dam has been corroborated: [16] During the Middle Kingdom, the Gadallah Dam and a continuation to the east (?) blocked the flow of the Bahr Yusuf towards the north; however, a sluice, most likely at the modern village of El-Lahun (in Ptolemaic and Roman times *Ptolemais Hormou, The Harbour of the Ptolemies*) may have directed this flow either to the west and into the Fayum depression (when the sluice was locked), or towards the north (when the sluice was open). It seems that the important step taken in the Middle Kingdom was to create means of controlling the water at the Lahun Gap with dykes, a sluice, *and* an *overflow* towards the north, the canal which is nowadays called the Magnûna Canal. The Bahlawan Dam, instead, now seems to be a construction of the Ptolemaic period.[17] It may have worked as an additional tool to regulate the influx into the depression at this point.

The existence or non-existence of such a continuation of the Bahr Yusuf towards the north is decisive not only for the question of whether or not we may assume that water flowed backwards out of the depression, but also with respect to the problem whether and to what degree the Fayum was used as an overflow for excess water from Upper and Middle Egypt during the Middle Kingdom. It seems that also excessive sudden floods could be controlled to a certain degree with the measures taken in the Middle Kingdom.

2.1 Appendix: Was there an underground exit of the lake?

It is rather unlikely that the lake ever had an underground exit, as Herodotus had been told by the locals (II 150)[18], and Ball continued to believe. Ball based

15 For this Middle Kingdom dam see GARBRECHT/JARITZ, 1990, p. 140–149; BALL, 1939, p. 212–213 still considered at least part of this wall the work of the early Ptolemies, corroborated by HASSAN/TASSIE, 2006, p. 40.

16 See HASSAN/TASSIE, 2006, p. 40.

17 Ibid.

18 ἔλεγον δὲ οἱ ἐπιχώριοι καὶ ὡϲ ἐϲ τὴν Σύρτην τὴν ἐϲ Λιβύην ἐκδιδοῖ ἡ λίμνη αὕτη ὑπὸ γῆν, τετραμμένη τὸ πρὸϲ ἑϲπέρην ἐϲ τὴν μεϲόγαιαν παρὰ τὸ ὄροϲ τὸ ὑπὲρ Μέμφιοϲ. “The locals told me also that the lake has an underground outlet to the Libyan Syrtis, turning into western direction and into the continent and running parallel to the mountains which are above Memphis”.

that assumption mostly on the salinity of the lake, which he considered to be too low for being a result of evaporation only, however, admitting that such an underground exit had most likely not existed anymore from the middle of the nineteenth century.[19] Such an exit would, according to Ball, have been level with the Nubian aquifer and at least part of the water would have gone to the Wadi Rayan by percolation. Shafei Bey's calculation of the salinity did not lead to such an assumption; he called the lake "water proof as if lined with Indian rubber".[20]

It has to be said that Ball's assumption rests on incorrect measures of the lake in Ptolemaic and Roman times.[21] Ball does not take into consideration that, by the dramatic enlargement of the arable areas at that time, much more water was distributed and used before it reached the lake. We now know that at that time the lake stood more or less at the same level as today, at –45 m; the well dug in the area to the east of the lake (north of Karanis) in the early Ptolemaic period,[22] obviously drawing from an underground basin and reaching down to –4.6 m, is therefore obviously not connected to the level of the lake, but to the surrounding canals. The same applies to a rectangular pit which was connected to a 34 m long underground tunnel, and which reached to +5 m. Beside the fact that the levels of these two installations do not match the level of the lake at that time, it is unlikely that there was an aquifer (coming underground from the lake) somewhere here, because that would have been used more extensively. However, this area was never impressively fertile.[23] The same observation should be applied to a well found by the American excavators in Karanis which reaches down to -7 m. This well cannot have been connected to the lake, but was certainly drawing from the nearby Bahr Wardan.[24]

19 Ball, 1939, p. 285–289.

20 Shafei Bey, 1940, p. 293.

21 Ball, 1939, p. 286.

22 Ibid., fig. 24 on p. 211, and fig. 25 on p. 212, both after Caton-Thompson/Gardner, 1934, p. 17–18; Ball, 1939, insists that the entire system of channels up here to the north of Karanis must have been fed by underground water from the lake (p. 217). However, as is the case today, this part of the Fayum most likely received its water supply from the Bahr Wardan, of which a side arm branches off before arriving at Karanis. See also note 23.

23 Cook, 2011.

24 With a different interpretation Ball, 1939, p. 218–219.

3. The Second Golden Age of the Fayum: From the Early Ptolemaic into the Roman Period

The first Ptolemies recognised the opportunities to create new fields for the thousands of new settlers who flocked into the land on the Nile during the third century B.C.; they used the potential of this landscape by enforcing measures to reduce the level of the lake more extensively than before.

When Ptolemy I, son of Lagos, had himself crowned as Pharaoh of Egypt in Memphis in 306 B.C., he must already have had the plan to make Egypt heir and representative of the Greek culture, which until then had flourished in mainland Greece and particularly in Athens. Ptolemy initiated the Museum and the Library in Alexandria; his son, Ptolemy II, continued in his footsteps, intensifying the work of his father. According to the teachings of Aristotle that all visible phenomena in the world can be investigated and compared with each other to draw the right conclusions about their existence and their meaning, the natural sciences were also fostered in Alexandria, and flourished: medicine, astronomy, geography, hydrology and agriculture.[25]

The Greeks felt enormous admiration for the age and depth of Egyptian culture; they had been educated for this admiration after Herodotus had visited the country on the Nile; however, taking the stimulus of the teachings of Aristotle's school, the Greeks occupied Egypt with new ideas about how to use its natural resources, how to improve the crops to be planted and how to gain more land for agricultural use. On the estate of Apollonius, the financial minister of Ptolemy II, at Philadelphia in the eastern Fayum, experiments were carried out with new crops, for instance poppy seeds for the production of oil and irrigation facilities.[26]

Of course, for Ptolemy I and then Ptolemy II, the development of the Fayum was not just a test model. The Ptolemaic kings needed soldiers and administrators for their government; tens of thousands of Greek speaking settlers flocked into Egypt within the first half of the third century B.C., looking for a better and peaceful life. They all expected fields and housing to be given to them. The Ptolemies tried to solve that problem in a civil manner: they did not take land away from the indigenous people of the occupied territory on a grand scale, but

25 For the Greek culture, which gave so many new impulses to the Egyptian culture, see Fraser, 1972.

26 A short introduction to the activities on that estate and its manager Zenon is given by Clarysse/Vandorpe, 1995, in particular p. 93; for innovations in agriculture in the Fayum at this time see Thompson, 1999, p. 123–138.

sought to develop new areas for settlements and agriculture. The Fayum became the centre of their efforts, but reclamation programmes were initiated also in the Nile valley and the Delta land.

Looking at the situation in the Fayum at that time, it must have been clear to the Ptolemaic kings that it would not be an easy task to transform that swampy depression west of the Nile valley into a fertile landscape.

During the time of the Ptolemies and the Romans, the shore line of that lake was more or less at the same level as it is now, namely at -45 m.[27] This is corroborated by three archaeological sites in the east which lie on that level close to the lake, one of them being Qaret Rusas; they are unpublished until now.[28]

Thus, the big drop of the lake level from +20 to -45 m most likely occurred only step by step when the first Ptolemies took an interest in that landscape, because there are no settlements registered anywhere in the Fayum foundet in the period between the Middle Kingdom and the Ptolemaic period.

What did the Ptolemies do, and why were they so much more successful than the Pharaohs of the Middle Kingdom?

The Ptolemies added to the installations at the Fayum Gap by constructing a dam west of the Bahr Yusuf before it enters the gap (see above with footnote 15 and figure 3). It is also clear that they maintained the northern extension of the Bahr Yusuf towards Memphis with its sluice at El-Lahun = *Ptolemais Hormou, The Harbour of the Ptolemies*.[29]

Here, the transshipment from larger onto smaller boats took place for those who wanted to go on further into the Fayum.[30] The northern extension of the

27 Hassan/Tassie, 2006, show the lake during the Roman period at -40 m, and comment: "although the lake fluctuated by +40 m and -13 m from this level". During the Roman period, they show the lake at -30 m, and comment: "although the lake fluctuated by +30 and -23 m from this level". The life-spans of the villages in the east of the lake, Qaret Rusas and the other two settlements, have not yet been published, but they all seem to have been inhabited through the Ptolemaic and Roman periods; this makes it very unlikely that such large fluctuations of the lake were possible. Ceramics and the rare stone objects on the site of Qaret Rusas show salt incrustations, so that the flooding of the village for a longer period at a time is certain.

28 I owe this information to Willeke Wendrich, who is working on these sites.

29 El-Lahun is from Egyptian "The mouth of the canal of Moeris"; the Greek literally means "Ptolemais of the anchorage".

30 Römer, 2010, p. 607–608.

Bahr Yusuf continued to offer a direct connection to Memphis. For the Ptolemaic and Roman periods, the Barrington Atlas of Greek and Roman World[31] shows – correctly, as I believe – the continuation of the Bahr Yusuf towards the north as a canal which passes by the harbour of Kerke, reaching Memphis and beyond. There is enough evidence to be sure that this canal existed during the Graeco-Roman period. A good piece of such evidence is P. Phrur. Diosk. 17, a letter from Dioskurides, the commander of the castle in Heracleopolis, to his father in Memphis from 5 Phaophi 151 or 140 B.C. In this letter, Dioskurides assures his father that a messenger has been put into a boat to bring him his monthly ration and a coat to Memphis. The planning of a trip by water from Herakleopolis, located on the Bahr Yusuf, to Memphis only makes sense if there was this continuation of the Bahr Yusuf to the north. Kerke, the harbour of Philadelphia in the eastern Fayum, which was to be reached from Philadelphia on land over a hilly area and approx. 10 km away, was most likely located between this canal on its western side and the Nile on its eastern side. P. Mich. Zen. 61, 16 – 20 (248 B.C.) renders the stops of a journey from Krokodilon Polis to Kerke as: "From Krokodilon Polis ... to Ptolemais (Hormou), and from Ptolemais to Kerke".[32]

The Greek engineers[33] who came to the Fayum after Alexander the Great had taken Egypt, will have observed that the main problem of the depression now consisted in two ravines (most likely visible only during the winter, when the waters were low), which had their roots close to the course of the Bahr Yusuf where it enters the Fayum; one ravine led to the north and then in a long western curve towards and into the lake (today called the Bats Drain); the other one departed in the south, and continued straight ahead to the north and also into the lake (today called the Wadi Drain)[34]. Without any special measures, enormous water masses found their way directly from the Bahr Yusuf into the ravines and further on to the lake, on the one hand being without any benefit for agriculture, on the other augmenting the lake. At both drains, the Ptolemies installed heavy dams to prevent the water from proceeding down these drains uncontrolled. At

31 TALBERT, 2000.

32 For the harbour at Kerke see CLARYSSE, 1980, p. 95–97.

33 Names of two of the early engineers are known from a papyrological archive: Kleon and Theodorus, whose files cover the time between 260 and 238/37 B.C.; there is a still unpublished PhD Dissertation on this archive at the Katholieke Universiteit Leuven by VAN BEEK, 2006 (forthcoming in Collectanea Hellenistica); for the archive see the Internet page of Trismegistos, Archives; cf. COOK, 2011, p. 45–47.

34 Also called the Wadi Nazlah after the main settlement on its course.

the starting point of the Bats Drain they threw up a dyke in the area of the village of Hawaret el-Maqta which guided the waters further to the north-east and around the northern fringe of the Fayum through a canal, which is called the Bahr Wardan today and which terminates in a blind end.[35] At the roots of the Wadi Drain, a 9 km long dam was built, which still stands in parts between the modern villages of Itsa and Abu el-Nur (Figure 4). This dam channelled the floods further to the west and into the area of the north-western Fayum, up to the villages of Theadelphia, Philoteris and Dionysias. Today that canal is called "Bahr Qasr el-Banat"; as all distributing waterways, this canal also terminates in a dead end.

There is no doubt that the two ravines had existed already for a long time. Gardner considered them to "have been initiated on the fall of the Neolithic Lake".[36] Their existence during the Graeco-Roman period is corroborated by the line of ancient villages following the Wadi Drain on its eastern fringe (from south to north: Kom el-Arka, Abu Dinqash, Tell el-Kinissa),[37] and by the clear dating of the Itsa-Abu el-Nur dam to the Ptolemaic-Roman period; this dam does not make any sense without having been connected to that ravine close to it in the north.

The dam has been studied thoroughly by Garbrecht and Jaritz, [38] who – in my view – did not always draw the correct conclusions. Nevertheless, they gave a thorough description of the dam and recognised various building phases:

In a first phase, an earthen dyke had been backfilled between Itsa and Abu el-Nur in the early Ptolemaic period.[39] In a second building phase, a solid wall reinforced the original dyke; the wall was made of limestone; according to the authors, this process goes back to the very early time of Roman government in Egypt.

35 For a thorough interpretation of their activity in this area see KRAEMER, 2010, p. 365–376; GARBRECHT/JARITZ, 1990, p. 153–164; they date the dams at Hawaret el-Maqta to the same period as the dam between Itsa and Abu el-Nur, i.e. to the early Roman period. However, most likely, also in this area as at Itsa, the early Roman dam replaced the Ptolemaic earthen dam.

36 CATON-THOMPSON/GARDNER, 1934, p. 17–18.

37 See RÖMER, The Fayum Survey Project, The Themistou Meris (forthcoming).

38 GARBRECHT/JARITZ, 1990.

39 The Ptolemaic date of a dam here has been corroborated now by drillings in the earthfill core of the limestone dam; see HASSAN/TASSIE, 2006, p. 38 with photo.

Figure 4. Fragment of the late-Ptolemaic limestone wall between Itsa and Abu el-Nur, (Photo by the author).

And finally, in the third building phase, the limestone wall was replaced by a wall made of fired bricks.[40] Garbrecht and Jaritz also observed evidence that the old wall had broken down. There must have been problems with this wall at some stage; such problems would have immediately led to problems with the water supply down to Dionysias.[41] I think it is pretty obvious that the problems with this dam could have given the final blow to the villages in the north-western Fayum during the fourth century A.D. If that dam broke, the water would not have reached very far to the north.

4. The north-western Fayum as a test case for the use of the river Nile in the Fayum

The area of the north-western Fayum is most distant from the Fayum's source of the water supply; the water reaching the fields here had to flow approx. 60 km (as the crow flies) from where the Bahr Yusuf enters the depression coming from the Nile valley.

This landscape was special in several respects:

40 Ibid., 2006, date the "majority of the brickwork" to the Ottoman period.

41 See RÖMER, 2013, p. 169–179.

It was created from the drawing board in the third century B.C. and was abandoned and taken back by the desert during the fourth century A.D. The archaeological and even more the rich written evidence shows a part of the Fayum oasis which flourished despite of its distance from the Lahun Gap for approx. 700 years.

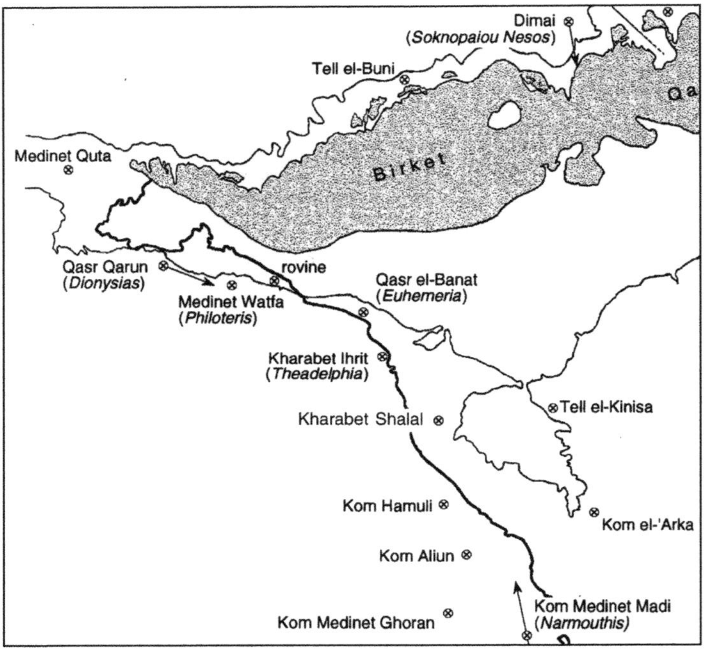

Figure 5. The north-western Fayum with the most important Graeco-Roman sites, see D*AVOLI* *1998, p. 367.*

It is this particular part of the Fayum, where we carried out an archaeological survey in the villages of Dionysias, Philoteris, Euhemeria and Theadelphia, all four settlements of the early Ptolemaic period, as their names and the written evidence from the sites tell; from the far end of the feeder canal upstream: Dionysias, the village of the Greek wine god, Philoteris named after one of the sisters of Ptolemy II, Euhemeria a speaking name of good omen to be translated as "the village where the day is nice", and finally Theadelphia, the village of the divine siblings, namely Ptolemy II and his sister-wife Arsinoe II. Despite their names, the main god here was the crocodile god Souchos, venerated under many different names such as Soknopaios, Psosnaus, Soxis etc. Temples here were built in purely Egyptian style and were erected side by side with Greek style public bath houses.

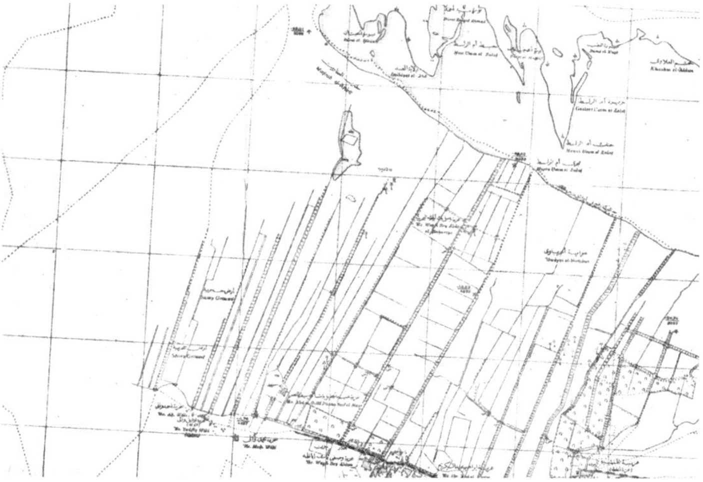

Figure 6. Survey of Egypt Map, 1989; canals descending from the main feeder canal towards the lake west of Dionysias.

During the Middle Kingdom, the entire area here had been under water, the villages having been located between 8 m above s.l. and the zero-line. Thanks to the abundant written evidence from papyri found in this area or in cartonnage from the Ptolemaic period, we know many details about the life in these villages. Euhemeria (at +7 m) is first mentioned in a papyrus dated to 243/242 B.C.[42] Since Dionysias lies on the zero-line, the lake must have been down to that measure by 229/228 B.C. at the latest when that village is first mentioned, but still called *the new village*.[43] At least 60 years, if not more, before that date, the work on the 9 km long dyke between Itsa and Abu el-Nur must have started. The entire Ptolemaic enterprise in the Fayum will have been already initiated around 300 B.C., when Ptolemy I was ruling.[44]

42 P. Petrie III 82, 8 (with BL 9, p. 211).

43 P. Lille Dem. 110, Vo col. 3, 13 "the new village"; Vo col. 1, 1 – 2 "the new village also called Dionysias"; De Cenival, 1980, p. 193–203.

44 For the calculation of the necessary timespan to bring the lake down from +20 to -45 m see Ball, 1939, p. 213–214; he calculates that this process cannot have taken more than 40 years and no less than 12 years; however, Ball bases this calculation

During the second half of the third century B.C., Dionysias had 732 tax-paying inhabitants, 391 male and 341 female, plus an estimated 250 soldiers with their families; two thirds of the population was Egyptian, one third Greek. That takes us to an overall population of approx. 1200 people who lived here. The numbers for Philoteris from the same document are nearly the same; [45] so it seems that, in the beginning, the villages were laid out for the same number of people. This detailed evidence comes from the tax-lists written on papyrus in Demotic and Greek.[46]

The prestigious names of these villages, which we do not find in any other part of the Fayum in such numbers, show the pride which Ptolemy II took in this development; and indeed, it was an extremely difficult task to bring water up to here on a regular basis in order to make these villages prosper.[47]

But once the dyke stood at the beginning of the main feeder canal, the Bahr Qasr el-Banat, the connected canals up to Dionysias and beyond could be activated. The task was now to bring the water from a level of approx. 16 m maximum, from where the canal started, to *line 0* at around Dionysias.

The construction of these ancient canals is clearly visible in some places in the farthest north-west of the Fayum: over long stretches, at least between Theadelphia and Dionysias, they were cut into the bedrock, as was the case also to the north of Karanis.[48]

As today, the canals had to be maintained; from the Roman period, we have abundant evidence of corvée work which had to be carried out by the locals to keep the canals clean. Every male inhabitant had to do 5 days of cleaning every year, mostly in June, before the flood, when the level of the water in the canals was at a minimum. We have more than 230 of the so called *penthemeros* certificates issued by local overseers to those who had been working for the 5 days. Most of these certificates date to the Roman period.[49]

---

mostly on the evaporation on the surface of the lake, not taking into consideration the extended and ever more expanding fields from which water also evaporated.

45 P. Count 11, Col. II 11–14 from 243–217 B.C.

46 A new edition of these tax lists and a thorough interpretation is given by Clarysse/Thompson, 2006.

47 For a thorough description of the dam and its location and significance see Römer, 2013, p. 169–179.

48 See Cook, 2011, Photos 63 and 64 e.g.

49 Sijpesteijn, 1964; for most recently published examples see Claytor, 2013, p. 49–75.

Methodologically it seems right to assume that the ancient canals provided the bed for the new canals built at the end of the nineteenth century over long stretches. The landscape itself offers an ideal line for the waterways; the ancient construction followed this ideal line as far away from the lake and as high above the lake as possible, to make the slope between canal and the lake as wide as possible; now, as during Ptolemaic and Roman periods, fields are irrigated by *gravity irrigation* via the vertical canals, which were branching off the main feeder canal and cutting through the slope (Figure 6). It is interesting to see very few waterwheels, or today diesel pumps, in this part of the Fayum. They are not needed here, for the natural flow of the water from the canal above provides sufficient irrigation.

A closer look into the location and the environment of one of the Ptolemaic villages reveals some further features which show the water management in this part of the Fayum.

The village of Philoteris as it shows itself now, does not stimulate much enthusiasm in archaeologists, I must admit. Remains of walls are scanty, single buildings difficult to identify. Most of the mud brick walls have been carried away by the farmers who installed new fields around here since the beginning of the twentieth century. But the site itself has been spared from agricultural activities.

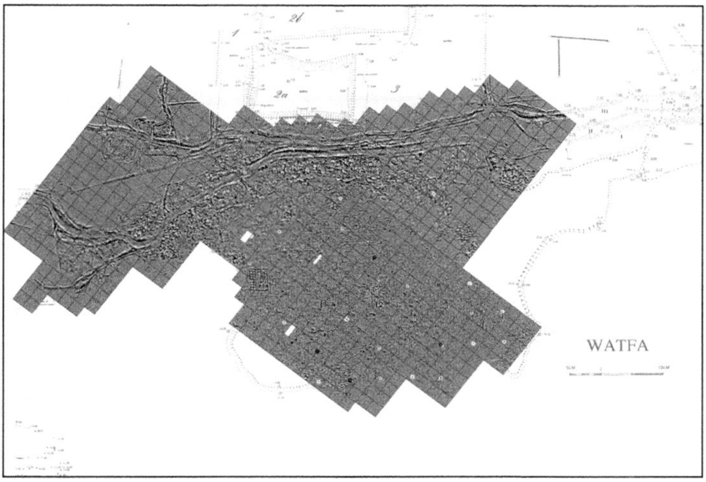

Figure 7. Geomagnetic map, produced by T. Herbich and his team 2011/2012.

During the archaeological survey[50] we already noticed that there were two canals approaching the village from the east, one flowing by directly, the other continuing along the village farther to the north and on a lower level. It seemed that the "upper canal" was built to transport the water further on to Dionysias – indeed, until 6 years ago we could follow it to that next village over a distance of 5.5 km – while the lower, receiving water from the upper, fed huge basins in the north (Figure 8). These basins are of different sizes, the largest measuring 28,000 m^2 with a capacity of 57,000 m^3, the smallest measuring 1,225 m^2. The basins are interconnected and were obviously supervised by a guard who resided on the ridge between the two largest basins. The purpose of these basins was clearly to store water, not to offer space for fields. Basin irrigation was practised in Graeco/Roman Egypt also in the Fayum, those basins being called περιχώματα or ὑδροςτάςια in the papyri;[51] however, where we have information about their sizes in the Fayum they were ten times larger than all the basins in Philoteris combined (P. Lille 1).

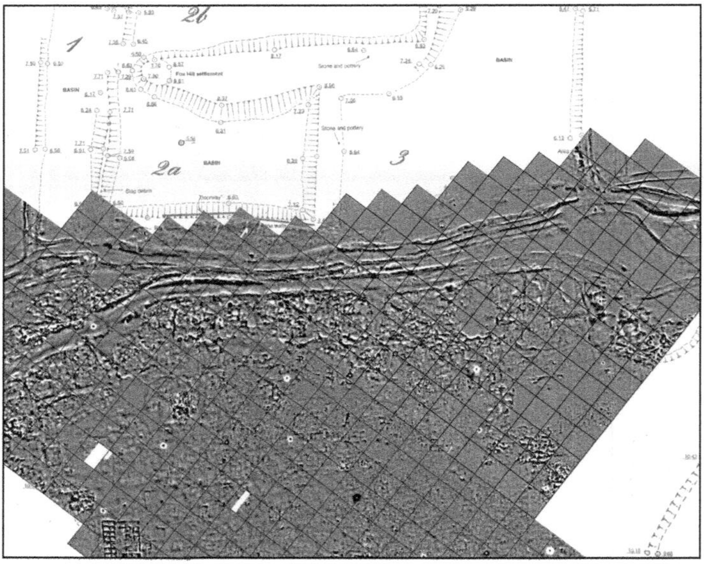

Figure 8. The basins in the north of Philoteris, canals and village; see Römer, 2004, Map in cover of volume, and Herbich in season 2011/2012.

50 Römer, 2004, p. 281–305.

51 Bonneau, 1993, p. 45–47. See now also Rapoport/Shahar, 2012; p. 1–31; basin irrigation was used in the Fayum in the area of Philadelphia, but not elsewhere.

It seems that the huge basins in the north were water storage basins meant to prevent the worst case scenario, which finally became true in the middle of the fourth century, when the entire area dried out.

The drillings and the small-scale excavation in November 2012 have given us new information about the levels between village, canals and basins. They have revealed another interesting feature of these basins. They were connected underground to a well above, which was built with carefully hewn lime stones on top of the bedrock. As in the cemetery to the south of the village, the bedrock is only approx. 60 cm thick here before it gives room to a hollow space which obviously extended largely under the village and the adjacent areas. The well was not fed by water from the canal above, but would fill from the hollow underneath which is positioned at the same level as the basins. The situation of the installations north of Karanis (see above) may the similar.

By drillings in the canals, their width and depth became clear. The upper and the lower canals were 7 to 8 m wide, the upper being at least 2.30 m deep. Here, boats may have travelled up to the landing places at the village.

In this area, more than 70 km away from the Nile, people used the water resources of the river for at least 700 years. In the Fayum, they celebrated the Katachyteria, the festival for the first arrival of the flood at Elephantine,[52] named their children Nilus or Nilammon,[53] and struggled with low or excessive floods.

5. Conclusion

Thanks to the new technical achievements of Google Earth, geomagnetic surveys, and intensive archaeological work it is now possible to understand better the measures taken to advance agriculture in the Fayum during the Middle Kingdom and the Graeco-Roman period. Today, the Fayum belongs to the poorest regions of Egypt and the interest in the water management of the past (in the Graeco-Roman period fields covered larger areas than today in some parts of the Fayum) is increasing.

52 P. Cairo Zenon II 59176, 39–40, from 255 B.C.

53 See the papyrological Database "Trismegistos", People.

Bibliography

BALL, JOHN, Contribution to the Geography of Egypt, Cairo 1939.

BONNEAU, DANIELLE, Le régime administratif de l'eau du Nil dans l'Égypte romaine et byzantine, Leiden 1993.

CATON-THOMPSON, GERTRUDE/GARDNER, ELINOR WIGHT, The Desert Fayum, London 1934.

COOK, RONALD JAMES, Landscapes of Irrigation in the Ptolemaic and Roman Fayum: Interdisciplinary Archaeological Survey and Excavation near Kom Aushim (Ancient Karanis), Egypt, University of Michigan PhD Dissertation 2011.

CLARYSSE, WILLY, Philadelphia and the Memphites in the Zenon Archive (Studia Hellenistica 24), Leuven 1980.

CLARYSSE, WILLY/VANDORPE, KATELIJN, Zénon, un homme d'affaires grec à l'ombre des pyramides, Leuven 1995.

CLARYSSE WILLY/THOMPSON, DOROTHY, Counting the People in Hellenistic Egypt, Vol. I (Population Register) and II (Historical Studies), Cambridge 2006.

CLAYTOR, GRAHAM, Penthemeros Certificates from the Granary C123, Karanis, Bulletin of the American Society of Papyrologists 50 (2013), p. 49-75.

DAVOLI, PAOLA, L'archeologia urbana del Fayyum di età ellenistica e romana, Napoli 1998.

DE CENIVAL, FRANCOISE, Compte de céréales de plusieurs villages du Fayoum. P. dém. Lille, in: Livre du centenaire (1980), p. 193-203.

FRASER, PETER MARSHALL, Ptolemaic Alexandria, Oxford 1972.

GARBRECHT, GÜNTHER/JARITZ, HORST, Untersuchung antiker Wasserspeicherung im Fayum/Ägypten, Braunschweig/Kairo 1990.

HASSAN, FIKHRI/TASSIE, GEOFFREY, Modelling environmental and settlement change in the Fayum, Egyptian Archaeology 29 (2006), p. 37-40.

HASSAN, FIKHRI, Lakes and Prehistoric Settlements of the Western Faiyum, Egypt, in: Journal of Archaeological Science 13 (1986), p. 483-501.

HAUG, BRENDAN, Ecology and the Administration of the Fayyūm. Paper Delivered at the Fayum Conference "Von der Pharaonenzeit bis zur Spätantike: Kulturelle Vielfalt im Fayum", Universität Leipzig, 29 May – 1 June 2013; on-line at Academia.edu.

KRAEMER, BRYAN, The Meandering Identity of a Fayum Canal. The Henet of Moeris/Dioryx Kleonos/Bahr Wardan/Abdul Wahbi, in: Proceedings of the 25th International Congress of Papyrology, ed. by TRAJANOS GAGOS, Ann Arbor 2010, p. 365-376.

RAPOPORT, YUSSEF/SHAHAR, IDO, Irrigation in the Middle Islamic Fayyum. Local Control in a Large-Scale Hydraulic System, Journal of the Economic and Social History of the Orient 55 (2012), p. 1-31.

RÖMER, CORNELIA, Philoteris in the Themistou Meris, in: Zeitschrift für Papyrologie und Epigraphik 147 (2004), p. 281-30.

—., Brief über das Umladen in einem Hafen, in: Honi soit qui mal y pense. Studien zum pharaonischen, griechisch-römischen und spätantiken Ägypten zu Ehren von H.-J. Thissen, ed. by HERMANN KNUF et al., Leuven 2010, p. 607-608.

—., Why did the Villages in the Themistou Meris die in the 4th Century AD? New Ideas about an Old Problem, in: Das Fayyûm in Hellenismus und Kaiserzeit. Fallstudien zu multikulturellem Leben in der Antike, ed. by CAROLYN ARLT/MARTIN STADLER, Wiesbaden 2013, p. 169-179.

SHAFEI BEY, ALY, Fayoum Irrigation as Described by Nabulsi in 1245 AD, in: Bulletin de la Société Royale de Géographie d' Égypte 20 (1940), p. 283-327 and map in front of article.

SIJPESTEIJN, PIETER JOHANNES, Penthemeros Certificates in Graeco-Roman Egypt, Leiden 1964.

ST. JOHN, JAMES AUGUSTUS, Egypt and Nubia, London 1845.

TALBERT, RICHARD, Barrington Atlas of the Greek and Roman World, Princeton 2000.

THOMPSON, DOROTHY, New and Old in the Ptolemaic Fayyum, in: Agriculture in Egypt from Pharaonic to Modern Times, ed. by ALAN KEIR BOWMAN/ EUGENE ROGAN, Oxford 1999.

Nilometers – or: Can You Measure Wealth?

SANDRA SANDRI

1. Introduction

In his 1982 article in the *Lexikon der Ägyptologie* Horst Jaritz mentioned, besides other types, nilometers consisting of:

> "Skalen [...] an freistehenden Säulen inmitten brunnenartig angelegter Räume, womit offenbar erstmals eine spezifische Bauform des N[ilmessers] entwickelt ist; röm. belegt nur nach Darstellungen, islam. neuerbaut auf Roda."[1]

The following article deals with images of such nilometers from Roman and late antique times and their iconographic interpretation.[2]

2. The nilometer motif – the documents

Doc. 1

In the lower part of two Coptic tissue medallions (Louvre, AF 3448) with an identical motif, a stonewalled, cylindrical structure, which is interpreted as a

1 JARITZ, 1982, p. 496.

2 See also for nilometers as an element of Nilotic landscapes: BALTY, 1984, p. 828. 831; BOISSEL, 2007, p. 368–373; BONNEAU, 1976; ID., 1991; HACHILI, 1998, p. 110–111; HAIRY, 2011; HERMANN, 1959, p. 62–63; ID., 1960; PFISTER, 1931–1932; RÖMER/ZANELLA, 2013, p. 906. 913; VERSLUYS, 2002, p. 271; MEYBOOM, 1995, p. 244–245, n. 77–78.

well by some authors,[3] is depicted (figure 1). At the bottom of the structure there are an arch and a staircase, which leads to the arch. In the middle of the structure there emerges a pointed stela. It is inscribed with the Greek numbers IZ and IH (17 and 18). A naked male child stands at the lefthand side of the structure and touches the stela with a chisel in his left hand near the number 18, while he strikes out with a hammer in his right hand. The child symbolises a cubit, the unit used to measure the rise of the annual flood on a nilometer scale. A second child with a water bird in his hands sits at the righthand side of the scene, perhaps in a papyrus boat. In the upper part of the medallions the male personification of the river Nile, Neilos, with a cornucopia in his left arm and a female counterpart, probably the personification of abundance, Euthenia, reclining in the middle of a Nilotic landscape marked by water birds and Indian lotus.

Date: seventh century A.D.
Provenance: said to be from Antinooupolis/Egypt
Bibliography: Du Bourguet, 1964, 132, D 36-37; Rutschowscaya, 1997; Id., 1999.

Figure 1. Coptic tissue medaillion (Rutschowscaya, 1997, p. 225); Figure 2. Mosaic Sarrin, province Osrhoene/Syria (Balty, 1990, pl. XXXIII, 1).

3 E.g. Meyboom, 1995, p. 244, n. 77.

Doc. 2

A scene in the Nilotic frame of a mosaic with Greek mythological motifs[4] shows a naked male cubit child with a chisel and a hammer in the same attitude as the child in doc. 1. He is shown standing in front of a rectangular structure positioned atop a slightly wider rectangle (figure 2). The upper shape is inscribed with the Greek numbers 17 and 18. A well from which a stela emerges as in doc. 1 is probably what is meant to be shown here.

Date: late fifth to mid sixth century A.D.
Provenance: a building of unknown character in Sarrin, province Osrhoene/Syria
Bibliography: BALTY, 1990, p. 65-67, pl. XXXIII; VERSLUYS, 2002, p. 474, cat. 44.

Doc. 3

A nilometer consisting of a column with five horizontal zones on an arched pedestal emerging from the river appears in the upper part of another late antique Nilotic mosaic (figure 3).[5] Three zones at the top of the column are inscribed with the Greek numbers from 15 to 17. A cubit child standing on the back of a bent second child seems to be marking the first figure of the number 17 with a chisel and a hammer. The nilometer is surrounded by further cubit children.

Date: second half of the sixth century A.D.
Provenance: a public building in Sepphoris/Palestine
Bibliography: HACHILI, 1998, p. 110-111, fig. 4; VERSLUYS, 2002, p. 233-235, cat. 130.

---

4 Like hunting Artemis, Dionysos and his Thiasos, the abduction of Europa, the birth of Aphrodite, Herakles and Auge, Meleager and Atalante, see BALTY, 1990, p. 24–57.

5 For the entire mosaic see VERSLUYS, 2002, p. 233–235. The nilometer and the cubit children are flanked by the reclining figures of Neilos and the female personification of Egypt, identified by a Greek inscription. Beneath the Nilotic scene there are other scenes with a city gate with the Greek inscription ‘Alexandria’, horsemen and fighting animals.

Figure 3. Mosaic Sepphoris/Palestine (Hairy, 2011, 109, fig. 23); Figure 4. Silver trulla, found in Perm/Russia (Bank, 1985, fig. 53).

Doc. 4

In the middle medallion of a silver *trulla* a nilometer well from which a pointed scale emerges is depicted. Only the letter Δ for the number four is clearly legible on the scale (fig. 4). On the lefthand side of the column, a cubit child standing on the back of a bent second child is engraving a figure at the top of the column. A reduced Nilotic landscape is indicated by fish, water plants and birds. The vessel was found in Perm/Russia, but it is unknown when and how it got there.

Date: beginning of the sixth century A.D.
Provenance: Perm/Russia
Bibliography: Bonneau, 1976, p. 9-10, fig. 14; Versluys, 2002, p. 216, cat. 115.

Doc. 5

In Kynopolis/Egypt, a headless marble statue of Neilos sitting on a throne was found (Alexandria, Graeco-Roman Museum 22173, figure 5). Along the left leg of the river god a rock is depicted as wells as two cubit children. One child, sitting on the shoulder of the second, raises his arms towards a stela. A scale or numbers are not visible, but there are remains of a Greek inscription.[6]

Date: second century A.D.
Provenance: Kynopolis/Egypt
Bibliography: ADRIANI, 1961, p. 57-58, no. 200, pl. 95, 311. 313; BAKHOUM, 2002, pl. 26, 2; BONNEAU, 1964, p. 282. 347-348; ID., 1995, p. 3212; JENTEL, 1992, p. 722, no. 35; PLATZ-HORSTER, 1992, p. 17.

Doc. 6

On a glass cup (London, British Museum 1868.5-1.919) with cut figured decoration, a beefy man with a helmet and a short cape stands with chisel and hammer in the typical position of a cubit child, facing a slim stela ending in a flat curve with several horizontal subdivisions (figure 6). Tatton-Brown recognises a *zeta* on the nilometer.[7] But the small, not very deep carving looking like a *zeta* is a little higher than the top of the chisel and above are other meaningless carvings. Thus, the alleged *zeta* could simply be another scratch in the glass. On the other side of the cup[8] a reclining woman, surrounded by a stylised Nilotic landscape with an Egyptian-style building, is holding a sistrum in her right and a cup in her left hand. Perhaps the woman is the Egyptian goddess Isis or the personification Euthenia.

Date: around 200 A.D.
Provenance: probably from Campania[9]
Bibliography: JENTEL, 1990; PLATZ-HORSTER, 1992, p. 13. 16, fig. 15; TATTON-BROWN, 1991; VERSLUYS, 2002, p. 169-171, no. 078, fig. 106; WALKER, 2006, p. 190, fig. 167.

---

6 SEG 20659. The inscription is very difficult to decipher. For divergent translations see FRASER, 1961, p. 141; BONNEAU 1964, p. 347–348.

7 "The principal scene shows a sturdy man with a chisel and a mallet, carving the Greek numeral 7 (the letter *zeta*) on a Nilometer, (...).", TATTON-BROWN, 1991, p. 87.

8 See Walker, 2006, p. 190, fig. 167.

9 See VERSLUYS 2002, p. 170. TATTON-BROWN, 1991, p. 87 suggests an Egyptian provenance because of the motif.

Figure 5. Marble Statue of Neilos (Bakhoum, 2002, pl. 26, 2); Figure 6. Glass cup with Nilotic scene (© British Museum).

Doc. 7

A nilometer is also depicted in various ways on several Roman coin series from Alexandria. Four series were engraved during the reign of Imperator Trajan. They show the female personification Euthenia reclining on a Sphinx or Neilos reclining on a hippopotamus. Above the legs of the gods, a child marking cubits stands in front of a nilometer. On the coins with Euthenia, the nilometer has the shape of an arch,[10] partly on a staircase with two steps,[11] but on the coins with Neilos the structure looks like a column.[12] The nilometer on the coins of Antoninus Pius is again an arch on a staircase with three steps (figures 7.1, 7.2). Sometimes the Greek letters *iota* and *sti* for the number 16 are legible inside the arch. On a series from the year A.D. 153/154, one child marking cubits stands inside the nilometer arch (figure 7.1),[13] on a second series from the

10 E.g. Geissen, 1974–1983, no. 607; Bakhoum, 2002, pl. 9, 11 (year 112/113 A.D.).

11 E.g. Geissen, 1974–1983, no. 578 (year 111/112 A.D.).

12 E.g. Ibid., no. 652 (year 113/114 A.D.) and 677–678 (year 114/115 A.D.).

13 E.g. Ibid., no. 1705–1708 (year 153/154 A.D.).

same year the child marking cubits sits on the shoulder of another (figure 7.2).[14] Here the two children are outside the arch. The coins of Annia Faustina[15] and Severus Alexander[16] from the first quarter of the third century A.D. adopt the child marking cubits on the shoulder of another, but the nilometer itself looks like a column or a pillar. Altogether, the nilometer motif was used on nine coin series over a century from the beginning of the 2nd century to the 3rd century.

Figure 7.1. Coin of Antoninus Pius (replica) (Roman Provincial Coinage Online, http://rpc.ashmus.ox.ac.uk/coins/4/13802/, accessed 18.05.2016); Figure 7.2. Coin of Antoninus Pius (replica) (Roman Provincial Coinage Online, http://rpc.ashmus.ox.ac.uk/coins/4/14929/, accessed 18.05.2016).

Doc. 8

Two reliefs on the opposite of a monumental stairway at the riverside of the island of Elepantine depict the same scene in mirror image (figure 8): [17] The reclining river god Neilos leans on a rock with a cornucopia in his left arm. Beside the cornucopia is a (water?) bird. Above the knee of the river god is a small arch. A tiny figure with raised arms stands on the top of the arch. In his article about the two reliefs, Hanz Günter Martin discusses whether the arch could be a nilometer because of the gesture of the figure on the arch, which is similar to the marking position of the

14 E.g. Ibid., no. 1709–1711 (year 153/154 A.D.).

15 E.g. Ibid., no. 2387 (year 221/222 A.D.).

16 E.g. Ibid., no. 2397 (year 221/222 A.D.).

17 MARTIN, 1986. One of these reliefs is still in situ, the other is located today in the Graeco-Roman Museum of Alexandria, see HÖLBL, 2004, p. 36, n. 114.

cubit child on the above mentioned silver *trulla* (doc. 4, figure 4).[18] He dismissed the idea, because there is no cylindrical construction on the Elephantine reliefs as in the nilometer well depicted in our document 4. However, there actually is a nilometer on these two reliefs: in the form of an arch and not in the form of a well. It seems not unlikely that one of the above-mentioned Alexandrian coins – with the reclining river god, a cubit child and a nilometer that looks like an arch (figure 7.1) – served as the model for the two reliefs in Elephantine. Because of their poor state of preservation, the reliefs in Elephantine can be dated only roughly to the second or third century A.D.[19] If this dating is correct, the reliefs were carved during the same period when the mentioned coins were in circulation. In contrast to a coin series of Antoninus Pius, the cubit child in the reliefs stands at the top of the arch and not inside it.[20] Perhaps it was difficult to cut the child with his widely extended arms inside the arch into the rough crystalline rock of Elephantine.[21] Coins are light and easily portable and in this way an Alexandrian motif may have reached the deep south of Egypt.

Date: second/third century A.D.
Provenance: Elephantine/Egypt
Bibliography: ADRIANI, 1961, p. 59, no. 203, pl. 96, 315; HÖLBL, 2004, p. 36-37, fig. 42; MARTIN, 1986.

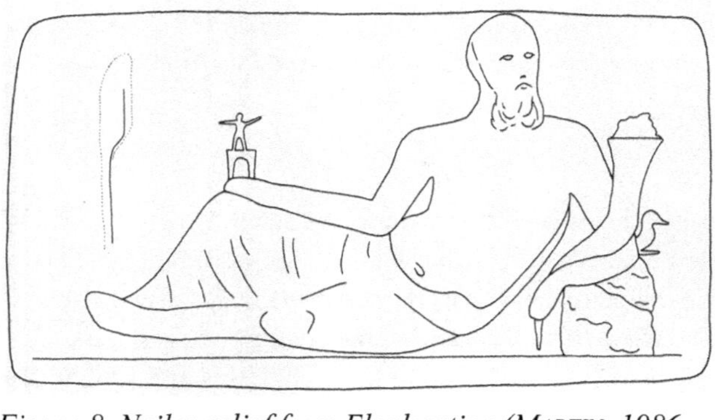

Figure 8. Neilos relief from Elephantine (MARTIN, 1986, pl. 23 b).

18 MARTIN, 1986, p. 190.

19 Ibid.

20 E.g. GEISSEN, 1974–1983, no. 1705–1708 (year 153/154 A.D.).

21 MARTIN, 1986, p. 188, n. 5.

Doc. 9

On a mosaic of a Christian church in Tabgha/Palestine, the upper part of a tower with a conical roof is depicted in a Nilotic landscape (figure 9). Although there is no child marking cubits, the interpretation as a nilometer is clear because of the scale with Greek numbers on the wall of the tower. Only the zone between the six- and ten-cubit marks is preserved, but it is likely that the scale originally started at 1.[22]

Date: fifth century A.D.
Provenance: Tabgha/Palestine.
Bibliography: Hachili, 1998, p. 110-111, fig. 2; Versluys, 2002, p. 228-230, no. 127.

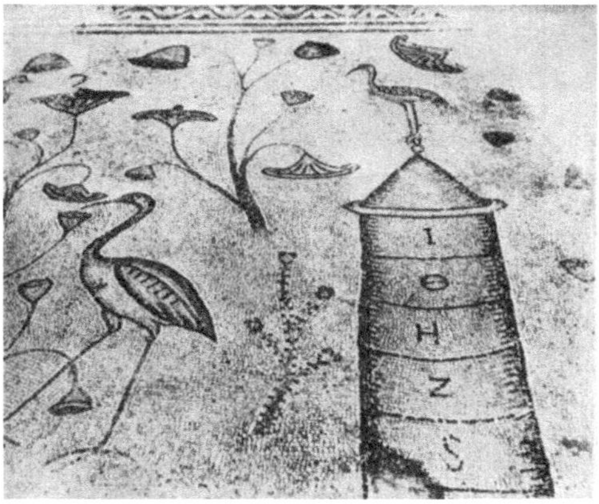

Figure 9: Mosaic in Tabgha/Palestine (Bonneau, 1976, 8, fig. 13); Figure 10: Mosaic in Beth Shean/Palestine (Balty, 1990, pl. XXXV, 1).

Doc. 10

A nilometer in the form of a tower and a city representation named Alexandria[23] stand near the reclining Neilos in a mosaic from Beth Shean/Palestine (figure 10). Here, the nilometer scale reaches from 10 to 16. The river god is shown holding a bird (?) in his right hand and a branch in his right arm. In addition, two

22 For the entire mosaic see Hachili, 1998, p. 109, fig. 2 c.

23 For late antique city pictograms see Decker, 1988; Duval, 2003.

fighting animals and a Nilotic landscape consisting of birds, fish, plants and a sailing boat are depicted.

Date: sixth century A.D.
Provenance: a guest room of a synagogue in Beth Shean/Palestine
Bibliography: HACHILI, 1998, p. 110-111, fig. 1; JENTEL, 1987, p. 211-212, pl. 8; NAUERTH, 1998, p. 194-196, fig. 3; VERLUYS, 2002, p. 226-227, no. 125.

Doc. 11

A Jordanian mosaic shows a similar composition as doc. 10. Unfortunately, only a sketch of the heavily damaged pavement has been published (figure 11).[24] According to an inscription, the reclining Neilos was depicted in the upper half. A nilometer in form of a column with a scale ranging from 11 to 18 on a rectangular pedestal stands beside him. In the lower part are a building labelled with 'ΕΓΥΠΤΩC' and a stretch of water with a fish and a sailing boat.

Date: sixth century A.D. (?)[25]
Provenance: Christian church in Umm al-Menābîa/Jordan
Bibliography: DECKER, 1988, p. 349; DUVAL, 2003, p. 223, fig. 7 b. 241; HERMANN, 1962, p. 79-86, fig. 10; PICCIRILLO, 1993, p. 341, fig. 752.

24 For the discovery and the further fate of the mosaic see AUGUSTINOVIĆ/BAGATTI, 1952, p. 286–288; GLUECK, 1951, p. 229–231; PICCIRILLO, 1993, p. 341.

25 AUGUSTINOVIĆ/BAGATTI, 1952, p. 288.

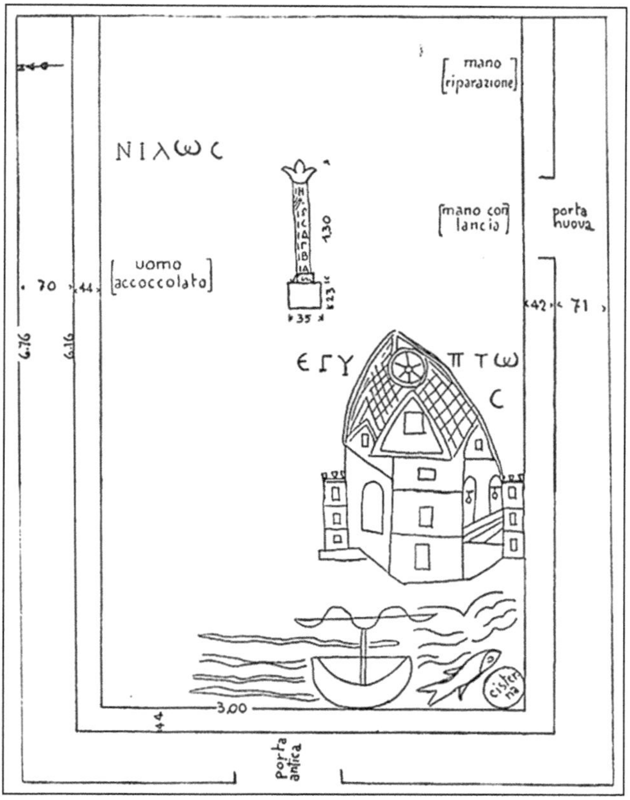

Figure 11. Sketch of the Nilotic mosaic in Umm al-Menābī/Jordan (HERMANN, 1952, 80, Abb. 10).

Doc. 12

On a famous Nilotic mosaic from the third century A.D., a procession of cubit children and two female musicians lead the river god Neilos, reclining on a hippopotamus, to a nilometer in the shape of an obelisk (figure 12). Two Egyptian priests await the procession sitting on the pedestal of the obelisk. Instead of a scale with numbers there is a Greek inscription at the top of the obelisk: ΑΓΑΘΗ ΤΥΧΗ "Good fortune".

Date: third century A.D.
Provenance: a bathroom of a villa in Leptis Magna/Libya

Bibliography: Boissel, 2007, p. 69, no. 61; Hermann, 1960, p. 37-41, pl. 2 a-b; Jentel, 1987, p. 210, pl. 5; Versluys, 2002, p. 185-186, no. 91.

Figure 12. Mosaic in Leptis magna/Libya (© Uwe Mahler).

3. Scholars and nilometers

In antiquity the annual cycle of the Nile and the devices for measuring this – the nilometers – were also of interest to Greek and Roman writers. Stephan Seidlmayer collected the statements of classical authors on this subject.[26] Most of them consider a flood with 14 to 16 cubits the optimum.[27] According to their records, a higher flood brings damage and a lower inundation causes famine. The famous description of the 16 fortunate cubits of Plinius the Elder belongs to these statements.[28] None of the authors mentioned a specific Egyptian nilometer where the optimum of 14 to 16 cubits was measured. Comparing the information given by Arabic authors and the known values of the nilometer in Roda we can establish that the nilometer of Memphis is meant.[29]

In the described nilometer images, the measured value or the maximum of the depicted scales are rarely the auspicious 16 cubits well known in antiquity (only on some Roman coins (doc. 7) and in doc. 10). Therefore, Alfred Hermann suggested that three nilometer images (doc. 1, 4, 12) reproduce particular nilometers in various areas of Egypt,[30] where different values were

26 Seidlmayer, 2001, p. 33–37.

27 Ibid., p. 34, table 2 (a).

28 For the Egyptian roots of the concept of the 16 fortunate cubits see Preys, 1999.

29 Seidlmayer, 2001, p. 33.

30 Hermann, 1960, p. 39.

measured because of the natural slope of the Nile from South to North. Hermann attempted to determine the real position of the depicted nilometers by means of the inscribed numbers. Contrary to most scholars, he assumed that Plinius' fortunate 16 cubits were measured in Herakleopolis Magna. In his opinion the nilometer of the Coptic tissue medallions (doc. 1) was located in Middle Egypt. Although only the letter *delta* on the scale of the silver *trulla* (doc. 4) can be read without doubt, Hermann adds an *iota* to every sector of the scale arguing that maximum values of fourteen and fifteen cubits are intended. In his opinion, this could refer to a nilometer in the area of Heliopolis or Memphis. Because of the presence of 12 cubit children in the mosaic of Leptis Magna (doc. 12) Hermann located the nilometer of the mosaic in the Egyptian Delta. However, there is no proof that the number of the cubit children can be interpreted as a measured value of the depicted nilometer.

Some of the nilometer images not mentioned by Hermann contain other references for their real place. The two exemplars of the Beth Shean mosaic (doc. 10) and the mosaic from Sepphoris[31] (doc. 3) stand near a city pictogram in form of a building with the Greek inscription Alexandria. This makes clear that a nilometer in this city is meant here. According to Jeanne Balty the nilometer besides a building similar to doc. 3 on the Syrian mosaic (doc. 2) could be an illustration of the Alexandrian nilometer, too.[32] The maximum scale values of the three probable depictions of the Alexandrian nilometer are surprisingly high (16 to 18 cubits). According to Egyptian records and classical authors, the peak of the flood could reach 28 cubits in Elephantine, but at the Mediterranean coast (Tell el-Balamûn) only 2 to 7 cubits.[33] If an inundation of 16 or even 18 cubits had really been measured in Alexandria, the rest of Egypt would have been under water.

Besides the mentioned mosaics (doc. 2, 3, 10), another exemplar includes a geographical identification: A building of the mosaic in Umm al-Menābîa/Jordan is labelled with ΕΓΥΠΤΩC. According to Hermann this indicates Babylon, the fortress near old-Cairo in late antiquity, but he does not provide an argument for this statement.[34] Subsequently, he identified the nilometer of the mosaic with the nilometer of Roda and regarded the entire mosaic as a more or less exact map of

31 See HACHILI, 1998, p. 111, fig. 4.

32 BALTY, 1990, p. 67.

33 SEIDLMAYER, 2001, p. 97, table 8. Aelius Aristides records only 2 cubits for Tell el-Balamûn (Orationes XXXVI, 115).

34 HERMANN, 1952, p. 82.

Old Cairo.[35] Other researchers did not share Hermann's opinion. They believed ΕΓΥΠΤΩC to be a designation of the whole of Egypt symbolised by a city pictogram.[36] In addition, DECKER proved that such pictograms in late antique mosaics never show actual views of certain locations or actual buildings.[37] They are exchangeable and can only be linked to a certain city by an appropriate inscription.[38]

In my opinion, it was never intended to show realistic values of different Egyptian nilometers in the mentioned illustrations. All indications to specific places (doc. 3, 10, perhaps 2) refer to Alexandria. But the given values are not likely for this location. Instead the number 16 of the Alexandrian coins (doc. 7) and in the Beth Shean mosaic (doc. 10) pick up the literary topos of the 16 fortunate cubits. Two scales feature lower values: on the silver trulla (doc. 4) 1 to 5 cubits and on the Tabgha mosaic (doc. 9) 1 (?) to 10 cubits. Perhaps the topos of the 16 cubits in these cases was not known to the designers and they created a scale which logically starts with 1 and ends arbitrarily or determined by the available space. The marked flood level of four nilometers (doc. 1, 2, 3, 11) is higher (17 and 18 cubits) than the literary optimum of 16 cubits. Perhaps the intention was here to show an inundation that is even a little better than the optimum.

35 Ibid., p. 86–92.

36 DECKER, 1988, p. 349; DUVAL, 2003, p. 241.

37 DECKER, 1988, p. 343–361.

38 This is an argument against Balty's opinion that the building near the nilometer of the Syrian mosaic symbolises Alexandria, see above and n. 25.

Figure 13. Remains of the nilometer of Alexandria (HAIRY, 2011, 107, fig. 21).

4. The Alexandrian nilometer

In Alexandria, the remains of a nilometer construction were found (fig. 13),[39] but it looks quite different to the mentioned nilometer images: A wide staircase leads down from the terrace of the Serapeum to a platform in an easterly direction. Another staircase with two turns connects the platform with a rock-cut chamber

39 SABOTTKA, 2008, p. 243–245.

(2.50 x 2.70 m). According to the excavator, Alan Rowe, a small corridor linked the chamber with an underground aqueduct supplied by a Nile canal south of Alexandria. Because of the indirect access to the Nile it is not clear whether the construction functioned like a real nilometer or only in a symbolic manner. During the modification of the Alexandrian Serapeum in Roman times the staircase was rebuilt at a larger scale. It is unknown whether the Ptolemaic nilometer remained accessible after these reconstruction works.[40]

Today, the nilometer is in a bad state of preservation, but one detail is very interesting: the staircase to the rock-cut chamber was open at its upper half and covered in antiquity at its lower half. The passage between the open and the covered staircase, which has now collapsed, was an arched door (figure 13).[41] A similar arch with steps is depicted on some of the coins mentioned above (doc. 7) and the Neilos relief from Elephantine (doc. 8), which is probably influenced by the coins. This leads to the assumption that this architectural detail of the Alexandrian nilometer is depicted on coins minted in Alexandria.

5. Provenance and date of the nilometer images

Only doc. 1, 5, 7 and 8 are of Egyptian origin, but seven documents were found outside: one in Syria (doc. 2), three in Palestine (doc. 3, 9, 10), one in Jordan (doc. 11), one in Libya (doc. 12), one probably in Italy (doc. 6) and one even in Russia (doc. 4). Both in antiquity and today it was necessary for the interpretation of the motif to know that the Nile floods its banks annually and that its flood is measured in cubits by nilometers. It must be assumed that those who initiated these non-Egyptian nilometer images were well-educated and highly interested in Egypt and that they – like the executing craftsmen – had probably never seen an actual nilometer with their own eyes.

Chronologically, two main phases can be recognised. Five documents (doc. 5-8, 12) date to the second and third century A.D. After a hiatus of about 150 years seven documents follow from mid-fifth to the seventh century A.D (doc. 1-4, 9-11).

40 Ibid., p. 245, n. 857.

41 Ibid., pl. 112–114; Hairy, 2011, p. 107, fig. 21.

6. Routes and means of transport of the nilometer motif around the Mediterranean

The oldest undubitable images of a nilometer[42] can be found on Alexandrian coins (doc. 7) from the beginning of the second century A.D. – objects which were produced in large quantities and were widely spread because of their material value and their small size. It can be assumed that these coins significantly influenced the development of the older group of nilometer images from the second and third centuries A.D. After the third century A.D., coins with a nilometer were no longer produced. Perhaps nilometer images have therefore been out-of-favour for 150 years, too. The reason why nilometer images emerged again is not quite clear. Also in the younger group of nilometer images from the fifth to the seventh century A.D. some elements of the mentioned coins of the second and third century A.D. were still used.

Depending on the coin (see figures 7.1, 7.2) which was used as model, one child marking cubits (doc. 1, 2, 6, 8) or one on the shoulder of a second child (doc. 3, 4) were depicted. The gesture of the child marking cubits does not really make sense. It does not measure a water level but gestures in the air: It seems to be engraving the highest number of the nilometer scale with hammer and chisel which would be in fact under water during the depicted surrounding flood. As a matter of fact, the scale was certainly fixed before its first use during the inundation season.[43] But because of this "unrealistic" depiction it is easier for the observer to recognise the motif.

Another problem is the design of the nilometer itself, because it is normally an underground construction which can only be used while being flooded. The artisans solved this problem by using the *pars-pro-toto* principle. They depicted particular details side by side, like a scale with horizontal divisions and numbers floating in the centre of a well (e.g. figures 1 and 4), although it has to be at the bottom of the well, not visible from the outside.

42 A well in front of an Egyptian temple on the famous Palestrina mosaic from the first century B.C. is usually interpreted as a nilometer (see MEYBOOM, 1995, pl. 15). But some scholars refuse this interpretation, because there is no measuring scale; so it could also be a common well (see Ibid., p. 244, n. 77).

43 After the installation of a nilometer scale only the so called "Nilstandsmarken" were placed: inscriptions which indicate the height of the flood in a certain year (see BORCHARDT, 1908, SEIDLMAYER, 2001, p. 53). But this was not done every year.

The selection and appearance of the particular details depend on the used model. The arch with stairs on the coins of Antoninus Pius was adopted on three documents (doc. 1, 3, 8). On the Coptic tissue medallions (doc. 1) the arch with stairs is located on the wall of the well. The arch could probably be meant as the end of a canal from the well to the Nile, which, being underground, could not be depicted. But the staircase which leads from the outside up to the arch makes no sense at this place. Presumably, the artisan regarded a staircase and an arch as basic elements of a nilometer, but he was not really aware of their actual function. On the Sepphoris mosaic (doc. 3) the arch seems to be the beginning of a vault in a cubic base standing in the Nile, whereas the nilometer scale is above the water level.

The appearance of the nilometer scale itself differs both on the coins (see description doc. 7) and in the other documents. It can look like a stela (doc. 1, 5, 6), a column (doc. 3, 4, 11), a tower (doc. 9, 10) or an obelisk (doc. 12).

Presumably, the coins were not the only source for the nilometer depictions. Several classical authors such as Plinius the Elder dealt not only with the unique phenomenon of the Nile flood but also with the technical aspects of nilometers and their function.[44] Two of them, Strabo (first century B.C. – first century A.D.) and Heliodorus (third century A.D.), describe in similar words the nilometer of the temple of Khnum in Elephantine as a well consisting of rectangular stones and a cubit-based scale with horizontal lines at the inside wall of the well.[45]

> "The nilometer is a well on the bank of the Nile constructed with close-fitting stones, in which are marks showing the greatest, least, and mean rises of the Nile; for the water in the well rises and lowers with the river. Accordingly, there are marks on the wall of the well, measures of the complete rises and of the others. So when watchers inspect these, they give out word to the rest of the people, so that they may know; (...). This is useful, not only for the farmers with regard to the water-distribution, embankments, canals, and other things of this kind, but also to the prefects, with regard to the revenues; for the greater rises indicate that the revenues also will be greater."[46]

The mentioned stonewalled well is clearly visible on the Coptic tissue medallions (doc. 1), the silver *trulla* (doc. 4) and in a stylised form on the Syrian mosaic (doc. 2). In contrast, it was not possible to copy a nilometer scale inside the well,

44 Postl, 1970, p. 36–48. 138–154.

45 Strabo XVII.1.48; Heliodorus IX.22.

46 Strabo XVII.1.48 (Translation: Jones, 1982, p. 128–129).

so it was depicted either on a stela floating in the well (doc. 1, 2, 4) or on the outside walls of a tower-like building (doc. 9-10).

7. Technical aspects

Based on several nilometer images, Horst Jaritz assumed in his lexicon article mentioned at the beginning that a specific Roman design for nilometers existed in the form of a well with a column inside, on which the height of the flood was read in Roman times. But none of the images (doc. 1, 3, 4) that show a nilometer of this form predates the sixth century A.D. No surviving building of this kind dates earlier than the seventh century A.D., like the nilometer structures of Schedia near Alexandria[47] and Roda.[48] Unlike the images of the nilometer columns, these sites stood at the bottom of a rectangular and not a round well.[49]

The nilometer images are not an accurate reproduction of a technical construction and its data. They show a conglomerate of single elements which seem to belong to a nilometer and were important for the comprehension of the motif. In Roman times, a standardised nilometer "design" did not exist. It rather depended on the source – image and/or text – used by the artisan. Therefore it seems difficult to reconstruct an actual structure based on these images.

8. Conclusion

Most nilometer images are part of Nilotic landscapes. They, like other Nilotic landscapes, appear in several contexts: in a private house (doc. 12), in a public building (doc. 3), in the periphery of an Egyptian temple (doc. 8), in Christian churches (doc. 9, 11) and even in a synagogue (doc. 10).

Today, it is not inconsistent for us to find sphinxes, obelisks and pyramids on Christian cemeteries. We immediately recognise these elements as "Egyptian", even if they are in a Hellenized form like female sphinxes. In the context of a modern grave-yard their original meaning has narrowed: the function of the sphinx as a manifestation of the king or the obelisk as a solar symbol is irrelevant here.

47 Daressy, 1900; Hairy, 2011, p. 108, fig. 18.

48 Popper, 1951.

49 See Daressy, 1900, p. 91; Popper, 1951, p. II.

For well-educated ancient viewers Nilotic images were presumably just as "Egyptian" as sphinxes, obelisks and pyramids are to us. Apparently it was no problem to integrate motifs of a foreign culture and religion in their own houses and chapels. The symbolism of these pictures – wealth, prosperity, abundance – was attractive for believers of all religions. And, via the depicted nilometers, it was actually possible to measure the expected wealth with numbers.

Bibliography

ADRIANI, ACHILLE, Repertorio d'arte dell'Egitto greco-romano, A/2, Palermo 1961.

AUGUSTINOVIĆ, AGOSTINO/BAGATTI, BELLARMINO, Escursioni nei dintorni di ʿAglûn, in: Liber Annuus 2 (1952), p. 227-314.

BAKHOUM, SOHEIR, Dieux égyptiens à Alexandrie sous les Antonins. Recherches numismatiques et historiques, Paris 2002.

BALTY, JEANNE, Thèmes nilotiques dans la mosaïque tardive du Proche-Orient, in: Alessandria e il mondo ellenistico-romano. Studi in onore di Achille Adriani, ed. by Nicola Bonacasa/Antonio di Vita, Rom 1984, p. 827-834.

ID., La mosaïque de Sarrîn (Osrhoène), Paris 1990.

BANK, ALICE, Byzantine Art in the Collections of Soviet Museums, Leningrad 1985.

BOISSEL, ISMÉRIE, L'Égypte dans les mosaïques de l'occident romain: images et representations (de la fin du IIème siècle avant J.-C. au IVème siècle après J.-C.), Reims 2007.

BONNEAU, DANIELLE, La crue du Nil, divinité égyptienne à travers mille ans d'histoire (332 av. – 641 ap. J.-C.) d'après les auteurs grecs et latins, et les documents des époques ptolemaïque, romaine et byzantine, Paris 1964.

ID., Le nilomètre: aspect architectural, in: Archeologia 27 (1976), p. 1-11.

ID., Nilometer, in: The Coptic Encyclopedia 6, ed. by AZIZ S. ATIYA, New York 1991, p. 1794-1795.

ID., La divinité du Nil sous le principat en Égypten, in: Aufstieg und Niedergang der römischen Welt II/18/5, ed. by WOLFGANG HAASE/HILDEGARD TEMPORINI, Berlin 1995, p. 3195-3215.

BORCHARDT, Ludwig, Nilmesser und Nilstandsmarken, Berlin 1906.

DU BOURGUET, PIERRE, Catalogue des étoffes coptes I. Musée National du Louvre, Paris 1964.

DARESSY, GEORGES, Le nilomètre de Kom el Gizeh, in: Annales du service des antiquités de l'Égypte 1, Cairo 1900, p. 91-96.

DATTARI, GIOVANNI, Numi augg. Alexandrini, Cairo 1901.

DECKER, JOHANNES, Tradition und Adaption. Bemerkungen zur Darstellung der christlichen Stadt, in: Mitteilungen des Deutschen Archäologischen Instituts in Rom 95 (1988), p. 303-382.

DUVAL, NOËL, Les représentations architecturales sur les mosaïques chrétiennes de Jordanie, in: Les églises de Jordanie et leurs mosaïques, ed. by NOËL DUVAL, Beirut 2003, p. 211-285.

FRASER, PETER MARSHALL, Bibliography: Graeco-Roman Egypt. Greek Inscriptions (1961), in: Journal of Egyptian Archaeology 47 (1961), p. 139-149.

GEISSEN, ANGELO, Katalog alexandrinischer Kaisermünzen der Sammlung des Instituts für Altertumskunde der Universität zu Köln, Opladen 1974-1983.

GLUECK, NELSON, Explorations in Eastern Palestine IV, New Haven 1951.

HACHLILI, RACHEL, Iconographic Elements of Nilotic Scenes on Byzantine Mosaic Pavements in Israel, in Palestine Exploration Quarterly 130 (1998), p. 106-120.

HAIRY, ISABELLE, Les nilomètres, outils de la mesure du Nil, in: Du Nil à Alexandrie. Histoires d'eaux, ed. by ISABELLE HAIRY, Alexandria 2011, 2nd ed., p. 98-111.

HERMANN, ALFRED, Der Nil und die Christen, in: Jahrbuch für Antike und Christentum 2 (1959), p. 30-69.

ID., Die Ankunft des Nils, in: Zeitschrift für ägyptische Sprache und Altertumskunde 85 (1960), p. 35-42.

Id., Ägyptologische Marginalien zur spätantiken Ikonographie, in: Jahrbuch fur Antike und Christentum 5, 1962, p. 60-92.

HÖLBL, GÜNTHER, Altägypten im Römischen Reich. Der römische Pharao und seine Tempel. 2. Die Tempel des römischen Nubien, Mainz 2004.

JARITZ, HORST, Nilmesser, in: Lexikon der Ägyptologie IV, Wiesbaden 1982, p. 496-498.

JENTEL, MARIE-ODILE, La représentation du dieu Nil sur les peintures et les mosaïques et leur contexte architectural, in: Echos du Monde Classique/ Classical Views 34, n.s. 6 (1987), p. 209-216.

ID., Euthenia, coudées et nilomètre, in: Echos du Monde Classique/Classical Views 34, n.s. 9 (1990), p. 173-179.

ID., Neilos, in: Lexicon iconographicum mythologiae classicae 6, ed. by LILLY KAHIL, Zürich 1992, 720-726.

JONES, HORACE LEONARD, The Geography of Strabo VIII, London 1982.

MARTIN, HANZ GÜNTER, Zwei Reliefs mit Flußgöttern auf Elephantine, in: Mitteilungen des Deutschen Archäologischen Instituts in Kairo 43 (1986), p. 189-194.

MEYBOOM, PAUL G.P., The Nile Mosaic of Palestrina, Leiden/New York/Köln 1995.

NAUERTH, CLAUDIA, Antike Hafenbilder. Das Beispiel Alexandria, in: Studien zur altägyptischen Kultur 26 (1998), p. 191-202.

PFISTER, RODOLPHE, Nil, nilomètres et l'orientalisation du paysage hellénistique, in: Revue des arts asiatiques 7 (1931-1932), p. 121-140.

PICCIRILLO, MICHELE, The Mosaics of Jordan, Amman 1993.

PLATZ-HORSTER, GERTRUD, Nil und Euthenia. Der Kalzitkameo im Antikenmuseum Berlin, Berlin 1992.

POPPER, WILLIAM, The Cairo Nilometer, Berkeley 1951.

POSTL, BRIGITTE, Die Bedeutung des Nil in der römischen Literatur. Mit besonderer Berücksichtigung der wichtigsten griechischen Autoren, Wien 1970.

PREYS, RENÉ, Hathor, maîtresse des seize et la fête de la navigation à Dendera, in: Revue d'Égyptologie 50 (1999), p. 259-268.

RÖMER, CORNELIA/ZANELLA, FRANCESO, Nil I, in: Reallexikon für Antike und Christentum, Lieferung 199, Stuttgart 2013, p. 898-915.

RUTSCHOWSCAYA, MARIE-HÉLÈNE, in: Ancient Egypt at the Louvre, ed. by GUILLEMETTE ANDREU et al., Paris 1997, p. 224-225, cat. 114.

ID., in: Keizers aan de Nijl, ed. by HARCO WILLEMS, Leuven 1999, p. 179, cat. 59.

SABOTTKA, MICHAEL, Das Serapeum in Alexandria. Untersuchungen zur Architektur und Baugeschichte des Heiligtums von der frühen ptolemäischen Zeit bis zur Zerstörung 391 n. Chr., Études alexandrines 15, Cairo 2008.

SEIDLMAYER, STEPHAN JOHANNES, Historische und moderne Nilstände. Untersuchungen zu den Pegelablesungen des Nils von der Frühzeit bis in die Gegenwart, Achet. Schriften zur Ägyptologie A 1, Berlin 2001.

TATTON-BROWN, VERONICA, The Roman Empire, in: Five Thousand Years of Glass, ed. by HUGH TAIT, London 1991, p. 62-97.

VERSLUYS, MIGUEL JOHN, Aegyptiaca Romana. Nilotic Scenes and the Roman Views of Egypt, Leiden/Boston 2002.

WALKER, Susan, Die Portlandvase, in: Kleopatra und die Caesaren. Eine Ausstellung des Bucerius Kunst Forums 28. Oktober 2006 bis 4. Februar 2007, ed. by BERNARD ANDREAE/ORTRUD WESTHEIDER, München 2006, p. 184-193.

In Search of a Future Companion

Digital and Field Survey Methods in the Western Nile Delta

JOSHUA TRAMPIER

1. Introduction

After reading the *Companion to Ancient Egypt*,[1] it occurred to me that despite its authoritative scholarship and hefty 1,352 pages, it contained virtually no discussion of regional settlement and paleoenvironment.[2] Consider that eleven chapters of the *Companion* are for the decorative arts, five for language, but just two for settlement. These two chapters focus on exemplars of urban form and planning in the Pharaonic and Classical eras, respectively.

Compare this work with the *Companion to the Archaeology of the Ancient Near East*[3] from the same publisher, which offers *seventeen* chapters (counting the introduction) on regional settlement, urbanism, and human-environmental dynamism. Turn to another region in the Mediterranean[4] or the Near East,[5] and one finds regional survey-derived archaeological data intertwined with historical, cartographic, remote sensing, and physical scientific evidence to describe the delicate interplay of *longue durée* settlement patterning and the natural world. I wondered why discussions such as these are missing from fundamental survey

1 LLOYD, 2010.

2 TRAMPIER, 2012.

3 POTTS, 2012.

4 E.g., CHERRY et al., 1991; BARKER/HODGES/CLARK, 1995; GILLINGS/SBONIAS, 1999; CHERRY, 2003.

5 ADAMS, 1981; AMMERMAN, 1981; BINTLIFF/DAVIDSON/GRANT, 1988; WILKINSON, 2003.

volumes like Lloyd's *Companion*: A wealth of epigraphic riches?[6] A lingering divide between anthropological archaeology and Egyptology?[7] The perceived and real logistical challenges that working in the Nile floodplain poses versus the low desert fringe?[8] Disciplinary momentum? Or is it simply that the semantics of the book titles – and thus content – differ? That is, would things have been different if the book were titled "Companion to the Archaeology of Ancient Egypt"?

Let's look at this a different way. Where is the discipline of Egyptology/Egyptian archaeology (using the terms interchangeably) focusing its collective gaze, such that the two *Companion* volumes might be so different in content? The SCA official website from 2010 offers a representative snapshot of archaeological activity in Egypt to address this question. I geocoded and parsed the list of 239 projects published on the website that year, noting each project's location and the nationality of its director's home institution. Here is an example of one project's record: "Qubbat Afafendina, Old Cairo. Agnieszka Dobrowolska, Netherlands-Flemish Institute, The Netherlands".

Twenty-four countries, including two projects with Egyptian co-directors, were represented in the SCA list (figure 1, left). Overall, France had the highest representation with fifty projects, a quarter of which were in Luxor/Thebes and 18 % in Saqqara. American directors chaired forty-two projects, about a third of them in Luxor. Great Britain had just one project in Luxor at KV 57, with five of its twenty-five projects in Saqqara. The remaining top ten countries had sizable investment in Luxor and Saqqara as well.

Figure 1 (right) shows the geographic distribution of projects. As the country breakdown suggests, Luxor/Thebes contained the most (n=60), followed by Saqqara (24) and Alexandria (18). Together with Aswan (9), Cairo (9), Giza (7), and Abydos (7), these seven places constitute almost 60 % of the whole of foreign-funded effort. Enlarging this sample to include the entire Memphite (+ Dahshur, Memphis, Abu Rawash, Ma'adi, Helwan, and Abusir) and Amarna clusters (+ Sheikh Ibabda, Deir el-Bersha, Kom el-Ahmar, el-Amarna), the total rises to 65 %. Well over half of the projects listed were concerned with recording and conservation of monuments, and half of the total focused primarily on mortuary practices, such as tomb epigraphy and cemetery excavation.

Certainly the SCA list, and so figure 1, has some slight omissions. One of the largest is Egyptian-led projects. While it would be nice to have public records of

6 MESKELL, 2006.

7 LUSTIG, 1997.

8 VAN DEN BRINK, 1988; VON DER WAY, 1991.

these, anyone working in the country knows they are not readily available. Yet even adding in such omissions or having an up-to-the-minute list from 2015, it is doubtful the picture would change drastically. Based on my limited experience, I suspect that more Egyptian-led projects would see better coverage of floodplain salvage work by Delta inspectorates around developing towns.

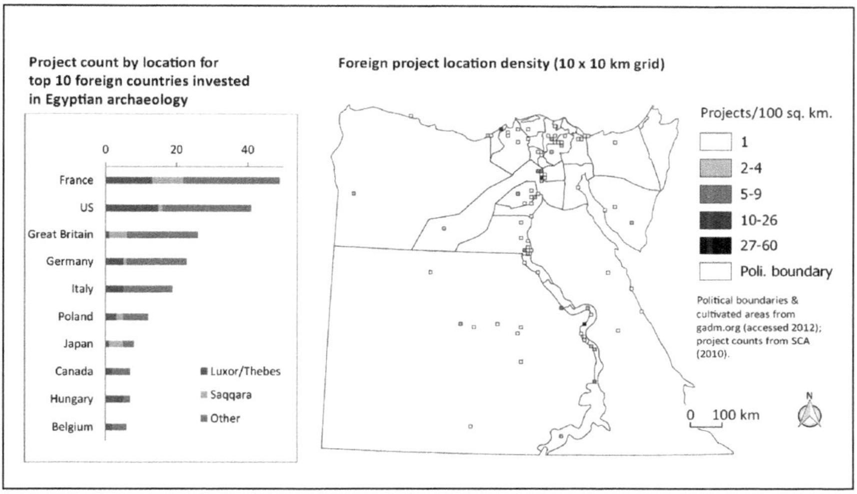

Figure 1. (left) Project count by country of origin and areas of interest for top foreign-institution-affiliated, active projects in 2010. (right) Geographic distribution of these projects.

2. The Delta floodplain in Egyptology

With this uneven geographic distribution, it is not surprising that less than 20 % of the projects on the list actively investigated settlement. Because our discipline concentrates its intellectual and financial resources largely on the landscapes of the powerful and the dead, our understanding of the places where people lived and relationship with their local environment remains primitive. The Nile Delta floodplain – a place where the majority of Egypt's population has lived and continues to live – remains largely overlooked, despite its vast potential and warnings about destruction of its archaeological contexts.[9]

9 VAN DEN BRINK, 1987; BIETAK, 2009.

Several projects in recent decades have worked to address this imbalance, conducting combined archaeological and geomorphological surveys around Qesna,[10] Tell el-Muqdam,[11] Tell Tebilla and Middle Egypt,[12] Kom Firin,[13] Kom Khawaled,[14] Memphis,[15] and along the shores of Lake Maryut,[16] to name a few. The Delta floodplain west of the Rosetta branch and north of the ancient confluence of the Rosetta and Canopic branches (hereafter, the "western Delta") has seen its share of survey too, particularly the Naukratis survey[17] and Wilson's Western Delta Survey[18] shown in figure 2.

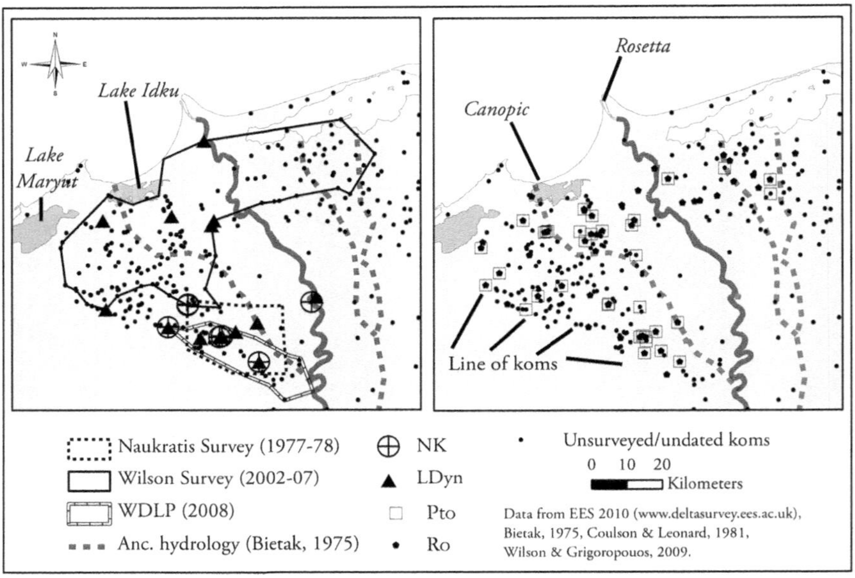

Figure 2. Location of EES Delta Survey koms *and previous surveys in the western Delta.*

10 Rowland/Wilson, 2006
11 Redmount/Friedman, 1997.
12 Parcak, 2005.
13 Spencer, N., 2008.
14 Sinclair, 2006
15 Lutley/Bunbury, 2008.
16 Blue/Khalil, 2011.
17 Coulson/Leonard, 1981; Id., 1982; Coulson/Leonard/Wilkie, 1982.
18 Wilson/Grigoropoulos, 2009.

The Egypt Exploration Society's Delta Survey has catalogued some 638 *kom*s and *tell*s in the floodplain by collating scholarly research and plotting points from British Survey of Egypt (hereafter, SEGY) maps.[19] New Kingdom (NK), Late Dynastic (LDyn), Ptolemaic (Pto), and Roman (Ro) period *kom*s as dated by previous survey work are shown in figure 2, as is the location of the Western Delta Landscape Project (WDLP) in the southwestern Delta discussed in this paper.

When I began investigating the western Delta, I geocoded the EES Delta Survey data using latitude/longitude coordinates published on its website, consulting the map for any emergent patterns (see figure 2). On my mind was Butzer's notion that ancient floodplain settlements nestled on elevated areas, such as river levees, to mitigate impacts from the annual inundation.[20] Through much of Egyptian history, the Nile and its distributaries also provided a vital transportation system for ferrying goods and people. Arguably the choicest spots for settlement in the Delta would have been elevated mounds or levees adjacent to artificial or natural waterways. Scrutinizing the *kom*s of the western Delta, I imagined the lines of ancient waterways by mentally connecting these islands of humanity.

Through this examination, it became clear that a series of previously unmapped channels connecting dozens of *kom*s lay far to the southwest of the Canopic branch. Deeper investigation of satellite imagery and historical maps provided stronger evidence, prompting fieldwork to confirm observations and collect additional information. It was the search for these channels and a desire to characterize and date settlement in this region that led to the WDLP. Through a case study of Kom Qamha, I offer this paper to summarize the interplay of remote sensing data, surface collection of artifacts, geomorphological study, and historical cartography to assess settlement and paleoenvironment in the first millennium B.C. I contend that, contrary to previous assessments and disciplinary momentum, surface (i.e., non-excavation) techniques that work well for other parts of the Mediterranean can also provide repeatable results for the Egyptian floodplain. The project was cultivated under the aegis of the Durham University mission to Sa el-Hagar, directed by Penelope Wilson, though any views voiced are my own.[21]

19 Spencer, A., 2008.

20 Butzer, 1975; Id., 1976.

21 For a fuller treatment that surveys several dozen *kom*s, the reader is referred to Trampier, 2014.

3. Kom Qamha: A case study of WDLP survey method

Work on the Western Delta Landscape proceeded generally in four phases from spatially coarse-grained to fine-grained information. In the first phase (digital source review), the EES Delta survey data and Shuttle Radar Topography Mission (SRTM) imagery provided sketchy details that prompted a more detailed inspection for channel traces and *kom* boundaries. In the second phase (initial GIS analysis), channel traces were fleshed out on Corona and Landsat satellite imagery and Survey of Egypt maps from the early twentieth century. Changing *kom* extents were mapped using these two sources, as well as Quickbird-2 satellite images. In the third phase (fieldwork), several methods – surface collection on and off-*kom*, topographic survey, and targeted coring across channel traces and historic *kom* extents – provided new information on soil character and material culture distribution. In the fourth phase (post-fieldwork analysis), laboratory and detailed GIS work analyzed diagnostic artifacts and sediments. These efforts provided novel insights into 1) the spatial disposition of human and natural activity from diagnostic ceramics and optically stimulated luminescence (OSL) samples on channel soils, 2) the sedimentological character and distribution of riverine and human-made soils, and 3) periods of *sebakh* removal and *kom* destruction.

3.1 Phases 1 and 2: Digital source review and initial GIS analysis

No one remote sensing image or historic map provides a complete solution to defining archaeological areas of interest or finding traces of buried channels. Rather, they are most effective when used in combination. In this study, for example, an examination of a coarse resolution, colorized SRTM image[22] and EES site locations prompted closer inspection in several areas. This led to mapping a hypothetical series of relict channels west of the Canopic branch through iterative visual examinations of the SEGY maps and Corona satellite images,[23] especially in those spots where the SRTM showed visible curvilinear mounding indicative of a former waterway.

22 For details, see Id., 2014, p. 69–71.

23 Corona satellite imagery have proven their worth many times over in archaeology because their temporal coverage in the 1960s and 70s predates changes that have fragmented the historical landscape (Galiatsatos, 2004). In our work they provided a level of detail sufficient to distinguish sandy mounds of *kom*s and cropmarks

Kom Qamha (EES Delta Survey No. 710, SCA el-Beheira Register No. 100263) is located between the el-Hagir and Firhash canals and about 2.5 km downstream of another large settlement mound, Kom el-Barud (see figure 3). Kom Qamha and Kom el-Barud lie tucked in between a bright sandy ridge to the northeast and stable dunes intermixed with fields to the southwest. In comparing a 1970 Corona with a Digital Globe Quickbird-2 image from the last decade (see figure 4), one notices that the mound of Kom Qamha has shrunk significantly in forty years. About half of the mound on the western side has been lost to clandestine field cutting by local farmers, shrinking its size from 4.6 to 2.5 ha.

The *kom*s are connected by a series of curvilinear pools on the SEGY map that have corresponding cropmarks in a declassified Corona satellite image.[24] These "Corona channels" appear as darkened, curvilinear fields varying in width from 110 to 130 m that wrap around the south and west of Kom Qamha (see figure 4, left). The Quickbird image (see figure 4, right) retains remnants of the cropmarks as two large ponds to the southwest of the *kom*.

To put these observations in context, a detailed, inland knowledge of the dynamic, Holocene Nile branches remains woefully incomplete. Based upon sedimentary sequences collected around Lake Maryut, Warne and Stanley posited that a fresh water channel or channels had flowed from the southeast and drained into the lake during the Classical and Medieval periods, or possibly earlier.[25] However, their core logs nor earlier cores published in during the Geological Survey of Egypt provided sufficient evidence to trace this channel upstream.[26] Subsequently, Flaux et al. support for the idea that inland channels southwest of the Canopic supplied historical Lake Mareotis with freshwater, sketching these channels to the southwest of the Canopic on their overview map.[27] Forays at

that have since been obliterated or obscured by human activity. Corona frames DS1111–1025DF013 to DF016 and other historical maps were georeferenced to a Landsat ETM+ image within ESRI ArcGIS 9.1. These imagery were graciously provided free of charge by the Center for Ancient Middle Eastern Landscapes (CAMEL) at the University of Chicago (http://oi.uchicago.edu/research/camel). Subsequent to this research, researchers at the University of Arkansas published the Corona Digital Atlas, enhancing portions of the CAMEL archive to offer georeferenced Corona coverage of much of the Middle East (http://corona.cast.uark.edu; CASANA/COTHREN, 2013).

24 WILSON, 2000.

25 WARNE/STANLEY, 1993.

26 MINISTRY OF COMMERCE AND INDUSTRY/ATTIA, 1954.

27 FLAUX et al., 2012, fig. 1.

Kom el-Hisn[28] and Kom Ge'if[29] have collected more detailed geomorphological information on inland channels in the western Delta. At Kom Firin, Spencer and Hughes independently recognized the importance of the series of curvilinear pools on the SEGY maps that preserved a seemingly older hydrological system.[30] Hamroush drew a similar conclusion working from Kom el-Hisn.[31] Wilson has identified portions of the Canopic/Agathodaemon branch through a combination of satellite imagery prospection and coring.[32]

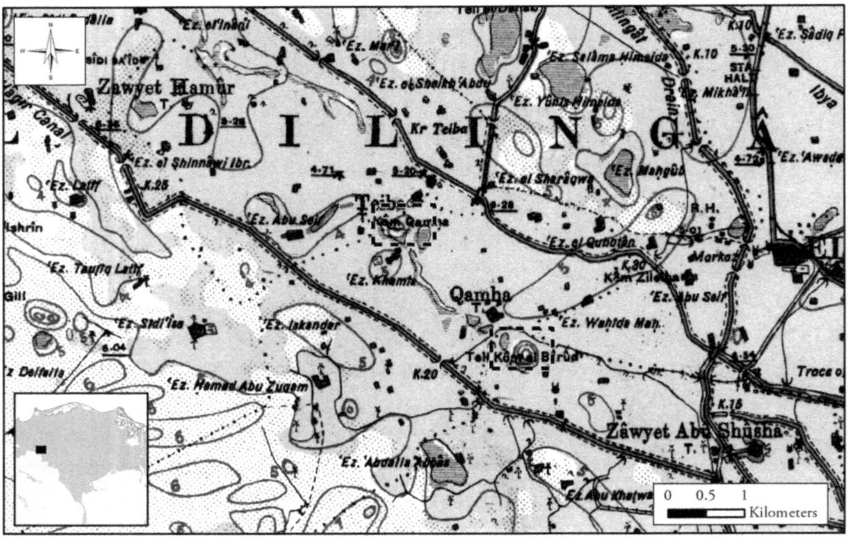

Figure 3. Kom Qamha, Kom el-Barud, and surrounding territory on the SEGY Kafr el-Ziyat sheet. Note the curvilinear pools that connect the two koms *and continue to the northeast (Coulson and Leonard, 1981).*

28 Buck, 1990.

29 Villas, 1996.

30 Spencer, N., 2007; Hughes, 2008.

31 Hamroush, 1987.

32 Wilson, 2006; Id., 2011; Wilson/Grigoropoulos, 2009.

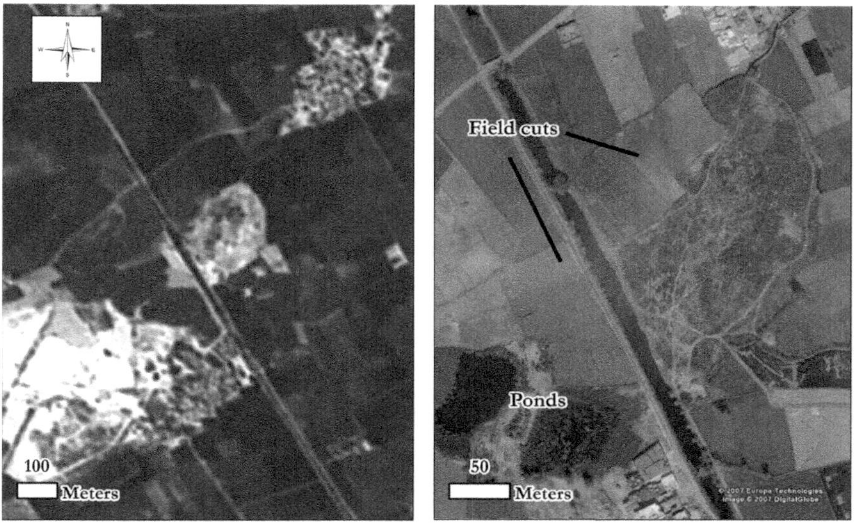

Figure 4. Kom Qamha on a Corona image from 1970 (left) and on a Quickbird-2 image from the early 2000s (right). The kom has shrunk to slightly over half its size from active field cutting. To the south and west of the kom on the Corona are curved field boundaries and dark cropmarks suggesting an ancient channel (USGS EROS. © Google © Digital Globe).

3.2 Phase 3: Fieldwork

Fieldwork included surface collection, coring transects that targeted Corona channels and koms, and topographic survey. Space is too limited here for a consideration of Kom Qamha's topography, though see Trampier, 2014.

Based upon trends observed in regional surveys within the Middle East and Mediterranean,[33] it was hypothesized that ceramic density would remain most concentrated within intensively occupied zones atop *kom*s, even if they had been disturbed. Furthermore, density was suspected to taper off at the edges of ancient human habitation/activity and then drop off rapidly to a relatively low, background ceramic scatter.[34] This work was in part designed to test Bailey's assertion that "large scale field surveys are in all probability not worthwhile in the cultivated areas of Egypt",[35] though it did not fully embrace the probabilistic

33 AMMERMAN, 1981; BINTLIFF, 1988; CHERRY, 2003; WILKINSON, 1998.

34 CHERRY et al., 1991.

35 BAILEY, 1999, p. 218.

or total coverage sampling method that may have been implied by his remarks. Surface collection units (CUs) were systematically gridded at 50 m intervals over a zone buffered at least 100 m beyond modern *kom* boundaries and often on the portions of the *kom* still visible on Corona satellite imagery (discussed below). A circular "dog-leash" CU with a diameter of 10 m was centered on each grid point with the aid of a Garmin V handheld GPS.[36] The size of the unit was chosen primarily because it would require fewer units to place overall. In retrospect, the sampling unit should have been much smaller (2–5 m radius) to allow for wider spatial coverage; too much time was spent counting hundreds of artifacts in the large sampling units. More often than not, the actual CU location was slightly shifted to accommodate mature or fragile crops and inaccessible areas. All cultural materials larger than 2 cm in length not embedded in the ground or in architecture (ceramic, wood, stone, etc.) were collected, counted, and weighed. Ground visibility and vegetation cover estimates provided a sense of each CU's effectiveness. Diagnostic sherds were retained for drawing and analysis of fabric, shape, date, and function, with all artifacts being repatriated at the end of the season. Surface collection was overseen by Jennifer Starbird.

Surface collection at Kom Qamha was challenging in that the mounds were thick with high halfa grass, camelthorn, and trash. All sorts of trash – plastic bottles and bags, paper, syringes, and even sewage – formed large piles on the northwest and north of the mound. These obstacles obscured visibility to an average of 50 % in all CUs. Density maps of surface ceramics were created in an effort to investigate their spatial patterning. The distribution and ceramic density (in kg/100 m^2) of the fifteen CUs at Kom Qamha are shown in figure 5 (left). In general, high sherd density (as measured by either weight or count) at Kom Qamha lay within the modern *kom* and/or Corona mound boundaries (i.e., the former edge suggested by the Corona). Mean density was 83.4 sherds or 1.6 kg/ 100m^2. The largest density was 354 sherds or 8 kg/100m^2, and the smallest had none. It would not appear to be a coincidence that the densest CU was located on a field cut into the Corona mound boundary, whereas the smallest density CU was about 100 m east of the mound's eastern edge. Two CUs to the northwest of the Corona mound boundary exhibited densities equivalent to those within the boundary, suggesting that in the past, the mound (or at least human activity) extended at least this far northwest.

36 BINFORD, 1964; SINCLAIR, 2006.

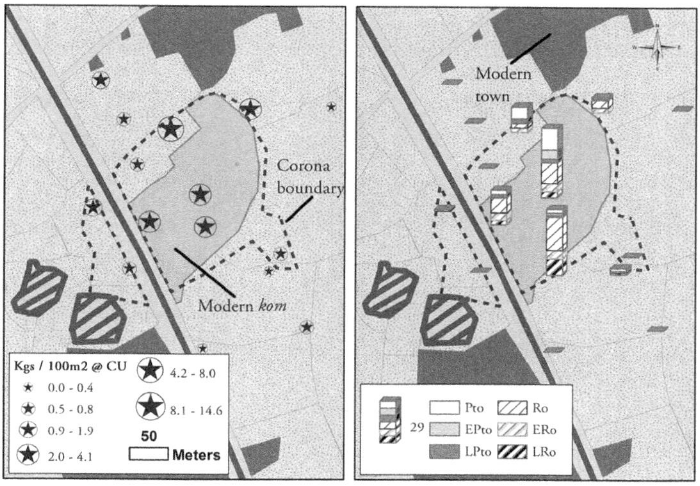

Figure 5. Distribution of CUs on Kom Qamha and their sherd densities in kg/100 m² (left). Distribution of dated ceramics on the kom *(right). Bar heights scaled to example shown in legend (29 units per bar height).*

Comparing these results with the other CUs collected at this *kom* and eight others studied in this project puts things into perspective. It was found that the ceramic densities of CUs placed on the modern *kom* as compared with the densities of CUs within the Corona-derived, historical *kom* extent were *not* significantly different according to an independent, two-sample t-test ($p > .05$). Yet there *was* a significant difference in density according to an independent, two-sample t-test ($p < .05$) between CUs made within the Corona and modern mound extent (n=72) and those made beyond it (n=33). Based on this information, it was theorized that the Corona historical *kom* extent often coincided with the limits of intensive human activity and occupation in antiquity. In this scenario, field cutting, *sebakh* removal, and other destructive activities had obscured but did not fully distort the archaeological record.

Geomorphologist Willem Toonen and the author planned coring transects (see figure 6) to investigate the components of hypothetical channels (e.g. distal floodplain, levees, channel bed deposits) and neighboring *kom*s to estimate the depth, extent, and nature of historical settlement and natural developments. A Garmin V GPS unit was used to position cores in the field. With the aid of an Edelman-style bucket auger provided by Penelope Wilson and the EES, cores were drilled up to a depth of 6.5 m in 10 cm intervals. Soil characteristics such as texture, organic remains, color, grain size, calcium/iron content, lamination, sorting, compaction, and cultural and natural inclusions were recorded.

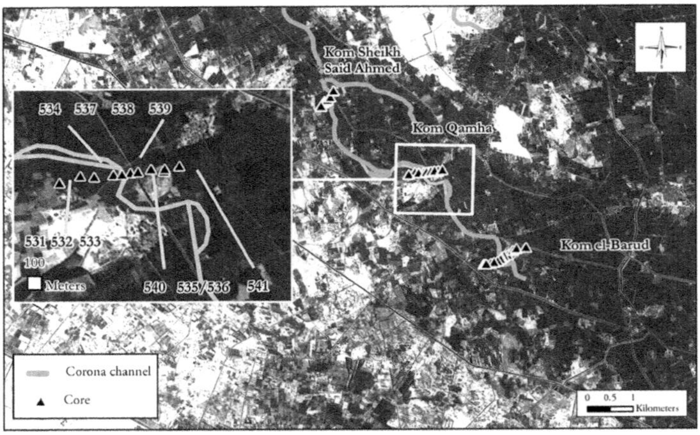

Figure 6. Coring transects at Koms Qamha, el-Barud, and Sheikh Said Ahmed.

3.3 Phase 4: Post-fieldwork analysis

After fieldwork, lab analysis of surface artifacts provided a sense of the historic periods and activities represented at Kom Qamha. Detailed examination of subsurface soils and artifacts provided new insights into the natural history of the *kom* and its hinterland, as well as *kom* destruction and *sebakh* removal.

Other authors working in agricultural lands have remarked on the relatively stable distribution of surface sherds with respect to subsurface remains. The distribution of artifacts in the plow zone is "smeared" by farming activity, for instance, but the bulk of sherds remain concentrated.[37] That being said, dated surface artifact distributions cannot always provide precise details about every activity at a site, including n-transforms such as erosion, weathering, sedimentation, and animal disturbance or c-transforms like structural recycling, ad-hoc sherd removal, and *sebakh* removal.[38] Surface observations are also just that – surficial – and merit subsequent testing by excavation.

37 DUNNELL, 1992.

38 SCHIFFER, 1975.

Abbreviation	Period or Subperiod Name	Time Period
Pto	Ptolemaic	Early third – Late first cent. B.C.
EPto	(Subperiod) Early Ptolemaic	Early third – Early second cent. B.C.
LPto	(Subperiod) Late Ptolemaic	Late second – Late first cent. B.C.
Ro	Roman	Late first cent. B.C. – Late seventh cent. A.D
ERo	(Subperiod) Early Roman	Late first cent. B.C. – second cent. A.D.
LRo	(Subperiod) Late Roman	5th – Late 7th cent. A.D.

Table 1: Portion of periodization adopted for ceramics in this study (Grigoropoulos 2009)

Kom Qamha produced 88 datable sherds. No New Kingdom, Third Intermediate Period, or Late Dynastic components were found (see figure 5, right). Rather, activity appeared to range from the Early Ptolemaic to Late Roman, following the ceramic periodization of Wilson and Grigoropoulos.[39] Diagnostics included imported and domestic fine wares such as Eastern Sigillata and African Red Slip Ware, Ptolemaic cooking vessels, and a bevy of imported and domestic transport amphorae.[40] One of the well-represented periods was Pto, the bulk of which occurred in the center of the extant mound, to the north, and to the southeast. A smattering of LPto occurred, though all but three were tentatively dated and/or dated to multiple periods. EPto did have a slightly lesser showing than LPto, and most of its firmly dated and multi-period diagnostics came from a CU in the center of the mound. ERo, Ro, and LRo ceramics clustered in the north half of the mound/Corona mound boundary. As ubiquitous as LRo ceramics may seem elsewhere in Egypt, it is important to note that it was only found in the center of the *kom* in limited quantities.

The two coring transects that intersected the Corona channels to the southwest of Kom Qamha and nearby Kom el-Barud (see figure 6) produced evidence for a dynamic channels and the ancient settlements' positions atop river levees, echoing Butzer's observations. In general, "[t]he entire cross-section [of Kom Qamha] show[ed] a fining upward trend, in which the coarse sands [were] first replaced by finer sands and finally transform[ed] into loam and clay dominated

39 Wilson/Grigoropoulos, 2009, Table1.

40 The reader is referred to Trampier, 2014 for the full ceramic analysis by Aude Simony.

deposits".[41] The geogenetic profile (see figure 7) summarizes probable human and natural origin and function of lithological groups.[42]

Given their depth, extremely poor sorting, and coarse fragments, the lowest meter or more of the sedimentological profile had probably originated from a Pleistocene braided channel. Its upper boundary was difficult to discern (as indicated by the dotted lines on the fluvial bed facies), and soils were highly calcareous. Core 532 provided some evidence of a boundary in the form of slight calcretion and plant remains from presumed growth on a stable soil. Above this layer, most deposits were Holocene-era, coarse and poorly sorted fluvial bed sands and moderate to poorly sorted levee sands and loams (located above the dotted lines). Both sudden abandonment and gradual channel siltation were present in the same location. For example, on either side of Core 534 one sees the saddle shape indicative of river levees, but in the center of the saddle where the channel bed would have been, there was almost no transition from coarse to fine sediments. Toonen suggested that the fine clays settled out of the pool that was left when the channel was abandoned (possibly from an upstream avulsion).[43] Just to this channel's east was a later phase of the channel, a saddle shape centered approximately on Core 538. This core showed a gradual transition toward gradually better sorting and finer sediments, suggesting that it had silted up over time. Accompanying drops in the water table at the low points of these saddles suggested differential patterns of water drainage based on the underlying soil matrix. Such differences in drainage would yield visual differences in crop health (depending upon the crop); the darker cropmarks on the Corona suggested that vegetation flourished here under such conditions.[44]

All of these developments preceded the formation of the *kom*. The mound seemed to have formed on the banks of an even later channel that had migrated east in a point bar system, as suggested by the coarse sandy body in Cores 539 and 540. The loams and medium sands of the channel levee provided a slight elevation as the base on which people first gathered. Coring had to be stopped in Core 535 due to thick, impenetrable clayey deposits (probably mudbrick). Just a foot over, material culture continued down in Core 536 to very top of the sandy clays. This channel was probably also operational during the Ptolemaic and Roman periods, given the scattered ceramics in its profile that may have resulted

41 TOONEN, n.d., p. 33.

42 For a detailed presentation of soil facies, the reader is referred to TRAMPIER, 2014, p. 185–93.

43 TOONEN, n.d., p. 35.

44 TRAMPIER, 2005.

from erosion of the *kom*'s southwest edge. Interestingly enough, preliminary analysis of OSL samples taken at the top of one of its levees and dated using optically stimulated luminescence indicate a minimum age of 800 b.p. and an average age of ~2200 years b.p.[45]

The geogenetic profile may also show evidence of sherd-laden *sebakh* being spread on fields, such as in the sherd-dense plow zones of Cores 537, 539, and 541. Core 540 was located at the northwestern edge of the Corona mound boundary; the repeated finding of sherds in this core and its location supported the argument that the channel had eroded the edge of the *kom*. Furthermore, there was an unusual concentration of ceramics in 538 *below* the level of the channel in 540. Since Core 538 had been placed adjacent to a modern canal, the disturbance caused by the continuous dredging of this canal may account for these ceramics and organics.

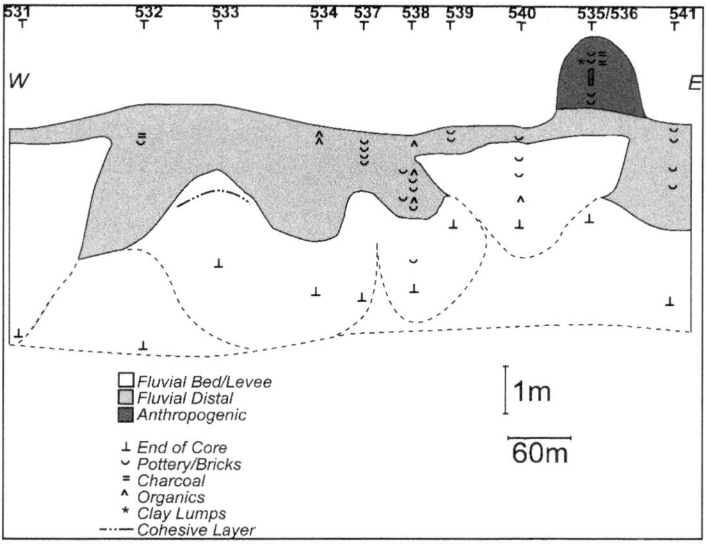

Figure 7. Geogenetic reconstruction of Kom Qamha (Toonen, n.d.).

45 Personal communication, Willem Toonen June 5, 2011.

4. Conclusion

Kom Qamha provides a good example of the complementarity of all of these datasets to hypothesize about former site extent while contextualizing this information within a broader study of the region's cultural and natural landscape. Simply put, much can be learned from sites in Egypt that would otherwise be considered "destroyed" or "lost." EES Delta Survey and SRTM imagery provided the basis for the first phase of hypothesizing and information collection. Comparing SEGY maps, Corona photos, and Quickbird-2 imagery proved an efficient second phase for collating historical changes to the landscape. In this instance, one could see on the Corona that the *kom* was probably once much larger than it is today. Surface collection and subsurface coring in the third phase corroborated this observation, suggesting that relative sherd densities of CUs within the Corona-derived mound extent were much higher than outside of it. In the fourth phase, mapping concentrations of surface ceramics by historical period also provided a sense of how small-scale activity here in the Early Ptolemaic subperiod grew to more widespread or intense occupation of the mound in the Early Roman subperiod, dwindling to smaller clusters in the Late Roman subperiod. Coring evidence corroborated observations made on the surface, as the transect revealed not only the *kom*'s development atop a Holocene river levee, but also deep cultural strata in areas where the mound had and had not been plowed under. Augering also provided subtle details on the development of several relict channels relative to the historical occupation. Based on survey conducted here and at other *kom*s, the WDLP has detailed several channels in the southwestern Delta which were active during the New Kingdom, Late Dynastic, Ptolemaic, and Roman periods.[46] Insights such as these for Kom Qamha and other *kom*s are intended as foundational data for new efforts into diachronic investigation of evolving political and economic institutions, settlement, and Nile distributaries through additional survey, testing, and data recovery in the Delta floodplain.

4.1 Writing a future Companion

The WDLP builds upon a durable trend in Egyptian archaeology of regional survey that integrates geoarchaeological, archaeological, textual, and GIScientific methods to arrive at a holistic study of a region.[47] Regional survey

46 TRAMPIER, 2014, p. 205–212.

47 E.g., PARCAK, 2005; ID., 2007.

in the Nile floodplain could have a bright future in Egyptian archaeology; a glance at figure 1 provides an idea of how much ground remains to be covered. The papers from this conference's own proceedings underscore the notion that targeted, interdisciplinary techniques can develop provocative, productive narratives of cultural and natural landscapes with minimal disturbance to archaeological contexts. One common thread is that satellite remote sensing and GIScience approaches are becoming almost indispensable for foundational information and organizational aids. Likewise, detailed understandings of local geomorphology can work hand-in-hand with artifact studies and philological insights. In that vein scholars working in the Delta have observed paleochannels of the Nile from Shuttle Radar Topography Mission (SRTM) data[48] and Landsat data.[49] It has become common practice to use remote sensing and historical cartography for prospection of ancient rivers, and to test their existence, explore their sediments, and date the channels through coring.[50]

Surface collection, while challenged by issues of artifact visibility, *kom* destruction and capping, and over-representation of later periods,[51] still offers a unique opportunity for Egyptian archaeologists to get the lay of the land before starting to dig. There is much to be learned in Egypt from combining surface work with targeted, small test excavations. By comparison, I spent several years in the U.S. Southwest and Missouri on an array of cultural resource management research projects. In one project in California in 2012, I directed several months' worth of surface collection, hand and machine-aided excavation, feature mapping, and soil sampling with complementary faunal, radiocarbon, obsidian sourcing, and several other lab studies. In that time we recorded on the order of *dozens* of artifacts (mostly stone tool fragments), fewer than *ten* datable cultural features, and encountered largely negative results in the lab work. By comparison the average Egyptian archaeologist encounters an embarrassment of riches in the sheer volume and range of expression in material culture, even when limited to the surface. In an eastern Nile Delta project, our team worked over a seven-week period to map features exposed by SCA salvage excavations; surface areas covered by the California and Egypt projects are roughly comparable. In that time we recorded *tens of thousands* of artifacts, *hundreds* of dated features, and produced over a *dozen* fruitful specialists' studies. This is not a value judgment on either country or its past cultures but a comparison of

48 USGS, 2003; STANLEY/JORSTAD, 2006.

49 WUNDERLICH, 1989; GLCF, 2004.

50 GRAHAM, 2010; HILLIER/BUNBURY/GRAHAM, 2007; ROWLAND/STRUT, 2007.

51 TRAMPIER, 2014.

scale. Even so large scale excavations have become the norm, not the exception in Egyptian archaeology, and regional survey until the past few decades has remained marginalized.[52]

The SCA 2010 list suggests that survey and settlement archaeology in the floodplain largely continue to take a back seat to studies of mortuary practices, epigraphy, and monuments. In no way does this statement question the quality of scholarship of current projects in Egypt, yet these circumstances continue to have direct consequences on how Egypt regards and administers its archaeological land. An October 2011 piece in *Ahram Online* quoted Dr. Mostafa Amin, then newly appointed Secretary General of the SCA, as saying that "lands declared free of monuments or artefacts will be offered for investment". He continued that "lands housing movable artefacts will be declared open for investment after all authentic objects have been removed to museums".[53] This may have been a pragmatic move to pay down the SCA's massive debts and a response to development pressures.[54]

Still, the language of this report is all too familiar, for it recalls the rhetoric of Ministerial Decree 43 made over a century ago.[55] This decree specified that once each and every antiquity was extracted from the *sebakh* comprising a mound, its *sebakh* could then be levied and sold as fertilizer or a source of saltpeter for gunpowder. The decree listed every *kom* or *tell* designated for *sebakh*, inspectorate by inspectorate, concentrating especially on the Delta. Some 545 mounds in all were sifted and carted away partially or completely, including some of the largest *kom*s in the western Delta. *Sebakh* extraction had no doubt happened earlier, but not previously on such a scale.[56] In some ways, the humbler manifestations of human cultures – broken and salty sherds, ashy deposits, windblown sand lenses, ancient earthworks – are denied legitimacy and a place in a culture's and a discipline's narrative when they are demoted to such a binary state: "antiquities" vs. "*sebakh*;" "archaeological land" vs. "lands declared free of monuments or artefacts."

The *Companion to Ancient Egypt* has some missing chapters, in large part because Egyptology as a field has missed some opportunities. Hopefully it is clear from the humble example of the WDLP and others working in the floodplain archaeology community that the cultivated lands offer numerous possibilities for

52 Id., 2014.

53 El-Aref, 2011b.

54 El Aref, 2011a.

55 Egypt, 1910.

56 Bailey, 1999.

piecing together past and present cultural and natural landscapes with minimal destruction for the sake of science. Perhaps after ten or twenty years, we can anticipate an edited update of the *Companion* that provides a few new chapters that move towards greater geographic and disciplinary balance. At the very least, a new chapter on diachronic settlement patterns and paleoenvironment would be a good start.

Bibliography

ADAMS, ROBERT, Heartland of Cities. Surveys of Ancient Settlement and Land Use on the Central Floodplain ot the Euphrates, Chicago/London 1981.

AMMERMAN, ALBERT, Surveys and Archaeological Research, in: Annual Review of Anthropology 10 (1981), p. 63-88.

BAILEY, DONALD, Sebakh, Sherds and Survey, in: Journal of Egyptian Archaeology 85 (1999), p. 211-18.

BARKER, GRAEME/HODGES, RICHARD/CLARK, GILL, A Mediterranean Valley. Landscape Archaeology and Annales History in the Biferno Valley, London/ New York 1995.

BIETAK, MANFRED, Archaeology in the Nile Delta, in: Egyptian Archaeology 35, p. 11.

BINFORD, LEWIS, A Consideration of Archaeological Research Design, in: American Antiquity 29, 4 (1964), p. 425-41.

BINTLIFF, JOHN, Off-Site Pottery Distributions. A Regional and Interregional Perspective, in: Current Anthropology 29, 3 (1988), p. 506-13.

BINTLIFF, JOHN/DAVIDSON, DONALD/GRANT, ERIC, Conceptual issues in environmental archaeology, Edinburgh 1988.

BLUE, LUCY/KHALIL, EMAD, A Multidisciplinary Approach to Alexandrias's Economic Past. The Lake Mareotis Resarch Project, Oxford 2011.

BUCK, PAUL., Structure and Content of Old Kingdom Archaeological Deposits in the Western Nile Delta Egypt. A Geoarchaeologica Example from Kom el-Hisn, Washington 1990.

BUTZER, KARL, Delta, in: Lexikon der Ägyptologie, Wiesbaden 1975, p. 1043-1052.

ID., Early Hydraulic Civilization in Egypt, Chicago 1976.

CASANA, JESSE/COTHREN, JACKSON, The CORONA Atlas Project. Orthorectification of CORONA Satelite Imagery and Regional-Scale Archaeological Exploration in the Near East, in: A Primer on Space Archaeology. In Observance of the 40th Anniversary of The World Heritage

Convention (Springer Brief in Archaeology 5), ed. by DOUG COMER, New York 2013, p. 31-41.

CHERRY, JOHN., Archaeology beyond the Site. Regional Survey and its Future, in: Theory and Practice in Mediterranean Archaeology. Old World and New World Perspectives, ed. by JOHN PAPADOPOULOS/RICHARD LEVENTHAL, Los Angeles 2003, p. 137-159.

CHERRY, JOHN,/DAVIS, JACK/MANTZOURANI, ELENI/WHITELAW, TOD, The Survey Methods, in Landscape Archaeology as Long-Term History, ed. by JOHN CHERRY/JACK DAVIS/ELENI MANTZOURANI, Los Angeles 1991, p. 14-31.

COULSON, WILLIAM.,/LEONARD JR, ALBERT, Cities of the Delta I. Naukratis, Malibu 1981,

ID., A., The Naukratis Survey, in: L'Égyptologie en 1979. Axes Prioritaires de Recherche I, ed. by MICHEL DEWACHTER/JEAN YOYOTTE, Paris 1982, p. 204-20.

COULSON, WILLIAM/LEONARD JR., ALBERT/WILKIE, NANCY, Three Seasons of Excavation and Survey at Naukratis and Environs, in: Journal of Egyptian Archaeology 79 (1982), p. 73-109.

DUNNEL, ROBERT., The Notion Site, in: Space, Time, and Archaeological Landscapes, ed. by JACQUELINE ROSSIGNOL/LUANNI WANDSNIDER, New York 1992, p. 21-41.

EGYPT, Arrête ministériel no. 43 S.A., Journal officiel du Gouvernement Égyptien, 1910, p. 313-8.

EL-AREF, NEVINE, No treasure in archaeologists' vaults, in: Al-Akhram Weekly Online, retrieved October 25th 2012, from http://weekly.ahram.org.eg/2011/1067/eg14.html, 2012a.

ID., Council of Antiquities survey paves way for land sell-off, in: Ahram Online, retrieved October 25th 2012, from http://english.ahram.org.eg/NewsContent/9/40/24287/Heritage/Ancient-Egypt/Council-of-Antiquities-survey-paves-way-for-land-s.aspx, 2012b.

FLAUX, CLÉMENT et al., Environmental changes in the Maryut lagoon (northwestern Nile delta) during the last ~ 2000 years, in: Journal of Archaeological Science 39, 12 (2012), p. 3493-3504.

GALIATSATOS, NIKOLAS, Assessment of the CORONA Series of Satelite Iamgery for Landscape Archaeology, unpublished Ph.D Thesis, Department of Geography University of Durham, 2004.

GILLINGS, MARK/SBONIAS, KOSTAS, Regional Survey and GIS. The Boeotia Project, in: Geographical Information Systems and Landscape Archaeology, ed. by MARK GILLINGS/DAVID MATTINGLY/JAN VAN DALEN, Oxbow 1999, p. 35-54.

GLCF, An Introductory Landsat Tutorial for Users of Geocover Data, 2004, http://www.landcover.org, 06.02.2007.

GRAHAM, ANGUS, Islands in the Nile. A geoarchaeological approach to settlement location in the Egyptian Nile valley and the case of Karnak, in: Cities and urbanism in ancient Egypt. Papers from a workshop in November 2006 at the Austrian Academy of Sciences, ed. by MANFRED BIETAK/ERNST ČERNY/IRENE FORSTNER-MÜLLER, Wien 2010, p. 125-144.

HAMROUSH, HANY, Geoarchaeology of the Kom el Hisn Area. Tracing Ancient Sites in the Western Nile Delta Egypt, Cairo 1987.

HILLIER, JOHN/BUNBURY, JUDITH/GRAHAM, ANGUS, Monuments on a Migrating Nile, in: Journal of Archaeological Science 34, 7 (2007), p. 1011-15.

HUGHES, ELLIE, Charting Ancient Waterways with the British Museum Mission to Kom Firin, unpublished M.Sc. thesis, University of Cambridge, 2008.

LUSTIG, JUDITH, Anthropology and Egyptology. A Developing dialogue, Sheffield 1997.

LUTLEY, KATY/BUNBURY, JUDITH, The Nile on the Move, in: Egyptian Archaeology 31, p. 3-5.

MESKELL, LYNN, The Practice and Politics of Archaeology in Egypt, in: Annals of the New York Academy of Sciences 925, 1 (2006), p. 146-169.

MINISTRY OF COMMERCE AND INDUSTRY/ATTIA, MAHMOUD IBRAHIM, Deposits in the Nile Valley and the Delta, Cairo 1954.

PARCAK, SARAH., Satelites and survey in Middle Egypt, in: Egyptian Archaeology 27 (2005), 8-11.

ID., Satelite Remote Sensind Methods for Monitoring Archaeological Tells in the Middle East, in: Journal of Field Archaeology 32, 1 (2007), p. 65-81,

REDMOUNT, CAROL/FRIEDMAN, RENEE., Tales of a Delta Site. The 1995 Field Season at Tell el-Muqdam, in: Journal of the American Research Center in Egypt 34 (1997), p. 57-83.

ROWLAND, JOANNE/STUTT, KRISTIAN, Minufiyeh. The Geophysical Survey at Quesna, in: Egyptian Archaeology 30 (2007), p. 33-5.

ROWLAND, JOANNE/WILSON, PENELOPE, The Delta Survey 2004-05, in: Journal of Egyptian Archaeology 92 (2006), p. 1-13.

SCHIFFER, MICHAEL, Archaeology as behavioral science, in: American Anthropologist 77, 4 (1975), p. 836-48.

SINCLAIR, PAUL, Kom el-Khawaled Field Report 2006-11-28, http://www.arkeologi.uu.se/digitalAssets/8/8060_KhawaledReport2006.pdf, 04.07.2012.

SPENCER, JEFFREY, New Ventures for the EES Delta Survey, in: Egyptian Archaeology 33 (2008), p. 2.

SPENCER, NEAL, The British Museum Expedition to Kom Firin. Report on the 2007 season, unpublished, retrieved October 25th 2012 from www. britismuseum.org/research/research_projects/kom_firin,_egypt.aspx.

ID., Kom Firin I, London 2008.

STANLEY, DANIEL./JORSTAD, THOMAS, Short contribution. Buried Canopic Channel Identified near Egypt's Nile Delta Coast with Radar (SRTM) Imagery, in: Geoarchaeology. An International Journal 21, 5 (2006), p. 503-14.

STANLEY, DANIEL./MCREA JR, JAMES/WALDRON, JOHN, Nile Delta Drill Core and Sample Database for 1985-1994. Mediterranean Basin (MEDIBA) Program, Washington D.C 1996.

STANLEY, DANIEL./WARNE, ANDREW/SCHNEPP, GERARD, Geoarchaeological Interpretation of the Canopic, Largest of the Relict Nile Delta Distributaries, in: Journal of Coastal Research 20, 3 (2004), p. 920-930.

SURVEY OF EGYPT, Rosetta 1: 100,000 Sheet 97/54, 1926.

ID., Kafr el-Ziyat 1: 100,000, Sheet 88/54.

ID., el-Ghayata 1: 100,1000 Sheet 88/48.

ID., Alexandria 1: 100,000 Sheet 92/54.

ID., Damanhur 1: 100,000 sheet 92/54.

TOONEN, WILLEM, Relationships between Paleo-environment and Ancient Settlement in the Western Nile Delta Egypt, University of Utrecht, unpublished document.

TRAMPIER, JOSHUA, Reconstructing the Desert and Sown Landscape of Abydos, in: Journal of the American Research Center in Egypt 42 (2005), p. 73-80.

ID., Expanding Archaeology in the Nile Floodplain. A Non-destructive, Remote Sensind-Assisted Survey in the Western Delta Landscape, in: Bulletin of the American Research Center in Egypt 194 (2009), p. 21-4.

ID., Review of: Alan B. Lloyd (ed.), A Companion to Ancient Egypt, Chichester UK 2010, in: Near Eastern Archaeology 75, 1 (2012), p. 60-2.

ID., Landscape Archaeology of the Western Nile Delta, Atlanta, Georgia (2014.)

USGS, United States Geological Survey, Shuttle Radar Topography Mission (SRTM), Fact Sheet 071-03, June 2004, http://egsc.usgs.gov/isb/pubs/factsheets/fs07103.html, 25.02.2011.

VAN DEN BRINK, EDWIN, A Geo-Archaeological Survey in the North-Eastern Nile Delta Egypt. The First Two Seasons. A Preliminaty Report, in: Mitteilungen des deutschen archäologischen Instituts, Abt. Kairo 43 (1987), p. 7-31.

ID., The Archaeology of the Nile Delta Egypt. Problems and Priorities, Amsterdam 1981.

VILLAS, CATHLEEN, Geological Investigations, in: The Survey at Naukratis and Environs, ed. by W. COULSON, Oxford 1996, p. 163-75.

VON DER WAY, THOMAS, Investigations Concerning the Pre- and Early Dynastic Periods in the Northern Delta of Egypt, in: The Near East in Antiquity. German Contributions to the Archaeology of Jordan, Palestine, Syria, Lebanon and Egypt, ed. by S. KERNER, Amman 1991, p. 47-61.

WARNE, ANDREW/STANLEY, DANIEL, Late Quaternary Evolution of the Northwest Nile Delta and Adjacent Coast in the Alexandria Region, Egypt, in: Journal of Coastal Research 9, 1 (1993), p. 26-62.

WILKINSON, TONY, Water and Human Settlement in the Balikh Valley, Syria, Investigations from 1992-1995, in: Journal of Field Archaeology 25, 1 (1998), p. 63-87.

ID., Archaeological Landscapes of the Near East, Tucson 2003.

WILSON, DAVID, Air Photo Interpretation for Archaeologists, 2nd ed., Stroud 2000.

WILSON, PENELOPE, The Survey of Sais (el-Hagar) 1997-2002, London 2006.

ID., Ramsîs. Ancient Memory, Archaeology and the Western Delta, in: Ramesside Studies in Honour of K.A. Kitchen, ed. by M. Collier/S. Snape, Bolton 2011.

ID., Sais I. The Ramesside-Third Intermediate Period at Kom Rebwa, London 2012.

WILSON, PENELOPE/GRIGOROPOULOS, Dimitrios., The West Nile Delta Regional Survey. Beheira and Kafr el-Sheikh provinces, London 2009.

WUNDERLICH, JÜRGEN, Untersuchungen zur Entwicklung des westlichen Nildeltas im Holozän, Marburg 1989.

The Dynamic Nature of the Transition from the Nile Floodplain to the Desert in Central Egypt since the Mid-Holocene

GERT VERSTRAETEN, IHAB MOHAMED,
BASTIAAN NOTEBAERT, HARCO WILLEMS

1. Introduction

The Nile and its floodplain played a crucial role in the development of the Egyptian civilizations. Being one of the most important allogenic rivers in the world, the Nile travels for more than 3000 km through an arid environment. The fertile floodplain, made up of silt and clay deposits through annual flooding events, and ranging in width between a few hundred meters and almost 20 km, thus stands in great contrast with the surrounding desert environment. The transition from the fertile floodplain to the sterile desert is not only an important geomorphologic and pedologic boundary, it also represented a religious and cultural limit in ancient Egypt between life and dead. However, this border between desert and floodplain is not a stable one. It changes through time through the interaction of fluvial processes operating in the floodplain of the Nile and geomorphic processes operating in the desert including aeolian processes and wadi activity. A correct understanding of the changing nature and location of the transition from the Nile floodplain to the desert is an important element in the reconstruction of the natural and cultural landscape of ancient Egypt. It not only determines the width of the floodplain, and thus indirectly also the maximum cultivable area, but also the location of settlements, cemeteries and harbors.

Here, we present some of our preliminary results of the geomorphic study of the desert-floodplain transition in Central Egypt (see figure 1). In this region, the floodplain is almost at its maximal width (± 15-20 km). The River Nile is flowing in the Eastern part of the floodplain and at several locations the river is actually eroding the eastern desert cliff, thus forming the border between floodplain and desert. The contemporary eastern course of the Nile has been suggested, for instance by Butzer,[1] to be the result of a gradual shift from west to east throughout the Holocene and this for the entire Egyptian Nile Valley. The dynastic Nile is thus suggested to have run more towards the centre of the floodplain. Also typical for this part of the floodplain is the Baḥr Yūsif, a side branch running parallel to the Nile for about 200 km before it enters the Fayyūm depression. Baḥr Yūsif runs in the western part of the floodplain, mostly at 2-4 km from the western desert edge, however, at a few locations it also comes directly in contact with the desert. The current morphology of the western and eastern deserts adjoining the floodplain is also different. Typical for the eastern desert edge is the limestone cliff with relative heights of 60-100 m: this cliff separates the limestone plateau (150-200 m a.s.l.) or higher desert from the pediments and wadi alluvial fans that form the lower desert (50-60 m a.s.l.) and the adjoining floodplain (45-50 m a.s.l.). At the western edge, the topography rises more gently: the width of the lower desert is much larger and the height of the plateau lower compared to the eastern side. Furthermore, not only limestone hills and cliffs form the western edge, also remnants of Pleistocene Nile terraces (gravel) as well as dunes from the Southern Rayyān Dune Field (SRDF) are present. In particular the SRDF plays a prominent role in the natural landscape of the western desert edge between Daljā in the south, and Sandafa in the north.

1 BUTZER, 1976, p. 134.

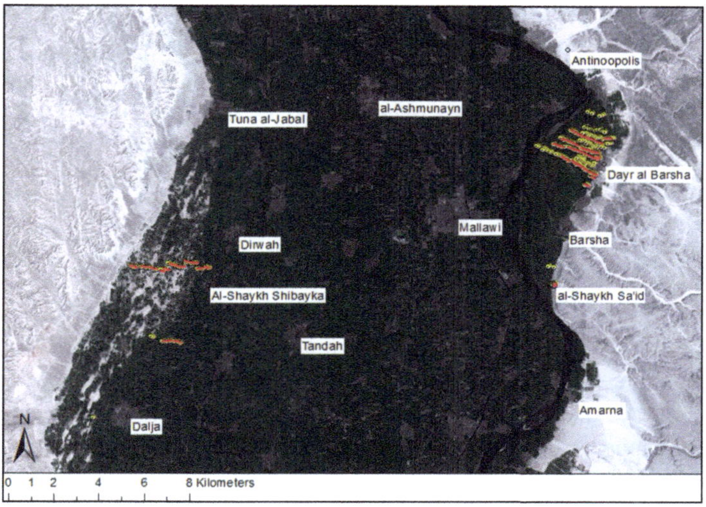

Figure 1. Aster satellite image of the Nile Valley in Central Egypt around Mallawi with indication of the main sites referred to in the text. The river channel in the western part of the valley is Baḥr Yūsif, whilst the River Nile is running near the eastern edge of the floodplain. Yellow dots indicate soil corings, whilst red lines show the location of the electrical resistivity imaging profiles.

2. Methodology

A wide range of techniques is being applied to reconstruct the floodplain-desert transition, including the analysis of current and historic topographic maps, multi-temporal remote sensing imagery, GPS measurements, soil coring, sediment dating and geophysics. The first three methods were used to describe the current landscape and to decipher sub-recent fluvial and desert dynamics. The oldest available satellite image is a CORONA image dating from May 1968, and together with the topographic map of 1945 produced by the Egyptian General Survey Authority, it provides a reference for the landscape before the closure of the Aswan High Dam, i.e. before the annual floods disappeared. Landsat TM 5 images from 1984 and 2003 with a spatial resolution of 30 m were used to detect and map the dynamics of the SRDF. A Quickbird image from 2005,

GPS measurements taken in 2009 and Geoeye high satellite images available on Google Earth from 2011 further extended the time range.

More than 200 soil cores have been made with depths ranging in between 1 and 9 m. These cores were aligned on east-west running transects across the transition area and in the Nile floodplain. A genetic interpretation of the sediment stratigraphy was made mainly based on sediment texture and lithology. Coarse sands were interpreted as channel deposits, loamy sand and sandy loam deposits are typical for levees and crevasse splays, whereas silt loam, silty clay loam and clay deposits are representative for the more distal parts of the floodplain. Wadi material could be determined based on the presence of sub-angular limestone fragments and the high calcium carbonate content of the sediments. Discriminating dune sand from fluvial channel deposits was performed by combining sediment texture including sorting, grain morphometry determined under scanning electron microscope, geochemistry and mineralogy, as well as the presence of freshwater organisms (snails) typical for river channels. The genetic interpretation was further interpolated between coring locations using electrical resistivity imaging (ER). ER profiles were performed with a SAS 1000 Terrameter using a Wenner-Schlumberger protocol and electrode spacings of 2 to 5 m. Depth of the ER profiles ranges between 10 and 40 m, and profiles are oriented east-west with lengths varying between 200 and 1500 m. ER profile interpretations were validated using soil cores. High resistivity values were interpreted as coarse sand and gravel under groundwater level, and sand above groundwater, whereas low values were interpreted as silt loam to clay deposits. The combination of soil cores and ER images provide more insight into the longer-term dynamics of the transition area. At several locations, organic fragments within the sediment record were dated using AMS ^{14}C. Also a few optically stimulated luminescence dates were obtained on point bar deposits and dune sediments. Both radiocarbon and OSL ages provided an age control of the sediment stratigraphy.

3. The evolution of the eastern floodplain-desert transition

We analysed in particular the transition area between al-Shaykh Saʿīd (immediately north of Amarna) and Antinoopolis, with a focus on the surroundings of Dayr al-Barshā. Several now-inactive Nile channels can be seen on older topographic maps and CORONA and these are still visible in the field. These channels were still active before the closure of the Aswan High Dam but became inactive due to the reduced peak flow discharge (see figure 2). However, field observations as well as soil coring and ER mapping, did show that some of these channels have been bulldozed over the last few decades and turned into arable land. At many locations, wadi material is first being brought into the former channels. Next, former river banks are leveled and silt and sandy loam deposits are moved from the banks into the channel (see figure 3). The same operations take place in and along former irrigation and drainage canals running south to north along the border between the floodplain and the desert. These post-1970 leveling activities make it difficult to interpret the current landscape, be it through topographic analysis or through the use of remote sensing data, to infer palaeo-channels that became inactive well before the Aswan High Dam closure.

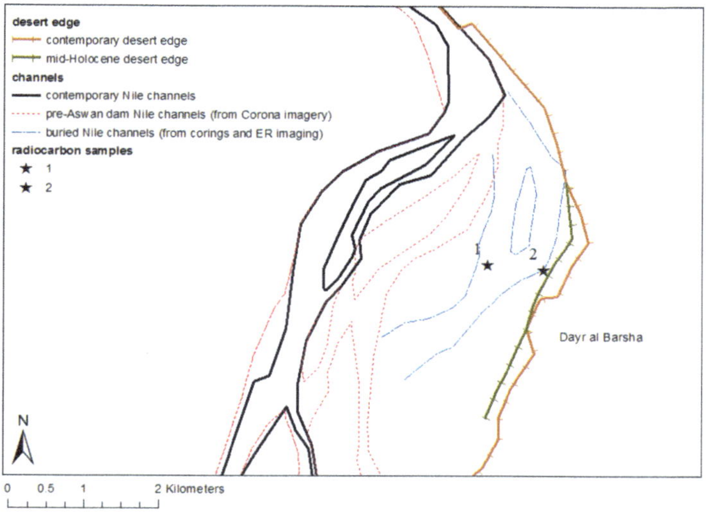

Figure 2. Indication of contemporary and ancient channels of the River Nile west of Dayr al Barshā based on Quickbird imagery (2011), Corona imagery, as well as coring and electrical resistivity imaging profiling. Contemporary and mid-Holocene desert edge (based on corings, see Figure 3) is also indicated. Stars denote sample location for AMS radiocarbon dating of fluvial point bar deposits (1: cal 650 – 780 A.D.; 2: cal B.C. 180 – 20 A.D.).

Figure 3. Wadi material is being used to level former Nile channels (left), whereas the banks of the former channels are being bulldozed (right) and the excavated Nile silts are being used as topsoil in the infilled channels. Pictures are taken 2 km south of Dayr al Barshā.

However, the field-based geomorphic approach provided evidence for ancient Nile channels that run further to the east compared to the twentieth century active channels. Both sediment corings and ER revealed the presence of a major channel in the form of meters thick sand deposits and high resistivity values, respectively. The width of the major palaeo-channel is approximately 250-300 m, which is similar to the main channel today. Further to the north, this channel bifurcates into two smaller channels of 100-150 m wide (see Figure 2). At the edge of these now buried channels, the top of the point bar deposits were characterized by high concentrations of micro-charcoal pointing towards deliberate firing activity along the banks of the river. Immediately on top of the charcoal, fine-grained sediment points to the infilling of an abandoned channel. At two locations, these micro-charcoal concentrations were used for AMS ^{14}C dating and returned an age of cal B.C. 180 – 20 A.D. and cal 660 – 780 A.D. This suggests that the activity of these Nile channels ceased around these dates. No indication for older or younger channels could be found. However, close to the desert edge near the village of Dayral-Barshā, no indications for channel activity could be found in the top 5 meters. At several locations close to the desert edge, charred plant material retrieved from the silt loam overbank sediments at 1.2 to 4.5 m depth returned radiocarbon ages ranging between cal B.C. 2890 – 2620 and cal B.C. 1400 – 1120. However, at a few locations, much more recent ages were obtained, i.e. cal 1210 – 1290 A.D. and cal 1420 – 1500 A.D. for material at 2.2 and 4.5 m depth, respectively. None of these dated samples were related to former channels. Hence, we conclude that at least for the last few thousand years, no major channel ran immediately next to the contemporary eastern floodplain-desert transition.

A more detailed coring dataset and ER profiling was made across the contemporary desert edge around the village of Dayral Barshā. Our results indicate that the location of the transition from desert to Nile floodplain did not change much over the last few thousand years, however, the morphology of this transition did change enormously. At present, there is a gentle grading of the alluvial fan at the mouth of the Wādī Nakhla towards the floodplain, with a small topographic step of ±1-2 m over a distance of 50-100 m. But the corings revealed a much more stepped and steeper desert edge. Over a distance of less than 40 m, the thickness of Nile silts covering wadi deposits increases from 0.5 m to more than 4.5 m, but locally probably much more (Figure 4 and 5). In dynastic times, the floodplain-desert transition thus must have been very abrupt. This is probably related to active undermining of the wadi pediment by the River Nile, which thus must have flown at the desert edge as is still the case between al-Shaykh Saʿīd and Amarna in the south and around Antinoopolis to the north of Dayr al Barshā (see Figure 6).

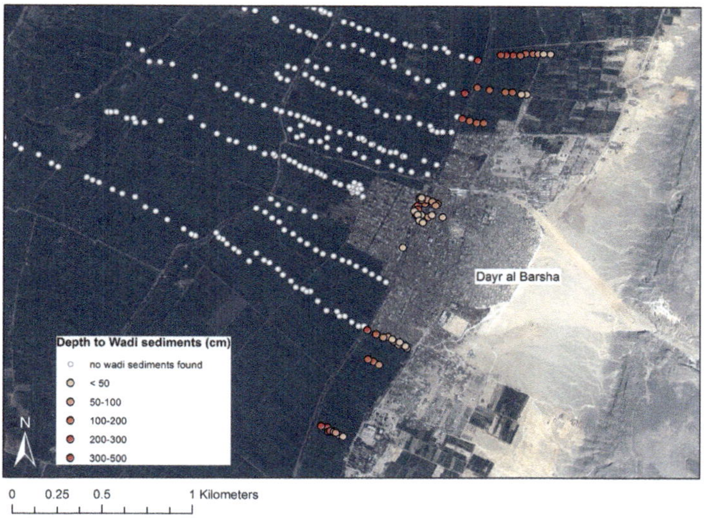

Figure 4. Depth to wadi sediments in the sediment corings performed in the surroundings of Dayr al Barshā. A gradual to abrupt thinning of the Nile flood sediments from west (> 5 m) to east (< 0.5 m) can be discerned over a distance of 100-200 m.

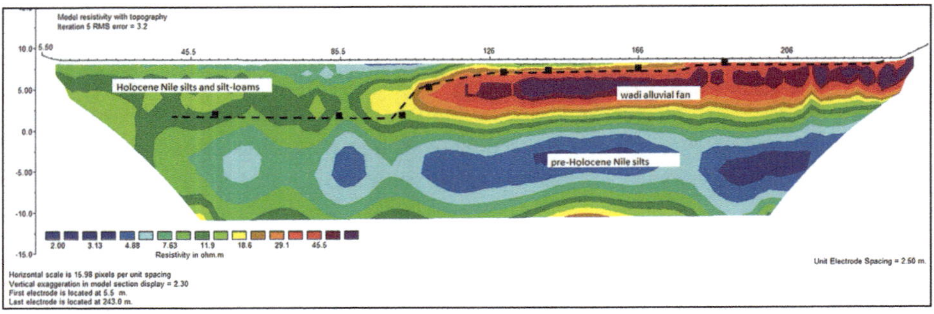

Figure 5. Results of the soil corings and electrical resistivity profile LRDB0813 located 300 m south of the village of Dayr al Barshā (southernmost set of corings indicated on Figure 4). High resistivity values correspond to relatively dry coarse-grained wadi deposits that can be related to the alluvial fans at the footslope of the limestone cliff. Low resistivity values refer to dry to wet alluvial Nile silts and silt loams. Black squares indicate the coring depth at which coarse-grained wadi material was found. The dotted black line corresponds to the interpreted topographic step in the landscape (height of approx. 5 m).

Figure 6. The left picture shows the sharp transition from the Nile channel to the desert near Antinoopolis. The height of the topographic step is approx. 8-10 meters. The right picture shows the gradual transition from the Nile floodplain (foreground) and the wadi fans whereby the boundary of the urbanized area corresponds to the desert edge. The profile shown in figure 5 is taken at this location. The major buried topographic step (5 m in height) shown on Figure 5 is located near the soil coring location in the foreground but is now completely buried under late-Holocene Nile silts.

For the eastern bank of the River Nile between Amarna and Antinoopolis, our data show that the Nile is shifting from east to west (see Figure 2), thus contrary to the generally accepted theory of Butzer.[2] This has important implications for a better understanding of the natural and cultural landscape of ancient Egypt as the position of the River Nile thus must not be looked for towards the central part of the floodplain, i.e. closer to al-Ashmūnayn. Furthermore, the important channel dynamics that occurred between the present-day course of the Nile and the eastern desert also implies that there was no continuous aggradation over time. Thus, sediment depth cannot be used for dating using the often used rule of thumb that 1 mm of silt corresponds to 1 year. Our results show a strong dependency of sedimentation rate to age. This effect, on geological timescales also known as the Saddler effect,[3] is typical when the lateral movement of channels is more important than the vertical aggradation.

4. The evolution of the western floodplain-desert transition

For the western study site, our main focus was the interaction area between the floodplain and the SRDF between Ṭūna al-Jabal in the north, and the city of Daljā in the south. It extended from the western desert to the channel of Baḥr Yūsif. In this region, dunes with heights above 20 m are separated by nearly flat and cultivated interdune areas. In this study, we were particularly interested in reconstructing the evolution of this part of the floodplain so as to analyse 1) the past changes in dune and floodplain areas, and 2) the possibility that the dunes influenced the Nile hydrology. Interactions between rivers and sand dunes can have far-reaching consequences on fluvial and aeolian geomorphology,[4] and thus also on the landscape inhabited by humans. Evidence for invading dunes blocking the Nile, thereby creating lacustrine environments and influencing settlement patterns (e.g. the Makhadma lake behind a dune field at Naj ʿḤammādī region) has been demonstrated for Upper Egypt in the Late Pleistocene (±22-14 kacal BP)[5] but never for central Egypt nor for the Holocene period.

The multi-temporal analysis of the SRDF using Landsat TM data for 1984 and 2003 showed that all the dunes moved in SSE direction, thus towards the

2 Butzer, 1976, p. 134.

3 Saddler, 1981, v. 89, p. 569–584.

4 See e.g. Liu, Coulthard, 2015 for a global overview.

5 Vermeersch et al., 2006; Vermeersch/Van Neer, 2015.

Nile floodplain.[6] However, their study also revealed that once the dunes invaded the floodplain – and are surrounded by vegetated cultivated areas – the migration rate is strongly reduced. Mohamed also calculated an average migration rate for 43 barchan dunes in the desert of 4.4 m/year from NNW to SSE.[7] Within the floodplain, dunes migrate only at a speed of approx. 1 m/year. Instead of moving at high rates, dunes tend to increase in height once they reach the floodplain. The combination of the migration rates for the analysed dunes with a map of the entire dunefield made it possible to calculate the total amount of sand that is annually being transported by migrating dunes towards the Nile floodplain. It is estimated that 0.34 million tons (Mt) of dune sand is transported towards Baḥr Yūsif on an annual basis, and this for the entire SRDF region between Samalūṭ in the north and Daljā in the south. Unfortunately, few data on the bedload transport of the Baḥr Yūsif are available. For the Nile at Banī Suwayf and Banī Mazār, this is estimated at 0.23 and 0.33 Mt/year, respectively.[8] These values are underestimating the transport capacity, however, as the Nile after the closure of the Aswan High Dam transports less sediment than it potentially can. Based on transport capacity equations,[9] we estimated the pre-dam transport capacity of the Nile and Baḥr Yūsif at 2.8 and 0.25 Mt/year, respectively. Thus, the rates at which sand is being flown into the floodplain is much less compared to the ability of the River Nile to transport sand as bedload. Under these conditions, it is very unlikely that the SRDF would have been able to influence Nile hydrology. However, the situation for Baḥr Yūsif is different. Sand influx and transport capacity are of an equal magnitude. Furthermore, for the dry periods which reduced the Nile discharge with up to 80 % towards the end of African Humid Period (approx. 5.5 ka BP),[10] we estimate the transport capacity of Baḥr Yūsif to 0.02-0.03 Mt/year. Hence, in these periods of major climatic and environmental change, the advancing dunes indeed could have blocked the flow of Baḥr Yūsif, which represents a major impact on the local hydrology of the floodplain, and thus also on the local population.

6 MOHAMED/VERSTRAETEN, 2012.

7 MOHAMED, 2012.

8 GAWEESH/VANRIJN, 1994; ABDEL-FATTAH et al., 2004.

9 e.g. KALINSKE, 1947.

10 KROM, 2002.

Figure 7. Nile silt and clay interfingering with dune sands in the eolian-fluvial interaction area between Dirwah and Dalja. The left picture shows thick clay-rich alluvial deposits at depth (> 3-4 m dune sand), whilst the right picture shows only cm thick Nile silts covered by 1-2 m of dune sand.

The long-term dynamics of the dune-floodplain interaction area was studied by ER and corings along a 3.5 km long west-east profile from the desert to Baḥr Yūsif near the village of Al-Shaykh Shibayka. The results show that close to the desert, the upper 5-10 meters are typical dune sands. However, at depth, some thin layers of clay deposits interfingered with the dune sands (see Figure 7). Often, these clay layers are only 1-5 cm thick. Further to the east, towards the Baḥr Yūsif, the thickness of the clay layers increases. In the most easterly part of the interaction area, the thickness of the dune sand deposits is less than 2 m and below it a continuous layer of several meters of Nile silts and clays was found (see Figure 7). In the central part of the interaction area, the clay layers interfingering with the dune sands are of variable thicknesses. However, this section is interesting for two other observations. First of all, remnants of a former river channel, nowadays buried under a few meters of dune sand, could be located, probably marking the location of an earlier channel of the Baḥr Yūsif. This indicates that the Baḥr Yūsif was forced further east by at least 900 m due to the advancing dunes. It also is in agreement with the quantification of the sand fluxes discussed above, which showed that the invading dunes were able to compete with the transport capacity of the Baḥr Yūsif. Next to the channel, we found in several detailed corings that the clay layers were mixed with the surrounding sand layers, with the presence of hoe marks. These indicate how these flooding deposits engaged agricultural practices in the interdune areas. Also at present, clay in the interdune areas is sometimes mixed with underlying sand to increase soil productivity (see Figure 8). One of the former cultivated layers at 7 m depth was dated at cal 2880-2620 B.C. (i.e. Early Dynastic Period to Old Kingdom) whereas the clay layer immediately below it is more than 2000 years older: this points towards a long period of stability which abruptly ended after the cultivation period by the invasion of the dunes. The cultivation period also falls shortly before the decrease in estimated Nile discharge[11] and a period of suggested contraction of Nile channels.[12] Although detailed datings are still missing, we interpret the advancement of the dunes after 2500 B.C. as the result of the ongoing mid-Holocene desertification typical for the Sahara.[13] Throughout the entire profile, the frequency and thickness of the clay layers also diminishes towards the top of the sequence, again showing how the influence of the fluvial processes is being reduced by the increasing intensity of aeolian processes. This increased intensity of aeolian processes during the last 4-5 ka is also reflected in the higher concentration of Saharan dust particles in Nile Delta sediments.[14]

Figure 8: Contemporary farming practices in the eolian-fluvial interaction area between Dalja and Dirwah (left). Farmers put silts and clays from Nile alluvial deposits atop new fields in the dune region (see soil heaps in the right picture) in order to make the sterile dune sand cultivable. Silts and clays are being mixed through farming practices: traces of such activities could be observed within the dune sands as well pointing to former agricultural practices in the area before the advancement of the dune system.

11 BERNHARDT et al., 2012.

12 MACKLIN et al., 2015.

13 KUPER/KRÖPELIN, 2006.

14 BOX et al., 2011.

For the interaction area between Ṭūna al-Jabal and Daljā, multi-temporal analysis of CORONA (1968), Landsat TM (1984 and 2003), and Geoeye (2011) showed a major reduction in dune areas. In 1968, more than 1700 of the ±2600 ha large study region was covered by dune sand. In 1984 this was already reduced to 1123 ha, and in 2003 only 256 ha of dunes were left, i.e. only 15 % of the original dune area (see Figure 9). For 2011 no quantitative estimate is yet available but our qualitative analysis shows that the reduction is still continuing. The intensity at which the dunes are disappearing is quite in contrast to the rate at which the dunes are invading the floodplain. It is clear that this is not related to natural processes but rather to anthropogenic processes of land reclamation through bulldozing activities and the expansion of arable land. During several field campaigns, the authors have witnessed that dunes are being levelled and that the older Nile silts and clays preserved below the dunes are being quarried and brought on top of the dune sand to create new fields.This rapid transformation of the landscape has important implications for the preservation of the rich cultural and natural heritage which is preserved below the dunes.

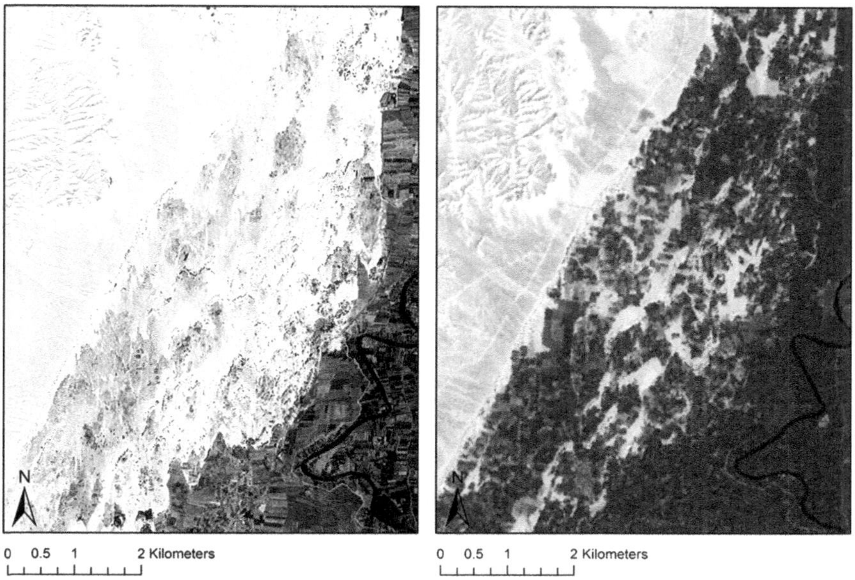

Figure 9: The Western floodplain-desert transition near Al-Shaykh Shibayka. Left: Corona image of 1968, right: ASTER image taken in 2003. Many dunes and interdune areas have been transformed to cultivated land.

5. Conclusions

The interaction area between the Nile floodplain and the desert is a highly dynamic and very variable environment. For the study area in central Egypt, the evolution along the western and eastern banks of the Nile is quite different. In the eastern part, the transition remained more or less at the same position throughout the last few thousand years, but the morphology drastically changed from a desert cliff eroded by the Nile to a more gently sloping surface. In the western part, the intensification of the dune migration into the floodplain following the mid-Holocene aridification of the Sahara meant that the areal extent of the floodplain decreased and that even the position of Baḥr Yūsif changed. Since the closure of the Aswan High Dam, human impact in the interaction area further increased whereby many of the dunes are being leveled thus destroying this unique interplay of dunes and Nile sediment dynamics. Also the cultural heritage preserved below the dunes in the floodplain deposits is nowadays threatened by the increase in cultivated land.

Acknowledgments

The authors would like to thank staff members from the Centre of Archaeological Sciences as well as the many Geography students and scholars from KU Leuven that contributed to this study through field work: Véronique De Laet, Koen D'Haen, Bert Dusar, David Kaniewski, Robin Jolley, An Steegen, Johan Schuermans, Wim Van Dessel, Jenny Segers, Marlon Pareijn, Pieter Van Turnhout, Jozefien Hermy, Jan Duhayon, Iris Deliever, Maarten Van Loo and Matthias Vanmaercke. The second author received funding from KU Leuven through a PhD-scholarship from the Interfaculty Council for Development Co-operation (IRO). The research fits within the APLADYN project funded by the STEREO II-program of the Belgian Science Policy-project SR/00/132, two projects funded by the Fund for Scientific Research-Flanders (FWO: G.0277.06 and G.0A94.14), and a research project funded by KU Leuven (OT/13/042).

Bibliography

ABDEL-FATTAH, S., AMIN, A., VAN RIJN, L.C., Sand transport in Nile River, in: Journal of Hydraulic Engineering (ASCE) 130 (6) (2004), p. 488–500.

BERNHARDT, CHRISTOPHER E., HORTON, BENJAMIN P., STANLEY, JEAN-DANIEL, Nile Delta vegetation response to Holocene climate variability, Geology 40 (2012), p. 615–618.

BOX, MATTHEW R. et. al., Response of the Nile and its catchment to millennial-scale climatic change since the LGM from Sr isotopes and major elements of East Mediterranean sediments, in: Quaternary Science Reviews 30 (2011), p. 431–442.

BUTZER, KARL W., 1976. Early hydraulic civilization in Egypt. A study in cultural ecology. The University of Chicago Press, Chicago 1976.

GAWEESH, M.T.K., VAN RIJN, L.C., Bed-load sampling in sand-bed rivers, in: Journal of Hydraulic Engineering (ASCE) 120 (12) (1994), p. 1364–1384.

KALINSKE, A.A., Movement of sediment as bedload in rivers. In: American Geophysical Union Transactions 28 (1947), p. 615–620.

KROM, MICHAEL D. et al., Nile River sediment fluctuations over the past 7000yr and their key role in sapropel development, in: Geology (January 2002), p. 71–74.

KUPER, RUDOLPH., KRÖPELIN, STEFAN, Climate-controlled Holocene occupation in the Sahara. Motor of Africa's Evolution, in: Science 313 (2006), p. 803–807.

LIU, BAOLI, COULTHARD, TOM J., Mapping the interactions between rivers and sand dunes. Implications for fluvial and Aeolian geomorphology, in: Geomorphology 231 (2015), p. 246–257.

MACKLIN, MARK G. et al., A new model of river dynamics, hydroclimatic change and human settlement in the Nile Valley derived from meta-analysis of the Holocene fluvial archive, in: Quaternary Science Reviews 130 (2015) , p. 109-123.

MOHAMED, IHAB, Evolution of South-Rayan Dune-Field (Central Egypt) and its interaction with the Nile fluvial system, unpublished Phd thesis, Faculty of Sciences – Geography, University of Leuven, Leuven 2012.

MOHAMED, IHAB, VERSTRAETEN, GERT, Analyzing dune dynamics at the dune-field scale based on multi-temporal analysis of Landsat-TM images, in: Remote Sensing of Environment 119 (2012), p. 105–117.

SADDLER, P.M., Sediment accumulation rates and the completeness of stratigraphic sections, in: Journal of Geology v. 89 (1981), p. 569–584.

VERMEERSCH, PIERRE, VAN NEER, WIM, GULLENTOPS, F., El Abadiya 3, Upper Egypt, a Late Palaeolithic site on the shore of a large Nile lake, ed. by K. KROEPER, M. CHLODNICKI, M. KOBUSIEWICZ, Archaeology of Early Northeastern Africa, Poznań Archaeological Museum, Poznan (2006), p. 375–424.

VERMEERSCH, PIERRE M., VAN NEER, WIM, Nile behaviour and Late Palaeolithic humans in Upper Egypt during the late Pleistocene, in: Quaternary Science Reviews 130 (2015), p.155-167.

The Analysis of Historical Maps as an Avenue to the Interpretation of Pre-Industrial Irrigation Practices in Egypt

HARCO WILLEMS, HANNE CREYLMAN,
VÉRONIQUE DE LAET, GERT VERSTRAETEN

> As the river retires, the fields are sown, which is sooner or later, according to their respective elevations; for some fields are not free of water till the month of December, and, in some temporary canals, it remains still longer. The canal Joseph is never dry, though at its beginning it is very shallow, and therefore soon loses the supply from the river.
> ANTES, 1800, p. 69

1. The Egyptological Debate on Early Irrigation Practices in Ancient Egypt.

The topic of how the Egyptian irrigation system worked has been amply studied. The Classical historians already referred to the remarkable fact that Egypt's prosperity depended on the silt annually deposited by the Nile flood ("Egypt is a gift of the Nile"). Classical and medieval Arabic authors also remark that different flood heights had a strong impact on harvests.[1] However, only few of these accounts provide detailed information on how the flood 'worked', and on

1 An overview of Classical and medieval Arabic sources on flood heights is given by SEIDLMAYER, 2001, p. 29-52. Several of the Classical sources are also dealt with by GARBRECHT/JARITZ, 1990, p. 195-206.

how humans intervened in the natural landscape to maximize the agricultural yield. Published Egyptological studies clarify these issues only in part.[2]

The first authors to express an informed opinion on flood irrigation in Egypt were not Egyptologists, but civil engineers working in Egypt in the latter part of the nineteenth and in the early twentieth century. At the time, Egypt was in reality, if not nominally, under British rule.[3] The colonial overlords had a keen interest in the territories under their governance. Partly, this was due to the curiosity for things exotic that so much characterized Victorian England, but partly it also reflected plainly economic concerns: the British were interested in maximizing agricultural yield. This led to the publication of a whole range of studies on the hydrology and agricultural use of the flood in Egypt and the Sudan at that time.[4] These highly detailed works provide accurate information on matters like the cyclic evolution of the Nile volume in the course of the year, the river slope, and the system of water distribution before, during and after the annual flood. They also provide insight in the operation of the artificial irrigation basins during the inundation as they observed it with their own eyes. Fig. 1 reproduces the simplified, and, in Egyptology, highly influential plan Willcocks produced of this irrigation system, which operated on the basis of the simple fact that water flows from higher to lower levels, no use being made of water-lifting devices. In the landscape of the Nile Valley, two complementary principles of water movement were in force, and determined the way irrigation worked. The first is a south-north water displacement from the higher river beds in Ethiopia and the Sudan to the mouths of the Nile branches in the Delta, which lie at sea level.The second is a lateral movement of water away from the river. Here it is important to realize that, during the annual floods, the largest and heaviest sand particles fall out first, generating high embankments, called levees, immediately beside the river. Farther away from the river, only smaller particles settle. This process generates a convex East-West cross-section of the floodplain. Floodwater would under natural conditions particularly flood the deepest area, which, on the

2 Some of the most important references are: BUTZER, 1976; SCHENKEL, 1978; GARBRECHT/JARITZ, 1990; EYRE, 1994, p. 57-80; MENU (ed.), 1994; HASSAN, 1997, p. 51-74; BOWMAN/ROGAN (eds.), 1999; LEHNER, 2000.

3 For the complex political structure of the Egyptian administration in the latter part of the nineteenth century, see DALY, 1998, p. 239-251; FERGUSON, 2003, p. 230-234.

4 E.g. BALL, 1938; HURST, 1957; LYONS, 1906; WILLCOCKS, 1889; WILLCOCKS, 1899, etc. For an overview of British interventions in Egyptian irrigation, see COOKSON-HILLS, 2013.

western bank of the Nile in Upper Egypt, lies closest to the edge of the Western Desert. These deeper areas lead to the emergence of natural drainage channels, and even though these can be quite significant waterways, they are never the main branch of the Nile.

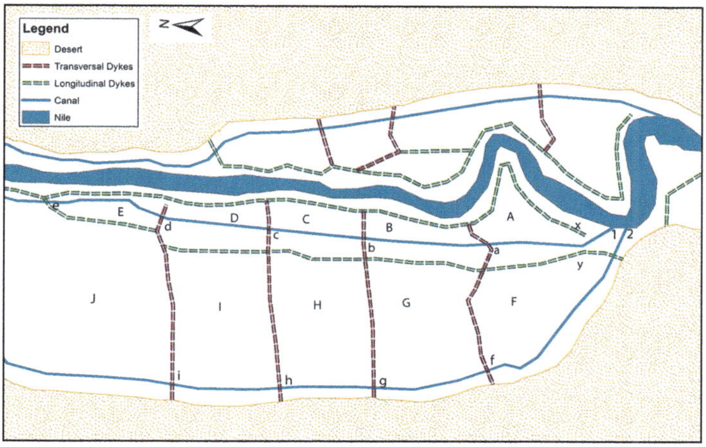

Figure 1. System of basin chains, after WILLCOCKS, 1899, pl. 14. N is to the left. Map Hanne Creylman.

Willcocks observed an irrigation system that cleverly exploited these two principles. We see that the floodplain west of the Nile is intersected by two systems of dykes.[5] The dykes running parallel to the Nile (here indicated by a double green dotted line) ensure that the water does not run off directly to the deeper areas surrounding the drainage channel dependent of canal head 2, but is partly kept in the higher basins A-E dependent on canal head 1 (see fig. 1). The transverse dykes a-i, which run perpendicular to the river, have more or less the same purpose. If these dykes would not be there, all water would continue its northward movement, with a smaller amount of silts settling. The transverse dykes ensured that the floodplain was parceled up stepwise in chains of basins, in which each basin would be allowed to be covered with c. 1-1.5 m of water. Fig. 1 visualizes a situation with two parallel chains of basins, one being located in the deeper parts of the floodplain near the western desert edge (basins F-J in

5 We will concentrate here on this western part of the floodplain, as, in general, the floodplain east of the Nile is very narrow. For the sake of brevity, we here leave out of consideration the so-called syphon canals discussed by WILLCOCKS, 1899, p. 58-60.

fig. 1), and one being located on the higher land closer to the levees fringing the Nile (basins A-E in fig. 1).

In this landscape, the flood was managed as follows. The higher and lower basin systems were fed with water from the canal heads at 1 and 2. When the floodwater in the main river channel had risen sufficiently, the dykes would be broken at these points, and the dependent systems flooded. However, before this moment, openings had to have been made already in the dykes further downstream. In the higher basin chain, for instance, dykes a, b, c and d were broken. Then, when the river dyke was broken at the canal head 1, the water ran through the feeder canal linking basins A, B, C, D, and E of course continuing to the deepest point, *i.a.* basin E downstream from the break in dyke d. Once enough water had collected in this basin, the hole in dyke was closed, and basin D began to fill. Next came basins C, B and A. The lower series of basins F-J would be filled following the same pattern. At the end of the flood season, the water would drain back into the Nile through an outlet at the end of the basin chain (not depicted in fig. 1).

Willcocks described this model in detail and with great clarity, occasionally suggesting in passing that his account concerned a very ancient system.[6] It is perhaps for this reason that several Egyptologists apparently assume that the same basin chain system was already operational in pharaonic days (cf. n. 13). If true, this idea has important consequences for ancient Egyptian society. The figure is of course only schematic, and it shows merely a small chain of basins. In reality, however, Willcocks observed far more extensive basin chains, which could extend over hundreds of kilometres. To operate such a chain, the persons monitoring dykes at the canal heads and other people at the end of the chain and everywhere in between would have to know exactly what the other was doing. This implies a well-structured and tight water administration with a firm grip over vast areas. The supposition that the same system was in use in pharaonic Egypt bears the implication that a similarly tight control over large areas was in place in that period, too. In fact, the hydraulic hypothesis of Karl Wittfogel, who formulated the idea that the emergence of early states generally depended on the capacity of the leadership to exert a tight control over the irrigation systems, and that the administration would have emerged from centrally controlled systems of irrigation management, is based to no mean extent on Willcocks' account.[7] For a long time, Egyptologists agreed that the success of the Egyptian state

---

6 WILLCOCKS/CRAIG, 1913, p. 299-300 attribute it to the time of king Menes at the beginning of the First Dynasty.

7 WITTFOGEL, 1957. This opinion is still found, for instance, in BROWN, 1997, p. 9.

depended largely on the same principle.[8] However, they initially hardly bothered about the extent to which their perceptions fitted information about the ancient environment in the first place.

This only changed in the 1970s, when the geomorphologist Karl Butzer published his seminal *Early Hydraulic Civilization in Egypt*. He argued that, before an artificial irrigation system making use of dykes and canals emerged, there must have been a simpler one that made use of natural irrigation basins. The meandering Nile had, over the millennia, generated many different levee systems, and these natural elevations would trap floodwater as the inundation receded. In the Neolithic the utility of such natural reservoirs for small-scale irrigation would have been first realized, leading to natural basin irrigation. After this, attempts to improve the natural basins by the construction of dykes and the digging of canals would gradually have improved the irrigation system, leading to more adequate forms and sizes of basins, with a stronger control over the flooding regime (artificial basin irrigation). Butzer placed the transition from natural to artificial basin irrigation in the Early Dynastic period, because the king Scorpion macehead depicts the king while digging a canal. According to Butzer, "This significant document leaves little doubt that the transition from natural to modified and, ultimately, artificially regulated irrigation had been *completed* by the end of the Predynastic."[9] This would show that, from this point in time, a new, more thoroughly regimented irrigation regime was in place.

Since the publication of Butzer's book there has been considerable debate over these issues. This, however, did not concern the principle of the development envisioned by Butzer, but only the detailed chronology of the evolution. Thus, W. Schenkel argued that the Scorpion macehead does not depict an irrigation canal, and that evidence for artificial irrigation only appears in the record in the First Intermediate Period. The famines referred to in texts from this period would in fact have triggered the introduction of artificial basin irrigation.[10] This can now no longer be considered an adequate interpretation, as there is reason to doubt that the floods in the early First Intermediate Period were substantially worse than in the preceding period.[11] Moreover, an early Fifth Dynasty scene in the so-called *Weltkammer* in the solar temple of Niuserre depicts a dyke enclosing

8 E.g. WILSON, 1951; KEES, 1955, p. 19-20 and *passim*; GRIMAL, 1988, p. 51; BONNEAU, 1993, p. XXI; LEHNER, 2000, p. 298-314 (we express our gratitude to Barry Kemp for this reference).

9 BUTZER, 1976, p. 20-21. (Italics: authors).

10 SCHENKEL, 1978, p. 37-49.

11 For an overview of the problems involved, see MOELLER, 2005, p. 153-167.

a flooded basin, showing in unmistakable fashion that artificial basins at least existed in the early Fifth Dynasty, and there is no reason to doubt that artificial irrigation emerged well before this point in time.[12] Even though this does not prove that artificial basin irrigation emerged in the time of king Scorpion, as Butzer had argued, this evidence does show it existed long before the First Intermediate Period.

A far more fundamental problem, however, is what scholars like Butzer, Schenkel, and others mean when they say Egypt deployed a system of artificial basin irrigation. None of them is very explicit about what they think the actual landscape looked like. Some scholars seem to believe that, as soon as artificial basin irrigation existed, it was immediately introduced across Egypt. When Butzer, for instance, states that the transformation had been "completed" in the Early Dynastic Period, he suggests it had been generally implemented, and many others plainly follow Willcocks by drawing an image of extensive chains of irrigation basins covering the whole of Upper Egypt.[13] However, the available early evidence for the existence of artificial basin irrigation neither proves that it was immediately adopted everywhere, nor that the same extensive chain systems existed that were observed in the nineteenth century. In fairness it must be admitted that Butzer does not explicitly state as much. He for instance assumes that in pharaonic times, irrigation was organized on a local basis. Still, the reader remains puzzled about what this exactly looks like to Butzer.

In 1992, Ghislaine Alleaume published cogent evidence against the hypothesis of large basin chain systems in premodern times, arguing that these were only created under Muḥammad ʿAlī Pasha.[14] She based her argument both on maps of the *Description de l'Egypte*,[15] which were drawn in 1798-1800 during Napoleon's

12 Willems, 2012, p. 1101-1107.

13 Ruf, 1994, p. 281-293; Bonneau, 1993, p. XXI; Lehner, 2000, p. 298-314 Grieshaber, 2004, p. 7-8; 11; fig. 8. This author remarks (without offering any real arguments) that "hinsichtlich der Dämme die Verhältnisse des 19. Jh., die sich besonders im Kartenwerk der Description und des Linant de Bellefonds widerspiegeln, cum grano salis die Verhältnisse der Antike zeigen."

14 For the agricultural policy of his time, see Lutfi al-Sayyid Marsot, 1984, p. 149-161.

15 *Description de l'Egypte ou recueil des observations et des recherches qui ont été faites en Egypte pendant l'Expedition de l'Armee française. Atlas Géographique* (Paris, 1826). All maps of this publication are easily accessible online (http://www.davidrumsey.com or www.descegy.bibalex.org), and the reader intent on following the details of the discussions of maps in this article is advised to consult

campaign in Egypt, and on written sources.[16] Focusing on the region between Luxor and Coptos, where the Nile Valley is narrow, she showed that basins were very small, and not connected by south-north feeder canals like those depicted in fig. 1. Rather, basins here had an east-west orientation perpendicular to the Nile, and each basin was fed by one or more east-west feeder canals that did not connect to adjacent basins. Clearly, irrigation was here a very local affair. The south-north feeder canal systems only emerged in the early nineteenth century. The system observed by Willcocks was hence not a survival from ancient Egypt, but a quite recent innovation, having been created earlier in the nineteenth century during a hydrological project that encompassed all of Egypt.

2. Aims of the Present Study

It is hard to overrate the importance of Alleaume's remarks for the issues we are interested in, as it shows that the nineteenth century agricultural landscape that has been so crucial in shaping our ideas on the situation in Antiquity distorts the picture rather than clarifying it. In this article we wish to show that historical maps produced before the interventions of the nineteenth century are essential to understand what the premodern landscape looked like. This does not mean that maps of more recent date are useless. Quite the contrary is in effect true. However, the earlier maps contain essential information that should not be discarded too lightly.

In what follows, we shall first go into the methodological issue of the extent to which the maps of the *Description de l'Egypte*can be relied upon. To this end, we will compare the record provided by these Napoleonic maps for different parts of Egypt. We will then focus on the area of Middle Egypt, where the authors have extensively worked in connection with the Dayr al-Barshā project of KU Leuven. To study this area, we will compare the Napoleonic maps with others, produced in the twentieth century, which provide details on surface relief which must have been of importance for the flooding regime both then and before.[17] Our analysis will not only show that in this part of Egypt, the irrigation regime of the

these sites (or the original publication). We have consulted the versions of the former website available on 01.12.2014.

16 ALLEAUME, 1992, p. 301-322; see also MICHEL, 2005.

17 KEMP, 2005 deployed very much the same method as we propose here, but we only became aware of the existence of his work after the present article had been finished. We express our gratitude to him for bringing his paper to our attention.

late eighteenth century displayed a much less thoroughly organized system of basin chains than the one that existed since the 1830s, but also that it was much more complex than the one in the Luxor-Coptos region that Alleaume analyzed. Also, some hitherto unobserved hydrological phenomena will be considered, which characterize not only the part of Middle Egypt we are most specifically interested in, but Upper Egypt north of Asyūṭ generally. Our new information is particularly intriguing when considered in connection with the landscape in the Banī Suwayf-al-Fayyūm area, which may have played a crucial role, not only as an important agricultural region in its own right, but also as an essential element in the water regime in Upper Egypt and the Delta generally. Finally, we will argue that our observations are not only relevant for eighteenth century Egypt, but also for ancient Egypt at least as early as the Middle Kingdom.

3. Methodology

3.1 The Accuracy of the Maps Published by the Description de l'Égypte

Alleaume's article clearly takes as its point of departure the assumption that the maps of the *Description de l'Égypte* provide information of sufficient accuracy to warrant their use in a study of the ancient landscape. However, a necessary preliminary question is whether these maps can really be relied upon. After all, they were produced in less than two years. In view of the vast surface covered, the short time span available, and the conditions under which the work had to be done (many of the maps indicate places where fights took place while the French troops were there), it would be naïve to suppose that the plans are error-free. In recent years, several scholars have in fact stressed their unreliability.

Thus, David Lorand has compared the French map for the area around al-Lisht with a Google Earth satellite image of the same area. He observed that the spatial relationships between sites show enormous differences between the two sources, which he attributes to the poor quality of the earlier map.[18] Similarly, a recent article by E. Subias, I. Fiz and R. Cuesta, aiming to reconstruct the middle Nile valley based on historical maps and satellite images, observed inaccuracies of such magnitude in the French maps that georeferencing them was considered quite impossible.[19] To Grieshaber, the unreliability of the *Description de l'Egypte*

18 LORAND, 2013, p. 139-143.

19 SUBIAS/FIZ/CUESTA, 2013, p. 29.

maps is such an entrenched conviction that he hardly takes seriously the dykes that are clearly drawn there.[20]

These authors are certainly right that the Napoleonic maps include numerous errors. A first problem (not highlighted in earlier studies) is that maps pertaining to certain parts of the country provide more or different kinds of details than others concerning other parts of the country. Thus, whereas the maps used by Alleaume depict numerous dykes in the area immediately north of Luxor, those concerning the area between Aswān and Luxor show no dykes at all, with two exceptions.[21] Similarly, the map of the *Description de l'Egypte*, illustrating the region of al-Lahūn at the mouth of the Fayyūm, duly indicates the presence of the Jisr al-Shaykh Jād-Allāh (in the *Description de l'Égypte* called "Gisr el-Sheikh Gadallah"), a large dyke linking the village to the desert edge near the pyramid of Senwosret II, but omits the second dyke (the Jisr al-Bahlawān") which runs southwest from al-Lahūn.[22] Since both dykes are of the same type and probably of the same age, and since they only make sense if both existed side by side, the cartographers clearly omitted to enter information on one of the two.

This confirms the impression that the maps produced by the *Description de l'Égypte* are not flawless. However, this does not mean they are useless, and perhaps they are not even as generally unreliable as the authors cited believe. One should reckon with the possibility that the teams that produced maps in different parts of Egypt may not always have consisted of the same people, that the time available may not always have been sufficient, or that other practical circumstances imposed a less strict work routine than was adopted elsewhere. Such circumstances could well account for the striking difference between the landscape as mapped by the French to the north and to the south of Luxor.[23]

20 GRIESHABER, 2004, p. 32-33.

21 *Description de l'Egypte. Atlas Géographique*, pl. 2-5. The exceptions are two adjoining dykes linking the town of Idfū to the Nile and one short dyke near Isnā. This issue has been addressed briefly by GRIESHABER, 2004, p. 9. It should be noted, however, that the apparent difference between the areas north and south of Luxor may not be due simply to errors in the maps. ALLEAUME, 1992 has argued, based on contemporary texts, that the elevation of land in the floodplain in the more southerly parts of Upper Egypt was so high that it was only to a very restricted degree reached by the flood in the eighteenth century.

22 *Description de l'Egypte. Atlas géographique*, pl. 19. We will discuss these dykes in more detail on p. 302-304.

23 The greater amount of details for the area north of Luxor may be related also to the fact that the advance of Napoleon's troops through Middle Egypt towards Luxor

Stated differently, before historical maps are used, their reliability should first be verified for each area investigated.

Figure 2. a. the Tirʿa al-Shaykh Hijāza ("Tora Cheik Hagazéh) near the village of Qulubba as drawn in the Description de l'Egypte, *map 14. b. the same channel today and crop marks betraying its former course (Google Earth).*

Figure 3. (Next page) Toponyms in the region of al-Ashmūnayn. All are given in Arabic phonetic transcription, except in a few cases (indicated by quotation marks) where the rendering in the Description de l'Égypte *is not entirely clear. "Tora" is undoubtedly the* Description*'s rendering of tirʿa. Map Hanne Creylman.*
Towns and villages: 1. Sabīl al-Khazindar; 2. Dairūṭ (al-Sharīf); 3. Banī Ḥarām; 4. Ismū; 5. Daljā; 6. Dayr Mawās; 7. Tānūf; 8. Kawm al-Sihāl; 9. Tandā; 10. Tūkh; 11. Nazlat al-Shaykh Ḥussayn; 12. Sinjirj; 13. Umm Qummuṣ; 14. Mallawī; 15. Al-Birka; 16. Rairamūn; 17. Qulubba; 18. Dayr al-Nuṣāra/Dayr al-Mallāk; 19. Al-Bayādīya; 20. Al-Ashmūnayn; 21. Ṭūna al-Jabal; 22. Al-ʿArin al-Baḥrī; 23. Nawāi; 24. Maḥraṣ; 25. Nazlat Abū Jāmī; 26. Kawm al-Riḥāla; 27. Iṭlīdim.
Waterways: a. Baḥr Yūsif; b. Tirʿa Tanūf; c. "Tora el Kiket"; d. Tirʿa al-Sanjāj; e. "Tora el Asarah"; f. "Tora Hoçein Cherkes"; g. Tirʿa Ḥasan Kāshif; h. Tirʿa al-Majnūn; i. Tirʿa al-Shaykh Hijāza; j. al-Sabakh.

---

was slower and more peaceful than their stay south of Luxor (STRATHERN, 2007, p. 275-306). For the impact of conflicts or the lack of appropriate geodetic equipment during some periods of the campaign, cf. GOMAÀ/MÜLLER-WOLLERMANN/SCHENKEL, 1991, p. 30-43.

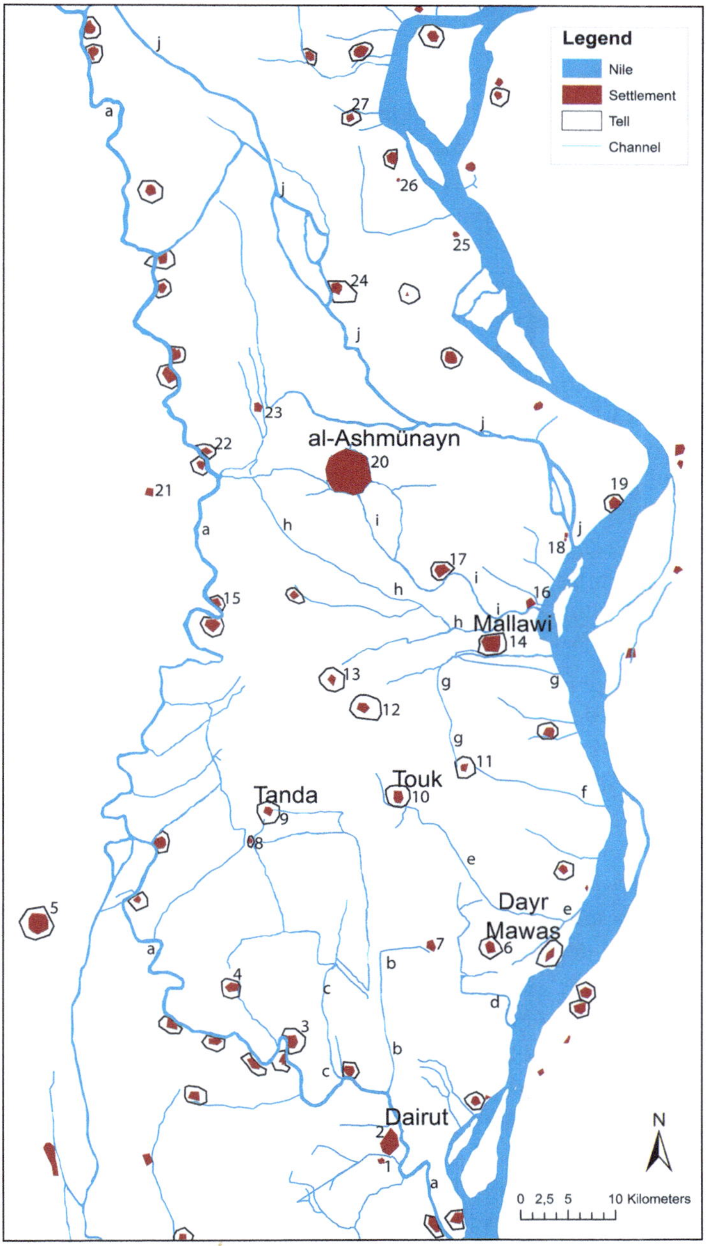
Legend
Nile
Settlement
Tell
Channel
al-Ashmūnayn
Mallawi
Tanda
Touk
Dayr Mawas
Dairut
N
0 2,5 5 10 Kilometers

Our study will be mostly concerned with the area around the town mounds of al-Ashmūnayn in Middle Egypt. Here, the maps of the *Description de l'Égypte*[24] were partially compared with information provided by more recent maps and satellite images, or observed by ourselves on the ground. Thus, the dyke linking al-Ashmūnayn to Ṭūna al-Jabal ("digue d'Achmouneïn"; our dyke 21[25]) still exists, being currently the provincial road linking the two places; the E-W dyke to the north of al-Bayadīya (our dyke 24) also still exists, as do parts of the waterway called al-Sabakh ("el Sabbak") on the Napoleonic map (west of al-Bayadīya) (channel j in fig. 3).[26] The wide channel linking al-Ashmūnayn to the Nile at the village of al-Rairamūn ("Tora Cheik Hagazéh"; i in fig. 3) still exists in part, and can be seen on the road from Mallawī to al-Ashmūnayn. In its more southerly parts, near the village of Qulubba, the channel disappears, but its contour is clearly visible in the cropmarks on satellite images (see fig. 2).

The isohypses in the fields to the west of the modern village of Dairūṭ as shown on fig. 5 follow the same contour as a dyke drawn there in the Napoleonic map ("Gisr Dairoût Chérif"; dyke south of our dyke 1, but not numbered in this article). Probably, this dyke itself no longer exists, but depositions of silt that had accumulated behind it led to the difference in height indicated in the isohypses. Finally, the course of a channel between "N. Bercheh" (*i.e.* Nazlat al-Barshā) and "Dêr en-Nakleh" (*i.e.* Dayr al-Barshā) on the French map corresponds well-nigh exactly to that of a canal (since 2006 filled in) shown in later maps and satellite imagery. These features are so clear and significant that the map of the *Description de l'Egypte* must be deemed very accurate at least for this part of Middle Egypt.

It should be pointed out that we have only infrequently made use of earlier maps than the *Description de l'Égypte*, because they are drawn at such a small scale as to be nearly useless. Exception must here be made for the maps published in 1753 by Robert de Vaugondy (fig. 7) and in 1765 by J.B.B. d'Anville.[27]

24 *Description de l'Egypte. Atlas géographique*, pl. 13-14. MICHEL, 2005, p. 256 has stressed the reliability of these maps.

25 For the numbering of the dykes, see figs. 5-6 and section 4.1.2 in this article.

26 The key to the location of settlements and waterways in this article can be found in fig. 3.

27 DE VAUGONDY, Carte de l'Egypte Ancienne et Moderne, dressée sur celle du R.P. Sicard et autres, assujetties aux observations astronomiques, par le Sr. Robert de Vaugondy Geographe ordinaire du Roy. Avec Privilege, 1753,' in : *Atlas Universel, Par M. Robert Geographe ordinaire du Roy, et Par M. Robert De Vaugondy son fils*

3.2 Georeferencing historical maps

Every map is a two-dimensional projection of a three-dimensional reality. The maps to be compared in this study were not made on the basis of the same projections, and therefore, they needed to be georeferenced. This means that the technical coordinates on the two-dimensional maps are transferred into geographical coordinates. For this study, the maps were superimposed in ArcMap, using the normal cylindrical Web-Mercator-Auxiliary-Sphere (WMAS) projection, making use of the geodetic datum WGS1984, as this is the projection system of Bing maps, which is the reference map used for this study. We used the georeferencing tool in ArcGIS 10.1 software using the AFFINE polynominal transformation function. This tool allowed us to link ground control points (GCPs) in each of the maps to the same points in an independently generated Bing Maps Aerial base map.[28] This approach avoids accumulations of errors in the transformation process.

As noticed before, Subias, Fiz, and Cuesta were of the opinion that georeferencing the maps of the *Description de l'Égypte* is not feasible for the area they investigated, and for this reason they did not incorporate it into their GIS.[29] The reasonable accuracy of the map for the region between Dairūṭ and al-Ashmūnayn suggests, however, that the situation may not be as hopeless here. Yet it should be pointed out that georeferencing always generates transformation errors (the root mean square error, RMSE). Bearing in mind the intrinsic inaccuracy, particularly of the maps of the *Description de l'Egypte*, a reasonably high inaccuracy

Geographe ord. du Roy, et de S. M. Polonoise, Duc de Lorraine et de Bar, et Associé de L'Academie Royale des Sciences et belles Lettres de Nancy, Avec Privilege Du Roy, 1757. A Paris, Chez Les Auteurs, Quay de l'Horloge du Palais, Boudet Libraire Imprimeur du Roi, rue St. Jacques; J.B.B. D'ANVILLE (1765 : Egypte, nomme dans le pays Missir. Par le Sr. d'Anville de l'Academie royale des Belles-Lettres, et de celle des Sciences de Petersbourg, Secretaire de S.A.S. Mgr. le Duc d'Orleans. MDCCLX). These maps can be conveniently consulted at the site http://www.davidrumsey.com. The D'ANVILLE map was frequently reproduced, e.g. in ANTES, 1800, RIPAUD, 1800; ARROWSMITH, 1812. It forms the basis of the Turkish *Cedid Atlas Tercümesi* map published in 1803. For this rare map, see the website of the Library of Congress: http://www.loc.gov/resource/g3200m.gct00235/?sp=63 (consulted 10 May 2015). For the Fayyūm region, an overview of the available historical maps has been published by GARBRECHT, JARITZ, 1990, p. 11-24.

28 See http://be.bing.com/maps/.

29 SUBIAS/FIZ/CUESTA, 2013, p. 29.

should be accepted. Considering the size of the area under investigation and the difficulty to identify identical points in both the *Description de l'Egypte* and the reference map, a RMSE of at most 200 m is considered to be acceptable.

Whenever georeferencing produced a larger margin of error, GCPs were added, or a different set of GCPs was chosen.

3.3 Analysing the maps

In this way an overlay was produced that combines different types of landscape features. The map of the *Description de l'Égypte* contains the following elements that are important for understanding the hydrology of the regions investigated: watercourses, settlements (including abandoned ones), roads, and dykes. The maps also include other elements, like field boundaries, wasteland, sand dunes, etc. However, it is not clear whether these were recorded accurately, and in mapping we have not systematically incorporated these elements, although we occasionally made use of them where other evidence was lacking.The maps also indicate marked differences in height, for instance in the case of town mounds. However, while the general aspect of surface relief is thus partly recognizable, it is impossible to quantify the difference in height, and less prominent elevations like levees are usually not indicated at all. Therefore, one can basically recognize the locations of some landscape features, but the result is still inadequate (see fig. 4). This is most clearly the case with the dykes, which are often interrupted. Clearly, such a dyke system can never have functioned properly.

It is likely that errors in the French map explain some of these anomalies: certainly the cartographers must have missed a number of dykes. It is impossible to assess the magnitude of such errors. Another problem is that dykes would of course not be built where there were natural elevations in the landscape, but since the *Description de l'Égypte* renders such elevations only in part, it is clear that the logic of the system will escape us unless we incorporate surface relief in our analysis. For this we have made use of the 1926 topographic maps (1: 100,000) of the topographic Survey of Egypt. These maps are superior to many others that have been published in that they indicate 1 m differences in elevation.[30] The integration of these maps with the Napoleonic ones is justifiable regardless of the time difference of about 125 years, because the accumulation

30 For Middle Egypt: the 1:100,000 topographic maps entitled *Minya* (Survey of Egypt map 26/402, sheets 56/54 and 56/60) and *Dairût* (Survey of Egypt 26/403, sheets 52/54 and 52/60). The 1957 map entitled *Mallawi* (52/263), sheet 140 offers closely similar information. Maps at a 1:25,000 scale, dated to 1940-1943, also

of alluvium on the surface would in this period have amounted to only about 10-20 cm.[31] This is much less than the 1 m intervals between the isohypses.

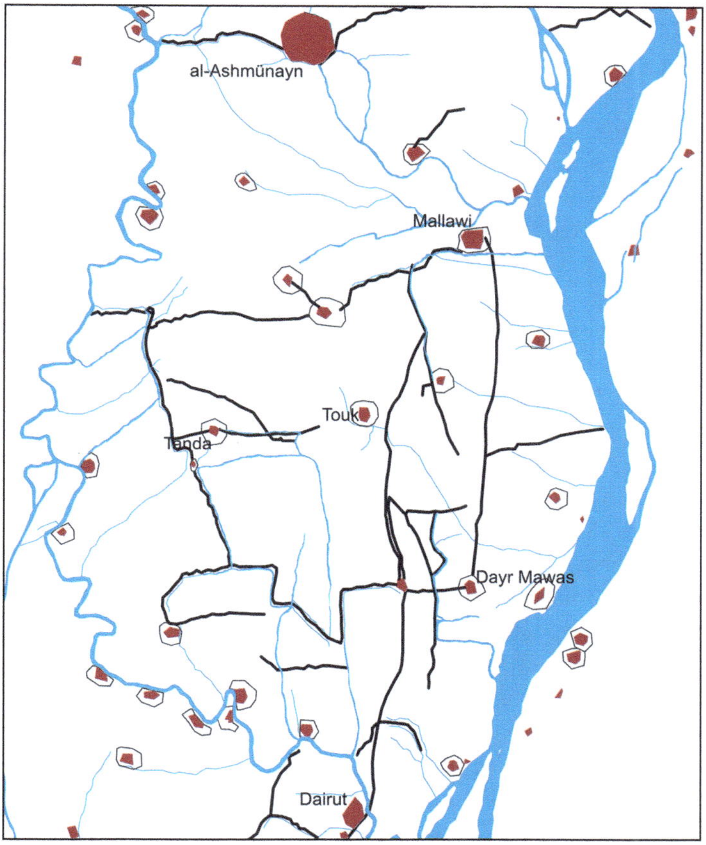

Figure 4. georeferenced map of a part of the area investigated in which the following information from the Description de l'Egypte *maps 13 and 14 has been included: settlements (red), tells (black contours around settlements), waterways (blue), and dykes (black). Map Hanne Creylman.*

exist, but their existence came to our notice too late for them being included in our analysis (see KEMP, 2005, p. 20, fig. 1.3).

31 As a rule of thumb, earth scientists often assume an average accumulation rate of 1 mm/y. The reality is of course more variable, and for this reason we here use a considerable margin of error.

In superimposing the two georeferenced maps, it becomes immediately clear that the interruptions in the dykes recorded by the *Description de l'Égypte* usually stop where the 1926 topographic map indicates a rise in surface level; we will offer ample illustration of this fact below (see fig. 5-6). A combined reading of surface relief and dyke contours leads to a meaningful subdivision of the floodplain in basins. Interestingly, many of these basins are defined by the combination of natural and man-made features that according to Butzer characterize early forms of artificial basin irrigation.

Another feature to be considered is that the Napoleonic maps often interrupt dykes shortly before they reach a settlement (for example near the village of "Touk" in fig. 4). A likely explanation is that these settlements stood on town mounds, and that the dykes stopped where the ground had risen to a sufficient height. Wherever town mounds were larger than the villages actually existing in the eighteenth century, the cartographic method deployed may suggest gaps in the hydrological system that did not actually exist.[32]

A further problem is that the maps of the *Description de l'Égypte* use topographic symbols that are not always easy to understand. In some cases, dykes are clearly recognizable, being drawn as a double line with an indication of the surface sloping down on both sides. In many cases, the map moreover includes the label "digue", or the Arabic "gesr" (i.e. jisr), which has the same meaning. Two problems have to be addressed here. In the first place, in some cases a dyke is characterized as an "ancienne digue." It is not clear to what extent this means the dyke was no longer in actual use. We have simply opted for accepting them as functional dykes. Secondly, some linear features in the landscape are also drawn by a double line, but without clear indications of surface relief. Are we not facing dykes here, but just roads at the level of the surrounding fields? This has a major impact on how the maps are interpreted. If road = dyke, then the number of basin demarcations will behigher than if we suppose that some roads crossed the basins without always following a dyke.[33]

32 That tells may have been larger than the settlements on top of them in the late eighteenth century is credible, due to a significant decrease in population in the preceding centuries (GARBRECHT/JARITZ, 1990, p. 188-189).

33 Schenkel has stated that many settlements in the floodplain lay simply within the basins, without any connection by dyke to neighbouring settlements (in: GOMAÀ/MÜLLER-WOLLERMANN/SCHENKEL, 1991, p. 25; 27). He does not clarify what this point of view is based upon. Although in the more southerly area we have been concentrating upon, settlements occasionally do lie in isolation in an area without dykes (thus Nawāy and Ḥūr northwest of al-Ashmūnayn), in most cases the

Since the map was apparently drawn in late autumn, when the flood had already mostly receded, the French troops might have marched through dry areas where such low-lying roads existed. It is, however, often assumed that most major roads in the floodplain followed dykes. The absence of graphic indications may indicate that some of the roads lay on top of dykes of only a small magnitude. The impossibility to determine with certainty whether all linear features were really dykes unfortunately introduces an element of randomness in our interpretation of the maps.[34] In our commentary, we will explain which considerations have induced us to accept or reject a linear feature as being a dyke.

Finally, even after the integration of the two maps, some dykes remain that seem to be utterly useless. Since it is as unlikely that the French geographers drew non-existing dykes as that the Egyptians took the effort of building dykes for no useful purpose, one has to assume in such cases that additional dykes existed that went unrecorded. Here, we have in a few cases (to be remarked upon later) inserted dykes not entered in the plans.

4. Analysis

4.1 The Hydrology of the area between Dairūṭ and al-Ashmūnayn

As noted before, a major feature of the hydrology of the Nile floodplain is the phenomenon of levees being deposited on the banks of the Nile, and the gradually decreasing elevation of the field surface as the distance to the river increases. The isohypses (height lines) in the maps reproduced in figs. 5-6 clearly indicate this phenomenon. Everywhere, the surface level immediately beside the Nile is on the west bank considerably higher than further west. Here, the surface level gradually decreases until the western desert edge is reached.

Although this longitudinal S-N depression can be observed in most places, the picture is blurred by secondary levee systems, of which the most important one follows the course of the Baḥr Yūsif, which branches off the Nile near the town of Dairūṭ, and follows a meandering course through the deepest part of

settlements lie on high levees or dykes according to the map of the *Description de l'Égypte*. Contrary to SCHENKEL, GRIESHABER, 2004, p. 29 seems to assume forthwith that the roads followed by the French army were dykes.

34 Some very significant dykes, such as parts of those near al-Lahūn, are in fact rendered as a mere double line in the *Description de l'Égypte*, map 19.

the S-N depression to the Fayyūm depression.[35] As explained elsewhere in this book,[36] the volume of water transported by this watercourse is much smaller than that of the Nile itself, and as a result, the levees in this area are generally much lower than those fringing the main river bed. But the Baḥr Yūsif levee system does lead to an interruption of the gradual decrease in elevation between the Nile and the western desert fringe. Here one can observe that, wherever the Baḥr Yūsif meanders away from the desert edge, depressions can be observed both east and west of it (see fig. 6). In some places, this even leads to the emergence of small lakes immediately east of the desert edge, but west of the Baḥr Yūsif. In the area investigated here, such a lake must have existed for instance near the town of al-Birka ("the Lake") near Ṭūna al-Jabal (no. 15 in fig. 3).[37] This situation suggests that the levees adjoining the Baḥr Yūsif are secondary, being deposited over the generally descending surface that existed in the western part of the Nile Valley.

The system observed here in some essential regards resembles the one in the area to the east of the Pelusiac branch of the Nile, in the eastern Nile Delta.[38] Here, too, the Nile is fringed by a gradually deepening depression, which functions as a natural drainage system. At its deepest point this has given rise to the emergence of a wide waterway locally known as the Baḥr al-Baqar, the "river of the cow". In width this waterway can reach some 50 m, not much less than that of the Baḥr Yūsif, although its depth probably is, and no accompanying levee system apparently exists here. Therefore the activity of the Baḥr al-Baqar is clearly substantially inferior to that of the Baḥr Yūsif, but to a large extent the two systems are nevertheless comparable. Both emerged in naturally grown depressions dependent on the Nile. However, the Baḥr al-Baqar is part of the Nile Delta, a vast expanse of low-lying land crosscut by two (and anciently as many as seven) Nile branches. As a result, the water volume spilling over the levees of the Pelusiac Nile branch must have been very much lower than in the Nile valley, and there were numerous drainage systems all of which transported only a relatively small part of the total flood volume.[39] By contrast, in Upper

35 This is what geomorphologists call a Yazoo river (Ward, S., 2004; Whittow, 2000).

36 Verstraeten et al., 2016, p. 248.

37 See p. 316 below. A better known example is the lake located due east of the pyramid of Amenemhat III at Dahshūr.

38 Bietak, 1975, p. 55-56.

39 Here one recalls that Classical authors recorded that in Elephantine the peak of the flood reached 28 cubits, but at the Mediterranean coast only 2 to 6 cubits. Compare also the remarks by Antes, 1800, p. 66.

Egypt, all water had to pass through the relatively narrow Nile valley, where the river slope is moreover steeper. Undoubtedly the much larger resultant local water volumes encountered here must have led to occasional breaks in the Nile levees, and ultimately to the (probably natural) creation of an open connection between the Nile and the Baḥr Yūsif drainage system. As a result, it would be correct to an extent to call the Baḥr Yūsif a 'branch' of the Nile, but yet this does not diminish the fact that it is not more than a secondary branch draining off Nile water to lower areas. In this regard, it differs completely from situations where the main bed of the Nile is split in two by islands, or where it bifurcates, as happens with the major Nile beds in the Delta.[40]

Today, the Baḥr Yūsif is the only remaining secondary system branching off the Nile, but the map of the Description de l'Egypte indicates several additional branches in existence in the late eighteenth century. One is the Tirʿa al-Shaykh Ḥijāza ("Tora Cheik Hagazéh") connecting the village of Rairamūn and the town of al-Ashmūnayn, another is al-Sabakh ("el Sabbak"), which crosses the Nile banks just south of the village of al-Bayaḍīya ("el Béïâdîéh") and following a winding course until it ends in the S-N depression just north of al-Ashmūnayn (see fig. 6).[41] For the location of these toponyms and names of waterways, see fig. 3. As figs. 5-6 show, these watercourses are also accompanied by their own, although less clearly expressed, levee systems.

Just as, to a lesser extent, the Baḥr Yūsif, these latter two waterways are secondary to the Nile itself. They are much narrower, and therefore undoubtedly also much shallower than the Nile. The obvious implication is that, when the Nile was low, an only small amount of water (or none at all) may have entered

40 The situation in the Baḥr Yūsif area reminds one of that around Memphis, where the current main branch of the Nile is accompanied on the west by the smaller, and less highly elevated Baḥr al-Libaynī. This waterway has been interpreted as the rudiment of an early dynastic Nile branch, the assumption being that the town then lay on the eastern Nile bank. This would only have changed as a result of a gradual shift of the Nile bed to the east (JEFFREYS, TAVARES,1994, p. 143-173). To us, it seems at least worth considering that the Baḥr al-Libaynī is a Yazoo-type river system (see n. 35) just like the Baḥr Yūsif. As a result, the perspective on the hydrology of ancient Memphis may have to be thoroughly reconsidered in the light of the present study.

41 *Description de l'Egypte. Atlas géographique*, pl. 14. At the mouths of these branches we have discovered remains of dams accompanied by brickwork water distributors built in the time of Muḥammad ʿĀlī. This suggests that the branches were blocked around 1830.

these waterways. They therefore mainly played their role in the hydrology of the floodplain during and after the Nile flood.

So far, all elevations along waterways have been termed levees, and we will continue to use this term below in this general sense. However, the landscape formation processes in the region studied here may bear witness not only to the result of sedimentation on embankments parallel to rivers. There is a second system, which evolves particularly inside river bends when large discharges of water erode small channels through an already existing levee. This results in large amounts of sediment being washed through the new gulley, and fanning out from there inside the basin ('crevasse splays'). Once formed, crevasse splay channels may be reactivated during subsequent floods, leading to the emergence of additional channels, and to sediment depositions on top of their embankments. Crevasse splays accordingly lead to the emergence of secondary levees bulging into the flood basins behind the main levees.[42] It will appear that this phenomenon also had an impact in the region here investigated.

4.1.1 Detailed Description of the Elements of the Hydrology of the Region

We will now discuss the hydrological elements (dykes, canals, channels, elevations in the landscape, etc.) in the region. This section is essential for this article, but for understanding its reasoning it is only of importance to those who might like to verify the detailed argumentation on which our assumptions are based. Readers not interested in these details are recommended not to read pp. 275-295.

The analysis to be presented here builds upon an account of the same issues published earlier in summary fashion by H. Willems.[43] This article was based on work carried out to prepare the construction of a functioning Nile model, which stood in the exhibition 'Keizers aan de Nijl' in the Gallo-Roman Museum in Tongres (1999-2000). For obvious reasons, this model had to be based on a highly simplified translation of reality. The base maps Willems used then were the same as are used in this article, but they were not properly georeferenced. Rather, the map of the *Description de l'Égypte* was somewhat impressionistically adapted to the topographical map of the survey of Egypt. Several errors were,

42 Bridge, 2003, p. 273; van Dinter/van Zijverden, 2010, p. 20-21 offer a good visualisation of the evolution of a crevasse splay. Note that the effects of crevasse splays in part of the area discussed here is already mentioned in Kemp, 2005, p. 23-24.

43 Willems, 2013, p. 346-352 and fig. 1.

however corrected in the georeferenced maps used in the present article, which were produced by Hanne Creylman and Véronique De Laet. We will present the relevant data from south to north, starting just south of the town of Dairūṭ, where Baḥr Yūsif branches off from the Nile.

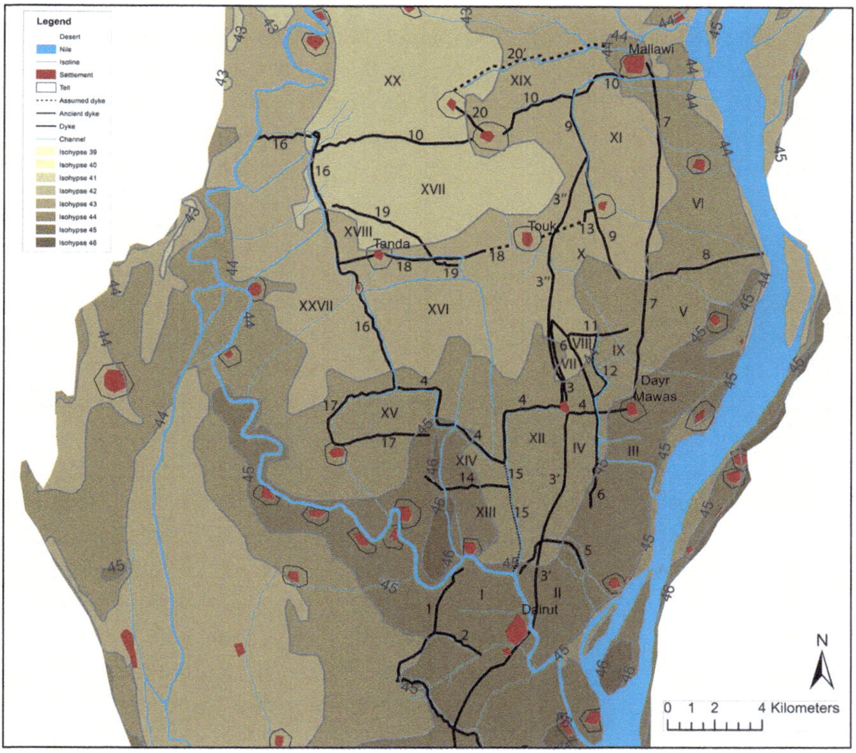

Figure 5. Georeferenced map of the northern part of Description de l'Égypte *plate 13, projected over a topographical map of the same area. Map Hanne Creylman.*

1. The Nile Levee near Dairūṭ (see fig. 5)

Dairūṭ lies immediately to the south of the Baḥr Yūsif, and accordingly on the levee of this Nile branch. The elevation of this levee (+ 45 m) is lower than that of the Nile levee southeast of Dairūṭ, the top of which lies above the 46 m contour line. Further north, the top of an island in the Nile reaches the same high level. The area near the beginning of Baḥr Yūsif, in between the two +46 m areas, is considerably lower (in the +44-45 m realm). This suggests that, at a certain point in time, Baḥr Yūsif cut through an existing +46 m Nile levee, perhaps caused by a bend in the course of the river Nile.

Events of this kind are only likely when a high flood generates such a water pressure that the levees break. The same exceptional circumstances may explain why the northernmost +46 m area developed into an island with, judging by the closeness of the isohypses, steep slopes, and why two small channels divert from the Nile NE of Dairūṭ.[44] This renders likely that a break in the Nile levee east of Dairūṭ evolved into the major Nile branch now known as Baḥr Yūsif. Possibly this waterway was at its origin a crevasse splay channel, but it has long since developed into a perennial river branch.

2. The dykes west of Dairūṭ (fig. 5)

The *Description de l'Égypte*, pl. 13, indicates two dykes west of Dairūṭ. *Dyke 1* branches off from the Baḥr Yūsif to the southwest. It is called Jisr Dairūṭ al-Sharīf (in the *Description de l'Égypte*: "gesr Dairoût el Cherîf", the "dyke of Dairūṭ al-Sharīf"). Due west of Dairūṭ, the unnamed *dyke 2* branches off eastwards from dyke 1 towards the town. Dykes 1 and 2 roughly demarcate the west and south of a basin, but the *Description de l'Égypte* does not indicate how this was closed off on the east and north. However, the northern part corresponds to the levee constituting the southern bank of the Baḥr Yūsif, even though the isohypses on the map do not show a rise in elevation in this area. On the east, there lay the town of Dairūṭ itself and, just south of that, a smaller settlement called Sabīl al-Khazindār ("Sibil el Kâzendâr"). It is likely that the elevations on which these settlements lay (and which are clearly indicated on the map of the *Description de l'Égypte*) constituted the eastern perimeter of the basin. This basin was fed by a channel perpendicular to the Baḥr Yūsif, which entered it just north of the town. We will call this *basin I*.[45] After the flood, the residual water from basin I must have flowed back into the Baḥr Yūsif.

44 Since on pl. 13 of the *Description de l'Égypte*, these channels have no name, they have received no number in fig. 3.

45 In this article we will not discuss a basin surrounded by dykes to the southwest of basin 1. However, it should be noted that its dykes as indicated by the *Description de l'Égypte* follow the same contour as the 45m isohypse in the topographical map, even though this does not indicate any roads or dykes here. Probably the eighteenth century dyke had been removed by the time the twentieth century map was drawn, although the silt deposition that had accrued behind it was still there. For such depositions to reach a height of this kind (the topographical map used iso-

3. The Depression in the Floodplain West of Dairūṭ

From dyke 1 as far as the edge of the Western Desert, not a single dyke is noted by the *Description de l'Égypte*. This implies that at the latitude of Dairūṭ the area within dykes is restricted to a narrow band west of the Nile. Further west, the topographical map indicates an extensive depression (elevation: + 43m), crosscut by three natural S-N drains, the central one of which further north breaks into the Baḥr Yūsif. This undoubtedly humid depression may have been less suitable for agriculture in antiquity[46]. West of this depression, a S-N elevation in the + 44 m range may be the levee of a former branch of Baḥr Yūsif. The town of Daljā lies on top of this elevation.

4. The Hydrology of the Region between Dairūṭ and Dayr Mawās

Between Dairūṭ and Dayr Mawās, north-east of the Baḥr Yūsif, a wide + 45 m levee fringes the Nile on the west, which must have bounded a somewhat lower and descending area further west. This area continued even N. of Dayr Mawās. Just north of the mouth of the Baḥr Yūsif, and northeast of Dairūṭ, the levee is intersected by two small channels which do not bear a name in the *Description de l'Egypte.* SE of Dayr Mawās, the levee was intersected by a small Nile branch called the the Tirʿa al-Sanjāj ("Tora el Sangág"; see fig. 3).

The village of Tānūf ("Tanoûf") lies due west of Dayr Mawās. The two settlements are connected by *dyke 4*, which also continues west of Tānūf (see 7. below). In Tānūf, dyke 4 seems to be crossed by a N-S *dyke 3* called the jisr Tānūf ("gésr Tanoûf"). In fact, the *Description de l'Egypte* writes the name Jisr Tānūf only on the stretch north of the village. In the area further south a double line continues, which could indicate a dyke, but also a road. However, in this area a dotted line indicates the marching route of the French army *beside* this double line, and in other places, such a dotted line alone often indicates a road. The fact that it here appears beside a double line suggests that the latter may be more than just a road. Moreover, further south, west of Jarf Sarḥān, the double line goes over another double line which is itself undoubtedly a dyke. Although the matter is not quite clear, we have assumed that the stretch south of Tanūf is a dyke, but because it is not entirely certain to be really the continuation of dyke 3, we will designate is as dyke 3'. This dyke continues south as far as the north bank of the Baḥr Yūsif opposite Dairūṭ. Thus, the Nile levees, the E-W dyke 4,

hypses at 1m intervals) the dyke must have been functional for a very considerable period of time.

46 Cf. the similarly low areas further north, to be discussed under 14, which, still in 1798, were only partly under cultivation.

the N-S dyke 3', and the levee on the north bank of Baḥr Yūsif circumscribe a roughly rectangular area which gently dips towards the west. This large area is intersected by two further dykes.

Dyke 5 branches off eastwards from dyke 3'in the direction of the Nile, but it does not reach it, undoubtedly because it runs up against the slope of the Nile levee. Dyke 5 separates two basins II and III. The area between the Nile levee and the Baḥr Yūsif, dykes 3' and 5 constitutes *basin II*, which is fed by a small channel. *Dyke 6* crosses dyke 4, continuing southwards through an area at an elevation of +44 m, but stopping where it meets the +45 m contour line. It seems likely that dykes 4 and 6 and the Nile levee surround a higher *basin III*, which mostly lies above the 45 m contour line. It is fed by the Tirʿa al-Sanjāj, and, in its southerly part, by a small channel already mentioned. This latter continues down into the +44 m depression east of dyke 3', which is here called *basin IV*.

Basins II and III mostly lie above the 45 m contour line, and will only have been flooded by a high inundation. The small, deeper areas of these basins, as well as basin IV, in the +44 m realm, were probably flooded almost every year. The *Description de l'Egypte* does not indicate in which direction floodwater drained off at the end of the inundation period, but small channels linking their deeper parts to the Nile exist on either side of dyke 5. It is likely that, at the end of the flood season, remaining floodwater was allowed to drain back into the Nile through these channels and the Tirʿa al-Sanjāj.

5. The Hydrology of the Region between Dayr Mawās and Mallawī, Eastern Part

This area is in many regards similar to the more southerly one we have just discussed. Immediately adjoining the Nile there is a levee, which gradually decreases northwards in elevation corresponding to the slope of the Nile itself.[47] In the south, the levee reaches a level of + 45 m, in the north of + 44 m. West of the levee, the fields slope down gently to the west.

Excursus: Concerning this Nile levee, a confusing element in the *Description de l'Egypte* must be discussed. Its map 13 indicates that, south of Rairamūn, the western bank of the Nile lay considerably farther west around 1720 than was the case in 1798 (and today). At that time, Mallawī would have lain immediately on the western Nile bank.[48] However, the extensive levee east of Mallawī, which we

47 According to WILLCOCKS, 1889, p. 8, the Nile slopes down 1 m per 12.9 km.

48 This is also reported in JOMARD, 1821, p. 316, and is occasionally accepted in the literature: KESSLER,1981, p. 85

have just discussed, renders this unlikely. Under normal processes of alluviation it can impossibly have accumulated between 1720 and 1798, or even 1926. The *Description de l'Égypte* does not clarify what its interpretation was based upon, but the indicated date ("vers 1720") renders likely that the source of information was a description by Claude Sicard dated to 1722, which Robert de Vaugondy used for the map he published in 1753.[49] However, on this map one can see that Mallawī did not lie on the river, but instead that the situation recorded by the French cartographers in 1798 already obtained then. We therefore assume that the course of the Nile did not change significantly between 1722 and today. Based on the remarks in 12.b and 12.f, it may even have lain further east at some point prior to Napoleon's campaign (End of Excursus).

West of the Nile levee, the N-S *dyke 7* leads south from Mallawī to Dayr Mawās. This long dyke is described in the *Description de l'Égypte* as an "Ancienne Digue".[50] About halfway between the two towns, *dyke 8*, called Jisr Khuzām ("gesr Kozâm"), runs perpendicular from this old dyke to the Nile. In this way, two basins are created. *Basin V* extends along the Nile from Dayr Mawās to the Jisr Khuzām, and this basin is provided with water from a small channel called "Tora el Asarah".[51] The adjoining *basin VI* fringes the Nile from the Jisr Khuzām to Mallawī, its northern limit probably being formed by the tell of Mallawī and the Nile levee. This basin is fed by three different channels. The contour lines show that, during low floods, only a narrow + 43 m area adjoining the ancient dyke 7 could be flooded in basin VI, but during higher floods, the covered area, in both basins V and VI, would increase significantly. In both basins, one channel continues into the adjoining basins on the west (basins IX and XI; see below).

6. The Hydrology of the Region between Dayr Mawās and Mallawī, Western Part

Under 4., we have already discussed the southern stretch 3' of dyke 3, extending from Dairūṭ to Tānūf, but this dyke also continues north of Tānūf. There is some uncertainty, however, as to where exactly it stops. Leaving from dyke 4 to the north, dyke 3, explicitly designated as Jisr Tānūf, certainly continues northwards until the point where it meets the northward extension of the more

---

49 See n. 27.

50 It is uncertain whether it was still functional in 1798. If not, dyke 9, to be discussed below, may have taken over its function. In his discussion of this dyke, GRIESHABER, 2004, p. 33 apparently confuses dykes 7 and 3''.

51 Arabic transcription unclear.

easterly dyke 6. The triangular area surrounded by dykes 3, 4 and 6 is here called *basin VII*. The map does not indicate any waterway connected to this basin, so that it cannot be ascertained how it was flooded, and in which direction the remaining water drained off after the flood. This basin cannot, therefore, be used in reconstructing the hydrology of the region.

The problem is whether dyke 3 continues northwards from the point discussed here. The map in the *Description de l'Égypte* suggests it does not. It also shows, however, that a road (indicated by two parallel lines) runs immediately beside dyke 3, to its west. Now dykes often serve as roads, and the dotted line indicating general Desaix' march route ascertains that dyke 3 was no exception (for it makes clear the French troups followed the top of dyke 3 near the northern end of basin VII, and not the road running immediately beside it). If we read the *Description de l'Égypte* map at face value, two roads seem to have run completely parallel from the village of Tānūf northwards. The lower of the two roads would be perfectly superfluous, and it would render a considerable surface of agricultural land useless.

Therefore it is assumed here that dyke 3 and the road immediately adjoining it on the west in fact were one and the same. This assumption, for which there unfortunately is no final proof, implies that dyke 3 continued further north to the village of Tūkh, where it turned northnortheast in the direction of the town of Mallawī, and reaching *dyke 9* near the cistern indicated on the French map. The somewhat hypothetical stretch of dyke 3 extending northwards beyond the northern end of basin VII until the cistern will here be called *dyke 3"*.[52] This dyke runs over (and probably dammed off) the "Tora el Asarah" just SE of the village of Tūkh.

Before discussing dyke 9, the hydrology on the southern fringe of the town of Mallawī must be discussed. Just southeast of the town, a Nile branch designated as Tirʿa Ḥasan Kāshif ("Tora Haçan Kâchef") crosses the Nile levee, passing the town on the south, and turning south a short distance after in the direction of Nazlat al-Shaykh Ḥussayn ("N. Chéïk Hoçein").

West of Mallawī, the map indicates a settlement called Sinjirj ("Singerg"), and a complex dyke system (*dyke 10*) connects the two. At halfway between the two settlements, *dyke 9* branches off south, enclosing the Tirʿa Ḥasan Kāshif on the west,[53] passing Nazlat al-Shaykh Ḥussayn on the west, and stopping not far

52 Linant de Bellefonds' map indicates a dyke following exactly this course (GRIESHABER, 2004, p. 33).

53 On the north, the map suggests this waterway was not completely surrounded by dykes. Not far from Mallawī, the waterway is shown to break through dyke 10,

west from the juncture of dykes 7 and 8. The isohypses show that dyke 9 stops on the ascent of the Nile levee, somewhere above 44 m above mean sea level.

The above, rather complex, description shows that a large, rectangular basin system is contained within dyke 7 (east), 10 (north), 4 (south), and 6, 3" and 9 on the west. This large area can be subdivided into further basins.

Basin VIII: This basin lies east of dyke 6. It is bounded on the north by the E-W *dyke 11*, which runs eastwards from dyke 6, crossing a +43 m depression and ending in the + 44 m realm on the Nile levee. Further south, a second *dyke 12* also branches off from dyke 6, meeting dyke 11 further north. Dykes 6, 11 and 12 surround basin VIII, which lies in the +44 m and the +43 m area. Like in the case of basin VII, no waterways are indicated that could explain how basin VIII was filled and emptied. This basin cannot, therefore, be used in reconstructing the hydrology of the region. The fact that the adjoining basins VII and VIII have to be disregarded does not pose a grave problem, as both are rather small.

Basin IX: East of basin VIII lies basin IX. It is bounded by dykes 4, 7, 11 and 12. In its deepest part, it lies in the lowest +44 m altitude range. The *Description de l'Égypte* indicates that this dyke system is open in the northeast, but this area lies well in the +44 m range, so that the raise in natural relief may have served to close this area off. Arguably, dyke 11 may also have continued as far as dyke 7. Basin IX receives its floodwater through the "Tora el Asarah", which enters it from basin V.

Basin X lies north of basins VII and VIII, and northwest of basin IX. It is bounded by dykes 3", 9, 13 (to be discussed below), 11, and the +44 m area forming part of the Nile levee, which, because it bulges out far from the river bank to the west, may be a crevasse splay. It can only have been flooded through the narrow "Tora el Asarah," which has its head in basin V, passing through basin IX into basin X. Thus, we here have a basin chain V – IX – X.

Basin XI lies north of the previous ones, being surrounded by dyke 7 (east), 10 (north), 9 (west) and the +44 m contour forming part of the Nile levee in the south. No waterways connect Basin XI with the more southerly basins VII, VIII, IX and X. Instead, it receives its water from the east, from the Tir'a Ḥasan Kāshif and, further south, the "Tora Hoceïn Cherkes".[54] This implies an E-W

and to return within it a short distance further downstream. The small part of the Tir'a Ḥasan Kāshif lying outside the dyke connected to a small channel leading deeper into the adjoining basin. It seems hard to find a rational explanation for the course taken by the channel. Perhaps the published map is erroneous here, or it may reflect a situation that emerged after a partial destruction of the dyke.

54 Transcription unclear.

basin chain VI-XI. There seems to be no clear connection between this basin chain and more westerly basins, so that residual floodwater is likely to have moved back into the Nile from the channels through which it had arrived.

7. The Hydrology of the Area between Dairūṭ, Tānūf, and Ismū

Two main natural features dominate the landscape in this area: the +45 m levee of the Baḥr Yūsif and two +46 m elevations north of the Baḥr Yūsif, which begin under and extend to the north of the village of Banī Ḥarām ("Beni Harâm"). Immediately east of this elevation, a rather large waterway branches off from the Baḥr Yūsif, running straight north: the "Tora el Kiket".[55] The outline of the + 46 m, +45 m, and +44 m contour lines just discussed strongly suggests that these are levees and thus that the "Tora el Kiket" was once a very active waterway, or that there is a crevasse splay here. Perhaps it is what remains of a precursor of the Baḥr Yūsif.

East of the "Tora el Kiket", a second waterway branches off to the north from the Baḥr Yūsif, leading to the village of Tānūf. This is the Tirʿa Tānūf.

These two waterways feed a series of basins further north, that are bounded on the south by the levee of the Baḥr Yūsif, on the east by dyke 3', and on the west by the + 46 m levee system just discussed. A major problem in interpreting the landscape in this area is a further landscape feature: *dyke 14*. This dyke, called Jisr al-Nāṣirīya in Arabic ("gesr el Nasriéh"), starts in the west at the northern tip of the southernmost +46 m elevation and runs straight eastwards to the Tirʿa Tānūf, where it stops. Since it is quite unlikely that a dyke would stop in mid-air, it must have continued somewhere, or it must have hit a higher surface not indicated on the map.

One possibility is that the Tirʿa Tānūf was fringed by a levee just like the "Tora el Kiket". Since the *Description de l'Égypte* generally disregards levees, the absence of the feature would be understandable. However, the Tirʿa Tānūf seems to have been quite small, and it is therefore unlikely to have created a levee of significant height. Another possibility is that such a levee may have been artificially heightened by earth dredged out of the channel. Such a dyke fringing the Tirʿa Tānūf is, in fact, visible further north. This is the westward extension of dyke 4 discussed already under 4. From Tānūf, the dyke first follows a westerly course for c. 2 km, then it sharply turns south for c. 2 km, where it fringes the Tirʿa Tānūf. We will assume here that this dyke continued all along the western

55 Transcription unclear.

side of the Tir'a Tānūf to the Baḥr Yūsif.[56] This hypothetical dyke (indicated by a dotted line) is here called *dyke 15*.

Basin XII: This is the basin just discussed. It is surrounded by the levee of the Baḥr Yūsif (south), dyke 3' (east), dyke 15 (west), and dyke 4 (north and west). This basin does not connect to any other, so that any remaining floodwater from it would have to flow back directly into the Baḥr Yūsif after the flood season. From there, it would be transported, not back to the Nile, but northwest into the Baḥr Yūsif depression.

Basin XIII: This is the basin enclosed on the west by the southernmost +46 m elevation, on the north by dyke 14, on the east by dyke 15, and on the south by the Baḥr Yūsif levee. It is fed exclusively by the "Tora el Kiket", which continued north of dyke 14, leading into

Basin XIV: This is enclosed on the south by dyke 14, which, on its western end, links the southernmost +46 m elevation to the northernmost one, which bounds basin XIV on the west. On its north, basin XIV is enclosed by the westward continuation of dyke 4, and on its east perhaps by the hypothetical dyke 15. At the northwest of the basin, there is a small gap between dyke 4 and the +46 m elevation, but this is closed off by a small mound indicated in the *Description de l'Egypte.*[57] The "Tora el Kiket" crosses basin XIV from S to N, continuing further into basin XVI. However, just south of dyke 4, two branches of the "Tora el Kiket" lead east[58] and west, the western one providing water to basin XV, and the eastern one hugging dyke 4 on the south and stopping short of the hypothetical dyke 15.

Basin XV: This basin is enclosed on the north by the westward continuation of dyke 4, which continues until where it touches *dyke 16*: the Jisr Badramān ("gesr

56 In the study cited in n. 43, H. WILLEMS assumed that such a dyke only ran from dyke 14 southwards, which is inconsistent. If it at all existed, it most likely continued all the way from dyke 4 to the Baḥr Yūsif, and hence north of the Jisr al-Nāṣirīya.

57 In the above, the hypothetical but likely dyke 15 has been added. If it did not exist (and dyke 14 was accordingly utterly useless), basins XII and XIII would in fact constitute one basin fed by two channels. For the rest, nothing in our interpretation would change.

58 The one turning to the east runs in the direction of the Tir'a Tānūf, but before reaching it, it makes a turn to the south. This could be explained by assuming that the waterway could not prolong further in a straight course because of a rise in surface elevation, as we have assumed when arguing there was a dyke (15) here.

Badraman").[59] Beyond this, dyke 4 continues further to the west, but since, as of this point, the *Description de l'Égypte* gives it the new name of Jisr al-Qashāsh ("gésr el Qachâch"), we have called it *dyke 17*. After continuing for a while to the west, it bends to the south, and then eastwards past the village of Ismū ("Esmoû") until touching the northernmost +46 m elevation. Where dyke 4 meets dyke 16, a channel leads into basin XVI, which lies mostly in the + 43 m realm, and thus lower than basin XV. Therefore it is likely that any floodwater remaining in basin XV at the end of the flood season would flow out of it this way.

8. The Hydrology of Basin XVI

This basin is enclosed on the south by dyke 4, on the east by dyke 3-3", and on the west by dyke 16. In its southern part, its floor lies in the +44 m range, further north in the +43 m range. Just after entering this basin, the "Tora el Kiket" gables, one branch continuing to the north, and a narrower one turning sharply to the east, hugging dyke 4. Then it, too, turns to the north. In the northern part of the basin, this waterway splits into a westward branch reaching dyke 16, and a northern one continuing into basin XVII.[60]

The most problematic issue is determining the northern demarcation of basin XVI, here indicated as *dyke 18*. This is the Nazlat al-Shaykh Ḥussayn –Tūkh – Tanda – Kawm al-Sihāl dyke system. Under 6. we have discussed dyke 9, running south from Mallawī to Nazlat al-Shaykh Hussayn. *a.* At the latter village, a short length of dyke has been drawn by the French cartographers, which roughly goes westwards in the direction of the village of Tūkh ("Touk"). The dyke as drawn is so short that it makes no sense, and the isohypses do not make things easier to understand. Probably the rest of the dyke has not been drawn. The crop

59 GRIESHABER, 2004, p. 32-33 considers dyke 16 (with an overall N-S orientation) a "Querdeich", i.e. an E-W dyke. This clearly does not do justice to the situation.

60 What we here call "basin XVI" might actually cover two distinct basins. Map 13 in the *Description de l'Egypte* indicates a road or dyke departing northeast from the northeastern corner of basin XV, and continuing in a northeasterly direction towards the village of Tūkh and dyke 3". It is indicated in the map in exactly the same way as dyke 3" and therefore it would be consistent to interpret this as a dyke as well. On the other hand, while in the case of dyke 3" other indications discussed above suggested that the feature represents a dyke, such supplementary indications are lacking in this case. Based on this uncertainty, we have not considered this a dyke; if it would be one after all, the only difference for the hydrology would the existence of an additional basin in the chain. For our interpretation of the irrigation the question of whether this dyke exists or not is immaterial.

boundaries indicated in the *Description de l'Egypte* seem to indicate an E-W line connecting the end of the dyke near Nazlat al-Shaykh Hussayn to Tūkh (dyke 13, see 6), and this has here been taken as being the easternmost part of dyke 18. It must be admitted that this interpretation is conjectural.[61] *b.* Between Tūkh and Tanda ("Tendéh"), no dyke is indicated immediately west of Tūkh, but the field boundaries between differently oriented crops exactly follows the expected trajectory of dyke18. For this reason, and because, otherwise, dyke 18 would be useless, we assume that there existed a continuous dyke here. *c.* At a short distance southwest of Tanda lies the town mound indicated as Kawm al-Sihāl (8 in fig. 3). The *Description de l'Égypte* shows no dyke between the two places, but since both are mounds one may perhaps assume that the intervening land was slightly more elevated, rendering dam construction superfluous.[62] *Dyke 18* constitutes the northern perimeter of basin XVI, which lies almost entirely in the +43 m range.

9. The Hydrology of the Area Southwest of Mallawī: Basins XVII and XVIII

a. About halfway between Tūkh and Tanda, the E-W dyke 18 is crossed by another roughly following a WNW-ESE course. This *dyke 19* is called Jisr Tanda ("gesr Tendéh"). It almost continues as far as the Jisr Badramān (dyke 16). In the remaining gap the drawing published in the *Description de l'Égypte* does show a field boundary in the extension of dyke 19. The contour lines indicate that the dyke may also stop before reaching dyke 16 because of a rise in surface level in this area. *b.* Dykes 16, 18 and 19 circumscribe a small triangular basin here designated as "basin XVIII". This basin lies in the + 42 m and the +43 m realm. South of the Tūkh – Tanda dyke, dyke 19 continues its course for a short distance, creating another triangular, but extremely small basin. This does not have a dyke at its eastern end, but the *Description de l'Egypte* here indicates the presence of

61 If this length of dyke did not exist, then basin X would simply be a bit larger. The option has no consequences for the interpretation of the hydrology of either this basin, or of basin XVI.

62 From Tanda, a double line continues further westwards to the village of Badramān. It crosses the Jisr Badramān dyke previously discussed (dyke 16). This double line seems to be deliberately less marked than the lines used for dykes, and we have therefore assumed this is not a dyke but a road. However, the possibility remains that this is a dyke after all, in which case the E-W dyke system here being discussed continued all the way to about the 44 m contour line of the Baḥr Yūsif levee.

a small kôm, which could have closed the basin from this side. In view of its miniature size, this basin will not be incorporated in the overall interpretation in this chapter. *c.* The vast "basin *XVII*" extends north of dykes 18 and 19, east of the northern end of dyke 16, and west of dykes 3" and 9. The northern perimeter is more difficult to reconstruct. Under 6. we have discussed dyke 10 leaving the town of Mallawī to the west. Beyond where dyke 9 branches off, dyke 10 continues its westerly course to the village of Sinjirj ("Singerg"). From this point the drawing becomes more problematic. From Sinjirj, a double dotted line indicates the location of an "Ancienne Digue",which continues westwards as far as the Baḥr Yūsif levee, and which we interpret as the continuation of dyke 10. Some confusion on the part of the printer is apparent, as at a certain point, the map indicates a bridge passing over the dyke. This of course cannot be correct. Moreover, the westernmost part of this ancient dyke is graphically rendered, not as a dyke, but as a channel[63] running in the direction of the Baḥr Yūsif. Probably, there was a canal on the inside (i.e., south) of dyke 10 surrounding basin XVII. Because the dyke was "ancient", and therefore perhaps not easily recognizable everywhere, the tract of the dyke may have been unclearly drawn by the cartographers, being misinterpreted by the printers as a canal.[64] These considerations do not alter the likelihood that Sinjirj was linked to the levee of the Baḥr Yūsif by dyke 10, which linked up to dyke 16.[65]

Basin XVII mostly lies in the +42 m range, and thus deeper than basin XVIII and all other basins discussed previously. Inside this basin, the map of the *Description de l'Egypte* indicates only few waterways, but we have seen that there probably was one on the south of dyke 10, and flowing in the direction of the Baḥr Yūsif. However, this waterway must under most conditions have been unsuitable to transport off water into the Baḥr Yūsif, which in this area lies higher. Perhaps, a waterway crossed dyke 10 in the direction of basin XX. One

63 Because the empty space between the lines is filled in in grey.

64 It is also conceivable that the "ancient" dyke 10 was no longer functional and did not exist everywhere anymore. In this case, basin XVII would be continuous with basin XX. The resulting basin would however be so much larger than any of the other ones on the map that this seems less likely.

65 Where dyke 16, i.e. the Jisr Badramān, turns west, the map of the *Description de l'Egypte* suggests the presence of some rather complex water works: the dyke does not just make a turn to the west, but there seems to be another dam making a turn around the corner of the dyke. Precisely at this point, numerous smaller and larger channels occur. We have not succeeded in finding an explanation for this system.

can see, however, that the continuation of the "Tora el Kiket" enters basin XVII from basin XVI, immediately turning west into the +43 m basin XVIII.

No waterway is visible through which the water may have drained off from basin XVIII after the inundation. However, it is impossible for this water to have drained off across the much higher Nile levee. The only realistic direction is through the low basins XVII and XX. It is therefore likely that a waterway was omitted in the *Description de l'Égypte*.

10. The Hydrology between Mallawī and al-Ashmūnayn (fig. 6)

Just northeast of Mallawī lies the village of Rairamūn. Still today, this is recognizable as a large tell, although it is now completely overbuilt. In the late eigthteenth century, it lay at the beginning of a Nile branch designated as the Tirʿa al-Shaykh Ḥajāza ("Tora Cheik Hagazéh"), which continues past Qulubba to al-Ashmūnayn. This situation is already seen in the map by de Vaugondy, which reflects the situation in 1720. Both north and south of the mouth of the Tirʿa al-Shaykh Ḥajāza, which itself lies in the + 43 m realm, the Nile levees lie in the + 44 m range. This suggests that, at an earlier point in time, a continuous + 44 m Nile levee was broken, leading to the emergence of the Tirʿa al-Shaykh Ḥajāza. The situation reminds one of the inlet of the Baḥr Yūsif to the east of Dairūṭ al-Sharīf.

The isohypses indicate a clearly expressed levee system here, which must have been deposited by an active Nile branch starting southeast of Mallawī. It is indicated by the + 44 m contour line on which the town of Mallawī lies, by the + 43 m contour line on which lies al-Ashmūnayn, and by the + 42 m contour northwest of that town. One also notes that, from a point close to the mouth of the Tirʿa al-Shaykh Ḥajāza, many channels fan out in different directions. This pattern is reminiscent of what would be found in a large crevasse splay, and this may explain the form of the landscape here.

Apart from dyke 22, to be discussed later, no dykes are indicated anywhere in this area, suggesting that the levees were sufficiently prominent to function as such.

West of Mallawī, in the village of Sinjirj, *dyke 20* branches off to the NW from dyke 10, forming a connection with the village of Umm Qummuṣ ("Moqommos"). Unexpectedly, no dyke continues from here according to the *Description de l'Egypte.* Nor do the isohypses offer an explanation as to why this dyke would stop in mid-air. One possibility is that a dyke simply was not observed by the French cartographers, or that the printers may have misinterpreted the information they were provided with, confusing a channel with a dyke. Since exactly the same kind of problem has been encountered elsewhere in the same region (see under *9c*), this is perhaps the best explanation. Here one should consider two points. On several occasions, we have already seen that canals/channels lie immediately beside a dyke. For canals this is self-evident: earth had to be dug up to build a dyke, and the easiest way of working would be to use the soil dug up for digging a canal for constructing the adjoining dyke. Research carried out in other parts of Egypt suggest that such canals normally lay just south, i.e. upstream, of the dyke they accompanied.[66] Therefore, if the *Description de l'Egypte* indicates a canal but no dyke, a dyke may nevertheless have existed (just as in the clearer case of the western end of dyke 10). Now a channel runs from Mallawī to Umm Qummuṣ, and if we tentatively assume the existence of a *dyke 20'* beside this channel, the problem is solved. This dyke would run up against the tell of Mallawī, constituting a *basin XIX* to the west of that town. This basin would be enclosed by dykes 10, 20, 20' and the Mallawī town mound.[67]

11. The Hydrology Southwest of al-Ashmūnayn[68]

a. Dyke 21 west of al-Ashmūnayn: this "digue d'Achmouneïn" runs from the western part of al-Ashmūnayn to the levee of the Baḥr Yūsif, and it thus constitutes the northern perimeter of the large *basin XX*. This basin is bounded on the south by dykes 10, 20 and 20', on the east by the +43 m levee of the Tirʿa al-Shaykh Ḥajāza, and on the west by the +43 m levee of the Baḥr Yūsif. *b.* From the western end of dyke 21, the *Description de l'Égypte* shows a double line turning south on the east bank of the Baḥr Yūsif to the village of al-Birka ("El Birkeh"), continuing all the way south to dyke 16. Since this element runs over the highest point of the levee of the Baḥr Yūsif, it could be just a road; it could also be an artificial elevation serving as a dyke. However, in that case

66 This has been described in more detail for dykes in more northerly parts of Middle Egypt. It was observed that, to the south (i.e. upstream) of dykes, a thickness of between 0.5-1 m had been removed of a wide surface south of the dyke. Within this already slightly deeper area, drains were then dug (GOMAÀ/MÜLLER-WOLLERMANN/SCHENKEL, 1991, p. 66-67).

67 This is really not more than a guess. Note that MICHEL, 2005 found no evidence in papyrological sources or in Ottoman dyke lists suggesting that canals were ever dug during this long period of time. If he is right, the evidence studied by GOMAÀ/MÜLLER-WOLLERMANN/SCHENKEL must be rather recent, and our reasoning leading to the assumed presence of dyke 20' would be invalid.

68 From this point, all areas discussed can be found in map 14 of the *Description de l'Égypte*; see our fig. 6.

such a dyke would also be expected on the Tir'a al-Shaykh Ḥajāza levee, which reaches the same elevation. Since this is not the case, the feature discussed may be a road rather than a dyke. The toponym al-Birka means "the lake". Although no lake is indicated in the Napoleonic map, the topographic map we used, which was made 125 years later, does show one west of the village in the +42 m realm. The name of the settlement leaves no room for doubt that the lake must have been a reality already in the 18th century. The fact that the 'lake' was not noted when the map was drawn in late December 1798, suggests this may have been an exceptionally dry year. This should be kept in mind in interpreting other elements of the hydrology as well. *c.* Basin XX was supplied with water through the Tir'a al-Shaykh Ḥajāza branching off from the Nile at Rairamūn. This waterway runs all the way to al-Ashmūnayn, which it bypasses on the south, continuing its course as a narrow channel in the direction of the Baḥr Yūsif. North of Mallawī, a smaller waterway called the Tir'a al-Majnūn ("Tora el Magnoun") branches off to the west from the Tir'a al-Shaykh Ḥajāza. The Tir'a al-Majnūn crosses the centre of basin XX from SE to NW. Near the centre of dyke 21, it crosses the Tir'a al-Shaykh Ḥajāza again, and the fact that it seems to continue through dyke 21 suggests that this dyke could be opened to drain remaining floodwater off to the north, into the deepest part of basin XXVI (+ 41 m).[69] In the southwestern corner of basin XX, two further waterways enter it from the Baḥr Yūsif, so that this very large basin seems to have been inundated (or: inundatable) from two sides.[70]

69 Waterways called al-majnūna and draining off to the deepest part of the floodplain existed in different parts of Upper Egypt (MICHEL, 2005, p. 259).

70 In the southwestern corner of the area discussed, the *Description de l'Égypte* indicates a somewhat more complex dyke system than anywhere else in the area investigated. Although the functioning of this dyke system is not clear to the authors, it might have something to do with the fact that basin XX was fed by two different channel systems.

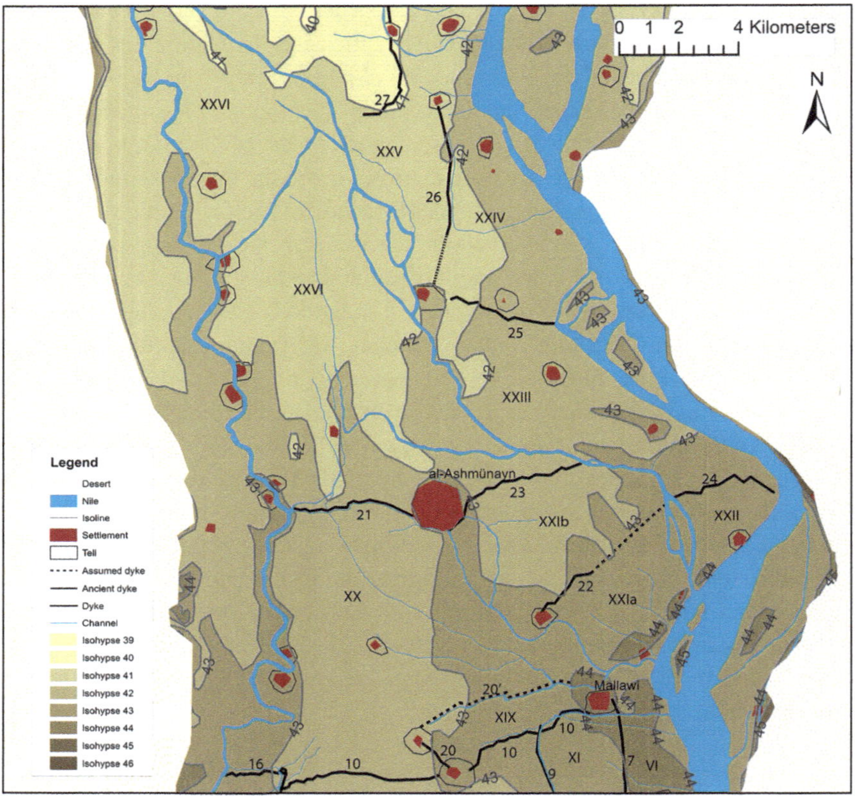

Figure 6. georeferenced map projecting information on dykes, waterways and settlements as provided by the Description de l'Égypte, *map 14 over a modern topographic map. Map Hanne Creylman.*

12. The Hydrology Northeast of al-Ashmūnayn

a. The main natural feature NE of al-Ashmūnayn is a watercourse branching off the western Nile bank at a place called "Deïr el Nosara" in the *Description de l'Égypte*, map 14. This village is today called Dayr al-Mallāk. This wide Nile branch, called al-Sabakh ("el Sabbak") may have been cut through the Nile levee by the force of the Nile waters in the bend in the Nile, in which case this levee system would, at least in origin, be a crevasse splay. From here, al-Sabakh followed a northerly course, turning west to the northwest of the village of al-Bayāḍīya ("El Béîâdîyeh"), bypassing al-Ashmūnayn on the north. As a result, this town lay in between two Nile branches: the Tirʿa al-Shaykh Ḥajāza and the much larger Sabakh. The isohypses indicate that the Sabakh is fringed by a +43 m levee system. However, this levee is expressed less markedly than the one accompanying the smaller Tirʿa al-Shaykh Ḥajāza, suggesting that, in the past, the latter may have been more active than al-Sabakh. *b.* Between Rairamūn and Dayr al-Mallāk, the +44 m Nile levee is interrupted. In this area lie not only the mouth of al-Sabakh, but also, slightly to the south, a much smaller, anonymous water inlet, which may also be due to erosion by the outside of the bend of the Nile.[71] West of this point, the contour of the isohypses is suggestive of a crevasse splay. Therefore it is likely that the Nile levee was broken here at some point in time.

Between Rairamūn and Dayr al-Mallāk, we have observed that the currently almost vertical profile of the western Nile bank cuts through Late Roman floor levels.[72] Clearly, a settlement has since the mid-first millennium A.D. been eroded here by a westward movement of the Nile.[73] Moreover, a core drilling (2012) at

71 Our plan in fig. 6 differs in two respects from the maps we used. Firstly, the *Description de l'Égypte* indicates the presence of a village called "K"(afr) el Reremoun" just beside this waterway. However, today there is no village here, but only farmland, with no traces of even a ruined village. Al-Rairamūn has enormously expanded, but even today, its northern fringe lies c. 1 km south of the channel here discussed. Therefore, the French map must contain an error, and we have therefore suppressed the village of "K. El Reremoun". Secondly, the Survey of Egypt map is erroneous, indicating that "el Malak monastery" (= the current Dayr al-Mallāk) lies to the east of, and closer to the Nile than, the + 44 m levee in this area. Our research there has shown that it lies very close to the river and on the highest point of the levee. This suggests that the + 44 m levee in this area has been somewhat misplaced. We have therefore adapted the map based on our knowledge of local conditions.

72 Clearly characterized as such by the presence of red brick and late Roman amphora sherds.

73 Above these archaeological depositions, the Nile bank has been heightened at some point in time as far as the level where brick waterworks were built in the time of Muḥammad ʿAlī to control al-Sabakh and the smaller water inlet. This suggests that the Dayr al-Mallāk-Rairamūn dyke topping the earlier levee was built prior to 1830, but after 1798, when no dyke was yet recorded in the map of the *Description de l'Égypte*. The likelihood of a westward migration of the main bed of the Nile in this area has been discussed elsewhere in this volume (see p. 242-247). This contradicts the widespread assumption that the Nile displayed a general eastward drift (Butzer, 1976, p. 35-36; followed by Grieshaber, 2004, p. 29; 34;

the site produced Late Period or Ptolemaic period ceramics. This movement may have caused the break of the levee between Rairamūn and al-Bayadīya. Farmers have told us that the smaller inlet in this area was used to supply the village of Tabūt, near al-Ashmūnayn, with water.[74] *c.* From Qulubba *dyke 22* runs NE in the direction of al-Sabakh, but it stops after only a short distance. Field boundaries drawn in *Description de l'Égypte* map 14 follow the line one would expect the dyke to have followed in the direction to the Sabakh. This renders likely that the dyke at some point in time had continued. Quite possibly, the event which caused the Nile levee to break between Rairamūn and Dayr al-Mallāk also swept away this dyke. *d.* From al-Ashmūnayn, *dyke 23*, called al-Jisr al-Ṣulṭānī ("gesr soultâni") runs eastward, also reaching the al-Sabakh levee, damming off *basin XXI* in between. This basin is bounded on the east by the Nile levee continuing northwards to the levee of al-Sabakh, on the west by the levee of the Tirʿa al-Shaykh Ḥajāza, and on the north by dyke 23. As noted under 11c, dyke 22 may once have continued and basin XXI may actually have consisted of two basins XXIa and XXIb. Basin XXIa may have taken its water directly from the Nile, forming a chain with XXIb, and draining off in the direction of al-Sabakh. It is here assumed, however, that the channel between Rairamūn and Dayr al-Mallāk was the result of the break of the levee discussed above. If this is correct, basin XXIa may have received its water from the westernmost branch of al-Sabakh, and basin XXIb from the easternmost branch of the Tirʿa al-Shaykh Ḥajāza, which in fact crosses the +43 m isohypse into the lower-lying basin. After XXIa and b had fused into a single basin, this was probably filled from al-Sabakh,[75] in which it may also have drained off again at the end of the flood season. *e.* North of al-Bayadīya ("el Béïâdîéh") a *dyke 24* runs E-W from the bank of the Nile to the al-Sabakh waterway. This dyke still exists today. Like the previous one, it probably formed the northern perimeter of a basin (*basin XXII*), of which the southeastern limit was formed by a levee (in fig. 6 partly indicated by the 44 m isohypse and the tell of al-Bayadīya) and the western one by the levee of al-Sabakh. As the latter is not indicated in the topographic map, it was probably not very high, and the same is likely for the dyke.This basin probably received its water from the easternmost branch of al-Sabakh at this point, with

ANTOINE, 2011, p. 18; BUNBURY/MALOUTA, 2012, p. 119-122; counterevidence is now also available for the situation at Karnak, see BORAIK/GABOLDE/GRAHAM, this volume). KEMP, 2005, p. 26-29 also argued that the current zone in which the Nile bed lies, differs barely from the situation since the Late Roman period.

74 This village lies in basin XXIb, to be discussed below.

75 Because the surface is higher there.

residual water flowing off that way again as well. *f.* North of dyke 24 the Nile banks show evidence of erosion due to high Niles. There are several remains of a +43 m levee which are crosscut by lower areas, and which in some cases have become detached from the river bank, constituting a whole series of high islands. This again suggests that the Nile has been moving westwards. Although the levee was degraded, and presumably reduced in height, it remains likely that further west the surface was still lower, although the difference cannot have been great. Along the northern bank of al-Sabakh, the isohypses also indicate a levee outline. North of al-Ashmūnayn, the village of Maḥraṣ lies on this levee. East of Maḥraṣ lies a "petite Digue" (*dyke 25*), continuing all the way to the river bank. The *Description de l'Égypte*s uggests this dyke started east of Maḥraṣ. The village is, however, drawn as lying on an elevation, and this is also recognizable in the isohypses. This elevation must have rendered unnecessary the construction of a dyke all the way to Maḥraṣ. The Nile levee, dyke 25, and the levee on the north bank of al-Sabakh enclose *basin XXIII*, which mostly lies in the +42 m realm, but in its northernmost part in the +41 m realm. Since the *Description de l'Égypte* indicates only one very small waterway entering basin XXIII from al-Sabakh, this basin was probably both filled up and emptied again at the end of the flood season from al-Sabakh.

13. The Hydrology of the Area North of Maḥraṣ

The amount of dykes in this area is very small.[76] *a.* North of Maḥraṣ as far north as the village of Itlīdim, *dyke 26* running parallel to the Nile, is clearly indicated, although according to the *Description de l'Égypte* it starts at quite a distance north of Maḥraṣ. Neither the *Description de l'Égypte* nor later topographic maps offer clear indications as to why this is the case, but as dyke 26 would otherwise be useless, we have assumed it continued further south than the *Description de l'Égypte* suggests. We have indicated this by a dotted line. The Nile levee, dyke 26 and dyke 25 surround *basin XXIV*. Immediately adjoining the Nile the levee lies in the +42 m range, just east of dyke 26 it is in the +41 m range. It is fed directly from the Nile by an anonymous waterway between the

76 The map of the *Description de l'Égypte* indicates the route followed by general Desaix through Middle Egypt as a double line. A part of this route, which branches off from dyke 27 south of at Itlīdim ("Atlidem") and passing east of al-Ashmūnayn, was earlier interpreted by Willems as a dyke (see n. 43). We have here assumed that this may not be a dyke. If it was one after all (we have seen that roads often followed dykes) this would not fundamentally change our interpretation, but just lead to the subdivision of some basins into more smaller units.

villages of Nazlat Abū Jāmiʿ ("N. Aboû Gâma") and Kawm al-Riḥāla ("Koûm el Rahâléh"). Two very small waterways crosscutting the Nile levee are also indicated near the village of Itlīdim. According to the *Description de l'Égypte* basin XXIV is fed through the already mentioned waterway directly from the Nile. As no waterways continue from basin XXIV further west, it is likely that, at the end of the flood season, residual water was channeled from this basin back into the Nile. *b.* Here, we will not discuss the dyke system north and northwest of of Itlīdim. Suffice it to say that, at the height of this village, only a very narrow strip along the Nile is confined within the Nile levee and a dyke (26 and 27). North of al-Ashmūnayn, the vast and low-lying floodplain area between the levees of Baḥr Yūsif and al-Sabakh (*basin XXVI*) and between the levees of al-Sabakh and dykes 26 and 27 (*basin XXV*) are entirely devoid of dykes. In this depression, the *Description de l'Egypte* indicates several large marshland areas that had not yet been converted into farmland. Presumably, the whole area had before been an area badly suitable for agriculture.

14. The Hydrology West and Northwest of al-Ashmūnayn

a. At the western end of dyke 21, a small road leads northwest to a village designated by the *Description de l'Égypte* as "el ʿArîn el Bahrî". This village lies on the east bank of the Baḥr Yūsif. A highly interesting point is that, between the village and the end of the al-Ashmūnayn dyke, a small and unnamed channel branches off to the northeast in the direction of the village of Nawāi, leading water into the large *basin XXVI* to the north and northwest of al-Ashmūnayn. This channel follows a winding course, and is joined along the way by the Tirʿa al-Majnūn from basin XX (see 11c). A distinguishing feature of the channel is that the map explicitly notes that the road from the al-Ashmūnayn dyke to al-ʿArîn al-Baḥrī passes over it by means of a bridge. Therefore, different from many channels in Egypt, this one was not contained within dykes. The only reason for this can be that an open connection existed between the Baḥr Yūsif and basin XXVI north of al-Ashmūnayn. Apparently, no system to control water existed here, and this probably correlates with the fact that the *Description de l'Égypte* does not indicate any dykes west of the line al-Ashmūnayn-Maḥraṣ (see 13). Also, as noted under 13, in this area there were in 1798 several large marshland areas. This probably means that this whole basin was ill suited for agricultural purposes.

15. The Hydrology along the Baḥr Yūsif (fig. 5)

We have already discussed some basins along the Baḥr Yūsif: basins I, II, and XIII. None of the more northerly basins we have discussed, except the last, lie immediately along that waterway, however. Basins XIV, XV, XVI, XVII and XVIII are all fringed on their western sides by natural elevations or dykes (16, 17) which lie at a considerable distance east of the Baḥr Yūsif. Along Baḥr Yūsif itself, this area is fringed by the + 44 m levee generated by that waterway and areas of lower altitude in the +43 and partly the + 42 m realm. The existence of dykes 16 and 17 shows that a conscious attempt was made to prevent floodwater entering freely from the west into the basins just listed. No dykes occur anywhere else in this area, so that it must have constituted a continuous basin XXVII. This basin is closed off at its northern end by the northernmost part of dyke 16, which here makes a sharp turning to the west, constituting a bar between basin XXVII and basin XX. However, precisely here lies the not entirely comprehensible dyke system crosscut by two waterways into basin XX (see 11). In several places, basin XXVII is crosscut by side branches of the Baḥr Yūsif, suggesting that during the flood period it would be entirely flooded. The existence of the dyke system suggests that any residual water that remained here at the end of the flood period was not allowed into the more easterly basins, but that it had to flow back into the Baḥr Yūsif. In basin XXVII, not a single settlement is marked on the map of the *Description de l'Égypte*, except on the levees of the waterways. This suggests the area must have been rather humid.

Lower areas also exist west of Baḥr Yūsif. These areas received all their water from there, which could only flow back again in the direction whence it had come. This simple system will not be commented upon here.

4.1.2 Analysis of the results

The detailed discussions above all suggest that in the way the basins were operated, the following varieties occurred: 1) Basins or basin chains receiving their water directly from the Nile and channeling residual water back to the river through the same channel at the end of the flood season; 2) basins or chains of basins receiving their water directly from Nile branches (the Baḥr Yūsif, the Tirʿa al-Shaykh Ḥajāza, or al-Sabakh) and channeling residual water back there through the same channel at the end of the flood season; and 3) basins or chains of basins taking their water from the Baḥr Yūsif or the Nile and channeling residual flood water to low-lying areas in the western part of the floodplain. Besides this, the information provided by the *Description de l'Égypte* is in the case of a few small basins insufficient to decide the kind of flooding regime that was deployed.[77]

1) Basins or basin chains receiving their water directly from the Nile and channeling residual water back to the river through the same channel at the end of the flood season. The following basins belong to this group: II, the basin chain III-IV, basin XXIV, and probably basins V and VI partly as well, since here the Nile is always close by and rather prominent waterways cross them in several places. In the case of basin V, one of these waterways continues into basins IX and X, and thence further into basins XVI and XVII, suggesting that some of the water in this chain of basins may have drained off to the northwest, and away from the Nile. Basin VI constitutes a chain with the lower basin XI. Both are crosscut by several waterways that probably served to channel water back into the Nile.

This means that most, but not all, of the basins immediately bordering on the Nile, returned their water back into the river at the end of the flood season, and that some small, dependent basins channeled some of their water back that way. In all other cases, however, the water did *not* move in that direction.

2) Basins or chains of basins receiving their water directly from Nile branches (the Baḥr Yūsif, the Tirʿa al-Shaykh Ḥajāza, or al-Sabakh) and channeling residual water back there through the same channel at the end of the flood season. This is the case in basins I, XII, and the very large basin XXVII linked to the Baḥr Yūsif. If our interpretation of basin XIX is correct, this received its water from a branch of the Tirʿa al-Shaykh Ḥajāza, through which the residual water was drained off again later. The situation in basin XXI is complex, but one reading of the facts could be that the same situation obtained here as in basin XIX. According to another interpretation basin XXI was fed with water from al-Sabakh, making a U-turn at the end of the flood season. Basin XXII was irrigated in the same way from al-Sabakh.

Even though in some of the cases here discussed, the Nile is close, the fact that the three Nile branches have a steeper slope than the river makes it likely that the water return did not move to the Nile, but away from it to the deeper parts of the floodplain further northwest.

3) basins or chains of basins taking their water from the Baḥr Yūsif or the Nile and channeling residual flood water to low-lying areas in the western part of the floodplain. The whole chain of basins XIII, XIV, XV, XVI, XVIII and XVII, the latter perhaps communicating with XX, is fed from the Baḥr Yūsif

77 This happens in the case of basins VII, VIII and some other, very small basins that were not attributed a number.

near Bānī Harām. At the entry point, the elevation is in the +45 m realm, at the end in the +42 m range. All residual water from the upper basins would, at the end of the flood season, naturally flow down to this latter level. Since, at this latitude, the basins along the Nile lie in the +44 m range, the only conceivable direction in which the water could flow was through the depression between the levees of the Nile and the Baḥr Yūsif, which must have developed into a very wet environment. North of basin XX, no dykes were even built; we have seen that, north of dyke 22 there was an open connection from the Baḥr Yūsif (with local levees at +43 m) to the +40 m[78] basin XXVI.

Based on the reconstruction of the irrigation regime by authors like Willcocks, Egyptologists often tacitly assume that at the end of the flood season, the residual floodwater receded into the Nile. In doing so, they project patterns observed in the late nineteenth century and later into the pharaonic past. However, the preceding pages have demonstrated that, even as late as the late eighteenth century, a completely different system was operative in the area between Dairūṭ and al-Ashmūnayn. Here, only a part of the floodwater drained back directly into the Nile. The rest flowed back into the Baḥr Yūsif, which transported the water northwards along the western fringe of the Nile Valley. Besides, a major amount of drainage water must have been transported via the deep depression between the Baḥr Yūsif and the Nile.[79]

It is still too early to estimate the ratio between the three groups. However, most of the basins from which water was channeled back into the Nile lie at a relatively high altitude, suggesting that during low Niles these received hardly any water at all, whereas the increased volume at periods of high Nile floods must still have been inferior to that collecting in the lower basins further west. Moreover, the lower areas which drained off either through Nile branches to the northwest, or through deeper-lying areas between the Nile and the Baḥr Yūsif, cover a substantially larger surface than the basins draining off directly to the Nile. This suggests not only that by far not all residual water flowed back into the Nile, but even that the latter water volume was substantially lower than the volume carried northward by Bahr Yūsif and the depression east of it. We will see below that this picture is confirmed by the situation further north in the Nile Valley.

78 West of Itlidim there is an area lower than the 40 m, isohypse.

79 The same point was already made by NICHOLAS, 2005, p. 259, without, however, going into the consequences of this observation, which will be studied in the present article.

4.2 The Hydrology of the area between al-Ashmūnayn and the Fayyūm

The first area to be crossed by these waters is the region between al-Ashmūnayn and the entrance to the Fayyūm depression. Most of this area was recently investigated by E. Subias, I. Fiz, and R. Cuesta.[80] The method they deployed much resembles ours, being based on a combined analysis of historical maps and remote sensing.

Their study leads to a reconstructed landscape that looks different in important regards to the results of the preceding pages. Based on the maps of the *Description de l'Égypte* combined with actually remaining traces of dykes, they showed that the basins in this area were, already in the late 18th century, separated by large transverse dykes linking the Nile to the Western Desert.[81] The roughly rectangular basins in this part of Middle Egypt were of a less varied size and shape than we have observed further south. However, Subias, Fiz, and Cuesta also found traces of earlier waterways, which they interpret as "crisscrossing" side branches of the Nile.[82] According to the authors, this irregular system predates a more recent stage when the dykes parceled up the landscape in rectangular basins. This earlier system is very similar to what we have observed near al-Ashmūnayn, where two large branches of this kind existed beside the Baḥr Yūsif.

They also suggest that a south-north watercourse ran through the depressions between the levee systems of the Nile and the Baḥr Yūsif.[83] Although they do

80 Subias/Fiz/Cuesta, 2013, p. 27-44. This article relies heavily on earlier research by Gomaà/Müller-Wollermann/Schenkel, 1991. Since Antoine's reconstruction of the ancient landscape in this region (2011, p. 9-27) does not take into consideration the reinterpretation of geography of the papyrus Wilbour proposed by Id., 1991, p. 105-168, we will not incorporate his views here. In his article in the present volume, he does consider the depression area between the Nile and the Baḥr Yūsif.

81 Subias/Fiz/Cuesta, 2013, p. 31-35. Unfortunately, their account is not as clear as it might have been. Partly this is due to a sometimes unclear and inaccurate use of terminology (thus it seems that they use the term 'levee' as a synonym for 'dyke'). Also, their reconstruction of the infrastructural improvements carried out in the course of the nineteenth century surprisingly suggests that the surface of the irrigated land was starkly reduced in the process (p. 32, fig. 4), which cannot be true.

82 Id., p. 38.

83 Id., p. 40. This situation was already pointed out by Kemp, 2005, p. 29.

not point this out, this waterway would be very similar to the natural drains existing between Nile branches in the Delta,[84] and to the area of basin XXVI studied above.

Figure 7. part of the map of Robert de Vaugondy, showing Middle Egypt between Mallawī and the Fayyūm.

84 BIETAK, 1975, p. 54-55.

In this connection they refer to the Robert de Vaugondy map dating back to 1753, which, as we have seen, was based on information collected in 1722 by Claude Sicard (see n. 27; fig. 7). This map is highly intriguing. It indicates a vast expanse of water thrice the width of the Nile, which lies about halfway between the river and the Baḥr Yūsif, and runs all the way from between the villages of Rawda ("Rouda") and al-Ashmūnayn, to an area just before the entrance of the Fayyūm depression. The same waterway is indicated on the d'Anville map (1765), where it is however shown to start farther to the north (see n. 27). This slight difference may reflect inaccuracies in the maps, or changes in the landscape that had occurred between 1720 and 1765, or different conditions occasioned by a difference in flood height in the two years when the maps were drawn. However, the fact that both maps indicate a very significant water volume between the Nile and the Baḥr Yūsif leaves no room for doubt that this very humid and extensive depression really existed.

Clearly, the maps do not show what Egypt looked like during the inundation (when most of the Nile valley would have been flooded). They can only be meant to visualize the situation after the end of the flood season, when the basins had been opened again, allowing residual water to drain off to the Fayyūm. This information must go back to observations made at a time the Nile was receding.

These maps confirm the hypothesis developed on the basis of the detailed analysis of the area further south, and it proves that vast masses of water plied their way through a depression between the Nile and the Baḥr Yūsif in the direction of the Fayyūm.

A very interesting point in the analysis of Subias, Fiz, and Cuesta is that their remote sensing images show anomalies reflecting an irregular pattern of waterways, whereas a more regularized pattern of artificial basins can be observed from the maps of the *Description de l'Égypte*. This is interpreted as evidence for an older (i.e. pre-Napoleonic), irregular pattern of basins that, by the time Napoleon's army arrived in Middle Egypt, had already been supplanted by a more regular system.

Interestingly, the *Description de l'Égypte* does not display any of the straight east-west dykes characterizing more northerly regions in the al-Ashmūnayn region. This suggests that Napoleon's cartographers recorded a hydrological system there that in other parts of Upper Egypt was already obsolete in 1798. From an Egyptological perspective this is important, as it suggests that the archaic al-Ashmūnayn hydrological system may be more similar to the pharaonic irrigation system than the ones that can be observed elsewhere.

4.3 The Hydrology of the Fayyūm and Lower Egypt

At the mouth of the Fayyūm, the Baḥr Yūsif turns to the west, carving its way through the narrow depression between al-Lahūn in the east and Hawwārat al-Maqṭā in the west (see fig. 8). From here, myriads of smaller waterways fan off from the Baḥr Yūsif, most of which eventually end up in the Fayyūm Lake. Several other natural phenomena in the region must have exerted great influence on the hydrology. These include:

1) Roughly between Banī Suwayf and Maidūm, the floodplain narrows significantly. Moreover, between Banī Suwayf and al-Lahūn,the Jabal Abū Ṣīr rises up from the surrounding fields. These two features result in a significant reduction of the surface of the area that could be flooded. Because in this way the northward flow of the Nile flood was restricted, this must have led to a significant local rise in flood height.[85] Precisely at this point, however, the Baḥr Yūsif branched off into the Fayyūm depression, enabling floodwater to drain away from the Nile valley.

2) The bottom of the Fayyūm depression today lies at about 45 m below mean sea level, whereas near al-Lahūn, it is at about +23 m.[86] This enormous difference in height carries the implication that water flowing into the Fayyūm

85 The extent of this difference can be roughly calculated. At the entrance of the Fayyūm there exists a dyke system to be discussed below. The distance from the dams to the nilometer at Rawdā is about 100 km. GARBRECHT/ JARITZ, 1990, p. 142 argue that the dams could have broken during major flood events such as took place in 1860. In that year, the peak flood level reached at Rawdā amounted to 19.5 m a..s.l. In view of the average river slope of 1 m per 12,9 km, one would expect that during the same flood event, the flood near al-Lahūn reached a height of about 7.75 m more, i.e. c. 27.25 m a.s.l. However, the crown of the dams was more than two 2 m higher (*op. cit.*, p. 144) than this. Since it is unlikely that the dam would really have been so much higher than the exceptionally high flood height of 27.25, there must be local factors complementing the average river slope to explain this. We assume that the reduced width of the floodplain must have played an important part in this. This also suggests that the general water (and groundwater) level south of the mouth of the Fayyūm depression must have been significantly higher (with reference to the surface level of the fields in the region) than elsewhere in Upper Egypt. This must have had its effect on agricultural yield in the region, and this may explain why the area south of the Fayyūm is generally poor in archaeological remains (cf. WILLEMS, 2014, p. 29, n. 94).

86 GARBRECHT/ JARITZ, 1990, p. 9; cf. also the article by RÖMER in this volume.

from the Nile Valley could do so with great force, transporting large quantities of alluvium into the depression. In fact, the town of Madīnat al-Fayyūm lies on a delta-shaped fan of alluvial mud, which must be the result of such processes. In Prehistory, when Man did not bar the Baḥr Yūsif from flowing into the depression, the water level was significantly higher than it is today.[87] This is often explained by referring to the greater flood height in that period. However, long after the flood levels had receded to the levels characteristic for the Pharaonic period, the Fayyūm remained poorly inhabited, which is easily explained by assuming that areas inhabited or cultivated later, were still covered by water.

It is generally accepted that irrigation measures taken in the Middle Kingdom and particularly in the Graeco-Roman period and later led to the area being made suitable for habitation and agriculture. However, the exact role of the dyke systems at the entrance of and inside the Fayyūm have never been fully explained, at least not with reference to the argument that has been developed on the preceding pages, to the effect that the flood regime in Middle Egypt had as its effect that residual flood water mostly did not flow back into the Nile, but that it drained off northwards along the edge of the Western Desert. During its course, huge water volumes inevitably ended up at the entrance to the Fayyūm depression. We will argue that this new idea is of immediate relevance to the interpretation of some major archaeological features in the region.

4.3.1 The al-Lahūn Dyke and Waterway System

The most remarkable feature of this system is formed by two vast dykes, which still exist today (fig. 8). One of the two, de Jisr al-Shaykh Jād-Allāh ("digue en pierres") starts in the northwestern outskirts of al-Lahūn, and follows a winding northwesterly course to the desert edge near the pyramid of Senwosret II (trajectory A-B in fig. 8). It is still largely intact.[88] It is a vast dam made of clay and provided with a stone casing on the northern side. On this side it is also buttressed. In some places it includes tunnels which must have served as sluices, allowing water to flow in or out. At its foot, this dyke measures c. 25 m in width, at its top 8 m. It still today stands about 3.25-4 m above the surrounding fields.[89]

87 As is shown by the fact that habitations of the Fayyūm culture lie at a level much above that of the current shore (Hassan/Tassie, 2006, p. 39).

88 Some of the authors inspected this dyke in March 2013 and March 2016.

89 See Garbrecht/Jaritz, 1990, p. 144 and Anlage 103. Gomaà/Müller-Wollermann/Schenkel, 1991, p. 62 give heights of between 3.4-4.2 m. The top width, these authors indicate as being between 12 and 15 m.

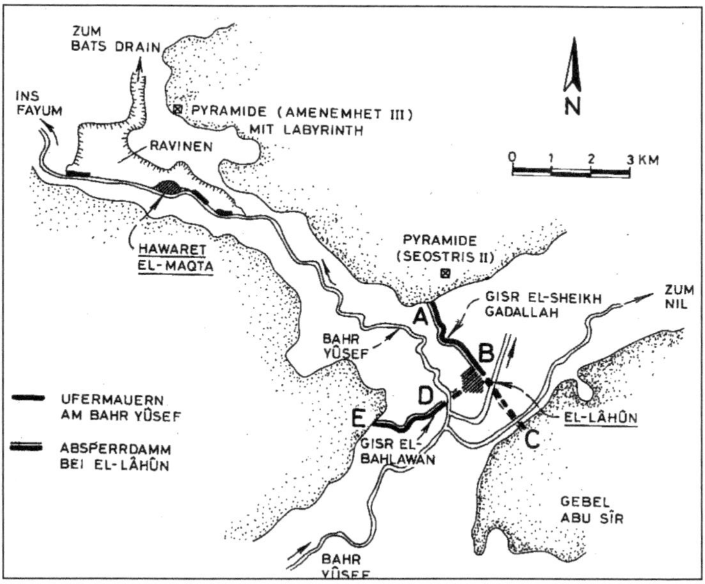

Figure 8. Waterworks at the entrance to the Fayyūm depression (after: GARBRECHT/JARITZ, 1992, p. 248, fig. 14).

A short distance south of al-Lahūn lies the village of Hawwārat al-ʿAdlān (in the *Description de l'Egypte*: "Howârah el-Kebîr"). From here, dyke D-E, the Jisr al-Bahlawān, runs west to the western desert edge. This dyke has been omitted on the Napoleonic map, but it is of the same type as the Jisr al-Shaykh Jād-Allāh, although it does not have a stone casing. It does, however, have the same bottom width of 25 m, a top width of 8 m, and a height of c. 3.25 m.[90] The top of both dykes lies at c. + 29.25 m a.s.l. On the inside of the dykes (i.e. in the direction of the Fayyūm) the field surface is c. 1 m lower than outside the dykes (i.e. in the Nile floodplain).[91]

Anciently, a third dyke seems to have existed running eastwards from al-Lahūn to the Jabal Abū Ṣīr (tract B-C in fig. 8). This dam is referred to in Arabic documents dating back to the eleventh century AD.[92] If, accordingly,

90 See GARBRECHT/JARITZ, 1990, p. 142 and Anlage 100. GOMAÀ/MÜLLER-WOLLERMANN/SCHENKEL, 1991, p. 63-64 indicate a bottom width of 19 m, its height is indicated as 4.2 m, and its top width as 12 m.

91 GOMAÀ/MÜLLER-WOLLERMANN/SCHENKEL, 1991, p. 64.

92 See GARBRECHT/JARITZ, 1990, p. 141 and Anlage 98b, referring to SHAFEI, 1957.

a continuous dam A-C would have blocked flood water advancing northwards through the depression between the Jabal Abū Ṣīr and the Western Desert, a substantial part of it would perforce move in the direction of the Fayyūm. This structure would accordingly have had the effect of diverting the vast volumes of water descending northwards away from the Nile valley, into the channel of the Baḥr Yūsif.

The Baḥr Yūsif passes between al-Lahūn and Hawwārat al-ʿAdlān. Today, an extensive sluice system, reconstructed in the early nineteenth century, exists here. It includes a series of large sluices between the two villages, which serve to control the water volume flowing into the Fayyūm depression. A second series of sluices exists at the head of a channel branching off from the Baḥr Yūsif to the north, which pursues its course through the depression between the western desert and the Jabal Abū Ṣīr, passing by dam B-C in fig. 8. By finetuning the two sluice systems, the amount of water entering into the Fayyūm could be decreased or increased, with the water moving north being increased or decreased. At least part of this system is ancient, as the sluice system between al-Lahūn and Hawwārat al-ʿAdlān is already indicated in the *Description de l'Égypte*, map 19. We will see below that it was in fact very much older than that. The possibility to block water entering the Fayyūm accordingly existed already at the time the premodern systems in Middle Egypt, discussed above, were operational.

4.3.2 The area between al-Lahūn and the Hawwāra pyramid

Between al-Lahūn and Hawwārat al-Maqṭā, the north bank of the Baḥr Yūsif is still today partly fringed by high stone walls. In several places, these dams were broken by high floods and subsequently repaired in the nineteenth century. These more recent dams, which are made exclusively of earthwork, also still partly exist.[93] The presence of these dams, as well as the fact they broke, suggests that water pressure in this area was repeatedly very significant. In fact, both east and west of Hawwārat al-Maqṭā, the north bank of the Baḥr Yūsif was badly eroded away, leaving a deep ravine.[94] Apparently, one or more flood catastrophes led to the erosion of the surface in this area, allowing vast quantities of water to flow away in the direction of Lake Fayyūm. The stone walls in this area must partly have been repair works after these events.

However, this is not the full story. Map 19 of the *Description de l'Egypte* depicts several walls/dykes in this area, which largely correspond to those still observed in 1988 by Garbrecht and Jaritz. The easternmost of these walls, the

93 GARBRECHT/JARITZ, 1990, p. 150-160.

94 Indicated by "Ravinen" in fig. 8.

Description de l'Égypte, however, does not merely designate as a dyke, but as a "déservoir", i.e. a facility serving to channel water away from the Baḥr Yūsif. This must correspond to structure 3 in the Hawwārat al-Maqṭā area, which Garbrecht and Jaritz date to a period between the thirteenth and the seventeenth/eighteenth century A.D., and which, according to them, was clearly the main sluice serving to channel water from the Baḥr Yūsif to the northern Fayyūm depression and Lake Fayyūm, through the easternmost of the channels (Bats drain) in the north of the depression.[95] It is quite possible that this sluice system is not very ancient. However, since the map of the *Description de l'Égypte* shows several very wide channels branching off Baḥr Yūsif to the north, it is clear that vast amounts of water must have carved their way to the lake from here, and undoubtedly other sluice systems like structure 3 must have existed earlier in this area. The draining of the Fayyūm in the Ptolemaic and Roman period is inconceivable without such devices. Even if they may have been washed away long since, they must have existed somewhere in this area.

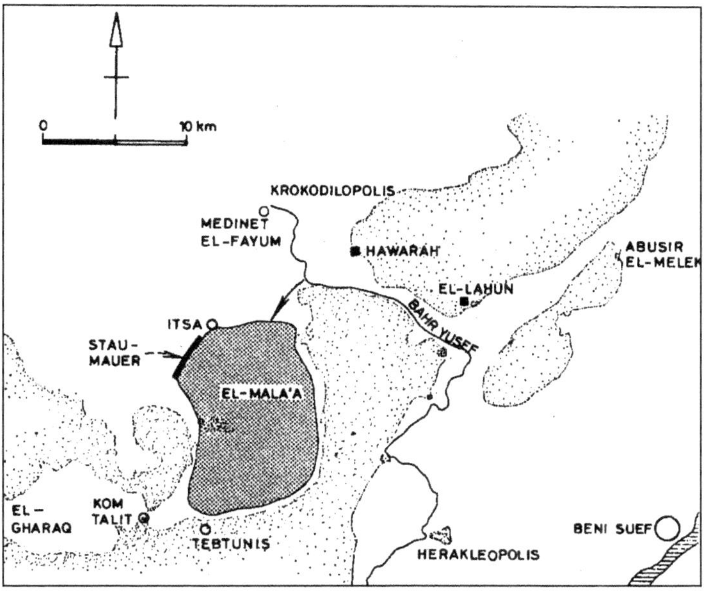

Figure 9. The setting of the al-Mala'a basin (after GARBRECHT/JARITZ, *1992, p. 248, fig. 13).*

95 ID., 1990, p. 152.

Shortly after the pyramid of Amenemhat III at Hawwāra, the Baḥr Yūsif branched into a large number of smaller channels, which allowed water to fan out all over the Fayyūm depression. It seems likely that these channels were once headed by sluice systems, but of these only the one near ʿIzbat Muṣṭafā al-Jindī now remains.[96] These waterworks must have controlled water entering into the artificial lake to be discussed next.

4.3.3 The Artificial al-Mala'a Lake

ʿIzbat Muṣṭafā al-Jindī lies southwest of the Hawwāra pyramid, at the juncture of the Baḥr Yūsif and a channel leading south into a depression called al-Mala'a (fig. 9). On its east and south, this depression is fringed by the desert, and on the north by the high alluvial deposits on which Madīnat al-Fayyūm lies. Towards the west, the surface slopes down into the Fayyūm, but here, there are remains of a vast dam running from Itsā in the north to the village of Shidmū in the south. The remains of this dam were documented in detail by Garbrecht and Jaritz.[97]

According to these authors, the top of the Itsā-Shidmū dam reached a height of about 17.50 m a.s.l.; the maximum flood level of the reservoir lay at about 16.50 a.s.l., the surface of the lake was 114 km^2, and it could hold c. 275 million m^3 of water.[98] Under optimal conditions, this would have allowed a surface of c. 260 km^2 to be supplied with additional water in the spring months, according to Garbrecht and Jaritz making a second harvest possible.[99]

4.3.4 Overall Interpretation of the Waterworks in the Fayyūm Region

The various dykes and drains discussed in the preceding paragraphs together constitute a well-conceived system serving to optimize the distribution of water in the whole area. Depending on how the sluices were used, the water volume flowing into the Fayyūm could be increased or decreased, leading to an inverse decrease or increase in water volume in the Nile valley proper.[100] Within the Fayyūm, the water could be distributed between Lake Fayyūm, the Mala'a reservoir, and the fields around Madīnat Fayyūm. We will show now in a probably simplified way how this functioned, discussing the consequences of

96 Id., 1990, p. 160-163; Id., 1992, p. 250-251.

97 Id., 1990, p. 38-134.

98 Id., 1990, p. 133.

99 Id., 1990, p. 170-173. That the lake was actually used to this effect is known from early eighteenth century sources (see Mikhail, 2010, p. 574-575).

100 What follows is in many regards similar to the shorter account presented by Römer elsewhere in this volume.

sluices in the dams being either closed or opened. Of course, in reality, there would be intermediate options as well, depending on the degree to which sluices were opened. However, by discussing the extreme options, it will be easier to conceive what the effects of the different options were:

Option 1: The dam system A-C (fig. 8) is closed. This will lead to large water volumes collecting in front of dykes D-E and the tell of the town of al-Lahūn, and this for prolonged periods after the end of the flood season. In this case, the sluice system between al-Lahūn and Hawwārat al-ʿAdlān will be opened. The Baḥr Yūsif will bring all its water volume into the Fayyūm depression.

Option 1a: The sluices east of Hawwārat al-Maqṭā are closed; no water flows through Bats drain into Lake Fayyūm. The ʿIzbat Muṣṭafā al-Jindī sluice is opened, and the al-Mala'a reservoir is filled with water.

Option 1b: The sluices east of Hawwārat al-Maqṭā are opened, so that water flows through Bats drain into Lake Fayyūm. The ʿIzbat Muṣṭafā al-Jindī sluice is closed, and the al-Mala'a basin receives no water.

In reality, options 1a and 1b were probably not mutually exclusive. The relatively small al-Mala'a reservoir would undoubtedly be filled every year, while the remainder of the water supply would be channeled into Bats drain and into the smaller channels around Madīnat al-Fayyūm. Moreover, in years when the water supply was particularly abundant, the sluices in the Itsā-Shidmū dam might remain open to permit water to flow out in that direction as well.

Option 2: Dam system A-C is opened, the sluices at al-Lahūn are closed. This will prevent water from the Nile valley spilling into the Fayyūm depression, allowing all water to continue its northward course through the depression between Jabal Abū Ṣīr in the direction of Memphis.

Again it is important to realize that we are here discussing an extreme option which was perhaps never put into practice. More realistically, options 1 and 2 were combined to different degrees. Let us envisage different possible situations:

Situation 1: the operation of the dyke and sluice system during a year with very low Nile floods. In this situation, only a small part of the fields could be reached by the flood, and opening the Baḥr Yūsif sluices at al-Lahūn would deprive the Nile valley of precious water. Closing dyke A-C, while having the effect of storing as much water as possible in Upper Egypt south of the Fayyūm, would also reduce water supply further downstream. To prevent this, the Baḥr Yūsif sluices near al-Lahūn could be closed and dyke A-C would be opened. Under these circumstances, only the Fayyūm would be deprived of water. To reduce this latter effect somewhat, the al-Lahūn sluices could be temporarily opened, but the sluices east of Hawwārat al-Maqṭā remained closed to prevent precious water spilling away into Lake Fayyūm. By opening the sluices at ʿIzbat

Muṣṭafā al-Jindī, however, the al-Mala'a reservoir could be filled, permitting a controlled flooding of certain areas further downstream. In this way, a substantial part of the Fayyūm would have access to water. After this, the al-Lahūn sluices could be closed to retain as much water as possible in the Nile valley.

Situation 2: A moderately high flood. In the Nile valley most, but not the highest basins would be flooded. Probably the sluices in dam system A-C and those at al-Lahūn would be open sufficiently long to flood both the fields in the Nile valley and in the Fayyūm. The al-Mala'a reservoir would be filled, but the amount of water draining off through Bats drain into Lake Fayyūm would be kept restricted at the Hawwārat al-Maqṭā sluice system.

Situation 3: A high Nile flood. In the Nile valley, even the highest basins on the levees would be flooded.[101] Probably the sluices in dam system A-C and those at al-Lahūn would be open sufficiently long to flood both the fields in the Nile valley and in the Fayyūm. The al-Mala'a reservoir would be filled, and the sluices at Hawwārat al-Maqṭā would be opened to drain off water into the Lake Fayyūm. This would have the effect of raising the water table in that lake. The sluices could be controlled to ensure that the volume of water flowing through the channels would not endanger the embankments and waterworks.

Situation 4: An exceptionally high flood. Under these conditions even the highest basins in the Nile floodplain would be inundated. The sluices in dam system A-C would be opened to allow as much water as possible to drain off to the north, while the sluices at al-Lahūn would be open, allowing a maximum volume of water to drain off into Lake Fayyūm. Quite conceivably, the sluices at ʿIzbat Muṣṭafā al-Jindī and in the Itsā-Shidmū dam would also be open so as to allow as much water as possible to drain off in a controlled way. Very high floods could, however, be of such magnitude that the waterworks were damaged, leading to a situation in which the amounts of water could no longer be kept in check. This explains the breaks in the Jisr al-Shaykh Jād-Allāh, in and behind the embankments of Baḥr Yūsif near Hawwārat al-Maqṭā, and in the Itsā-Shidmū dam.[102]

101 According to WILLCOCKS, 1899, p. 59, this would happen eight or nine times per century.

102 This has been associated to a very small number of catastrophically high floods listed by GARBRECHT/ JARITZ, 1990, p. 184-186. However, the cause of dams collapsing may have been the state of repair and bad maintenance as much as flood height, as the documentation from the early eighteenth century proves (MIKHAIL, 2010, p. 580-582). And the frequency of the collapses may accordingly have been significantly greater.

Under several of the scenarios described above, the water retained in the al-Mala'a lake could be stored for later use. It has been argued that this could lead to a second harvest in spring in part of the Fayyūm.[103] Taking into consideration a maximum volume of the lake of c. 275 million m^3 and an evaporation of 75 million m^3, the amount of available water would be in the range of 200 million m^3, sufficient for irrigating c. 150 km^2 for a second harvest, i.e. c. 12.5 % of the Fayyūm.[104]

Our study adds one, probably significant, element to this discussion. In the present section we have argued how the dykes and sluices at the entrance of and within the Fayyūm may have been used in the course of the Nile flood. However, we have also argued that, *after* the flood, the opening of the basin dykes in Upper Egypt must have led to the release of a very significant amount of water that flowed through the Baḥr Yūsif and through the depression between it and the Nile. Since this happened at a moment the flood waters had already been receding for some time, it must have had the effect of producing a second 'wave.' We put the word 'wave' between inverted commas, because this water transport through shallow basins must have been much slower than that of the inundation proper. Although we are unable to quantify the volume and speed of this water displacement, it is likely to have plied its way to the Fayyūm over a somewhat prolonged period of time, and the sluices at al-Lahūn and further west are likely to have been used in this period as well. This raises the possibility that the water in the al-Mala'a reservoir was fed by additional water, replacing water that had evaporated or that had already been channeled to the fields downstream. This suggests that the evaporation factor referred to in the above calculation may have been neutralized, and also that the reservoir may not have been filled just once, but may have been replenished continually over a prolonged period of time. This possibility, which has never before been mooted, has the consequence that a far greater part of the Fayyūm depression than the 12.5 % mentioned above may have been irrigated.

5. Chronological Aspects of the Waterworks Discussed

This article has thus far attempted to reconstruct the functioning of the hydrological systems in Middle Egypt and the Fayyūm in the late eighteenth

103 Id., 1990, p. 168.

104 Id., 1990, p. 170-173.

century, the aim being to understand their operation in the period before the major infrastructural works undertaken in the early nineteenth century by Muḥammad ʿAlī. In this way it has been possible to filter out a number of modern innovations that had hitherto often been assumed to be characteristic for far more ancient irrigation practices, and even for pharaonic Egypt. The result of our enquiry has been that this latter approach is unwarranted, since only thirty years before Muḥammad ʿAlī, the situation was already utterly different.[105] It is quite likely that the less advanced infrastructure of the *eighteenth* century brings us closer to the situation in the preceding millennia. However, it would be naïve to suppose without further ado that the Napoleonic maps record a system that would be directly transferable to the Roman or an even more distant past.[106] The present section has as its aim to study ways that might enable us to disclose with greater reliability which elements of the landscape as observed in Napoleon's day have ancient roots.

In doing so, attention will be paid to information from historical toponymy and ancient written sources, archaeological evidence, and information on sedimentation in the landscape recorded in the *Description de l'Egypte.* These aspects will first be discussed for Middle Egypt and then for the Fayyūm.

5.1 Middle Egypt

5.1.1 Historical Toponymy

In this section we will first compile the available evidence on place names in the region studied. We will proceed from south to north.

1. Dairūṭ (al-Sharīf). The etymology of this name is currently understood as *ỉw-rd* "plant island". It is assumed that the D in Dairūṭ goes back to the article

105 For the area between Ṣamalūṭ and the Fayyūm, however, GOMAÀ/MÜLLER-WOLLERMANN/SCHENKEL, 1991, p. 44, have argued that the dykes constructed in the nineteenth century lay more or less at the same locations as those rendered in the *Description de l'Egypte.*

106 ID., 1991, p. 52-53 have argued that the distances between the dykes observed by the Napoleonic mission between Ṣamalūṭ and the Fayyūm may reflect the pharaonic *itrw* measure, and thus that the basin system observed there dates back to at least the Ramesside period. Of course this single argument can hardly be accepted as proof. Moreover, in other supposedly ancient dykes, this distance was not observed (GRIESHABER, 2004, p. 20).

tȝ, Coptic ⲧ. For Dairūṭ, the Coptic variant ⲧⲉⲣⲱⲧ is in fact known.[107] Gardiner still linked *Ỉw-rd* to a place north of al-Ashmūnayn, because P. Harris I, 61b,3-7 features a town list arranged *Ḫmnw – Ḥw.t-wr.t – Ỉw-rd*.[108] The al-ʿArīsh list also features the town after *Ḥw.t-wr.t*. This might be taken as an argument against identifying *Ỉw-rd* with Dairūṭ. However, there are many places in Egypt with the name Dairūṭ, and therefore there were arguably as many different ancient places named *Ỉw-rd*.[109] Even though it is hard to attribute any one textual reference to *Ỉw-rd* to any one of the many sites with the name Dairūṭ, the name as such is well attested since the Old Kingdom. This is an argument for assuming that the Dairūṭ in the area we are concerned with, is of (probably early) pharaonic date. In the area we are here interested in, two villages of this name are mentioned in the *Description de l'Égypte*: the larger settlement at the beginning of the Baḥr Yūsif, (Dairūṭ al-Sharīf), and a smaller village on the eastern bank of Baḥr Yūsif due west of al-Ashmūnayn.

2. Ismū. In P. Duk. Inv. Miss 88, listing a number of settlements between al-Ashmūnayn and Ṣanabū, there is a mention of a village called Σόμου, which van Minnen identifies with Ismū. This village is also attested in other documents, some of them dating back to between the first and third centuries A.D.[110]

Maspero has suggested that the modern name Ismū goes back to Coptic ⲤⲘⲞⲨ or ⲤⲒⲘⲞⲨ, which is attested since the fourth or fifth century A.D.[111] Drew-Bear expressed doubts against this, since according to a papyrus in Munich, ⲤⲒⲘⲞⲨ would have been the place of birth of Apa Phib, whereas the Life of Apa Phib attributes his birth to a village named Ψινομουνις in the Antinoite nome. Since the Antinoite nome centres around Antinoopolis on the east bank of the Nile, she believes that ⲤⲒⲘⲞⲨ must have lain on the east bank of the Nile as well, and therefore could not have been identical with Ismū.[112] However, the etymology is very unconvincing. Ψινομουνις must go back to Egyptian *Pȝ š n Ỉmn* "The basin of Amūn". Drew-Bear's interpretation involves that both the genitive-*n* and the *n* in the name of Amūn would have been dropped, for which

107 ZIBELIUS, 1978, p. 26-28; DREW-BEAR, 1979, p. 289; TIMM,1984, p. 562-565; PEUST, 2010, p. 34-35.

108 GARDINER, 1947,II, p. 87*-88* (379A).

109 For this reason, JACQUET-GORDON's suggestion (1962, p. 115 and *passim*), that all Old Kingdom references to domains named *Ỉw rd* must refer to the place identified by GARDINER, is far from compelling.

110 VAN MINNEN, 1994, p. 85.

111 MASPERO, 1931, p. 96, n. 1.

112 DREW-BEAR, 1979, p. 328-329.

she fails to adduce any parallel, and which is singularly unlikely. Moreover, the Antinoite nome during certain periods may well have extended to the west bank of the Nile, and the Munich papyrus locates ⲥⲓⲙⲟⲩ in the nome of Hermopolis. The only argument is the purported place of birth of saint Phib. But as a glance at for instance Timm's toponymical list shows, literary accounts bristle with errors in accounts of this type. In my view, there can be no doubt that Ismū = ⲥⲙⲟⲩ, ⲥⲓⲙⲟⲩ = Σόμου, an etymology which would bring us back uninterruptedly from the present to the first century A.D.[113]

As we will see in the discussion of the toponym Sinjirj, the element Si- in place names is often interpreted as "place", but it might as well be connected to *š* 'basin". The element -ⲙⲟⲩ following this could go back to Egyptian *mw* "water", in which case the toponym would mean something like "basin of water". However, this sounds tautological. -ⲙⲟⲩ might also go back to *m3.wt* (ⲙⲟⲟⲩⲉ) "island", in which "island" might refer to a high place not normally flooded by the inundation. In this case the name would mean "basin of the island". It must have been commonplace for basins to be located in the vicinity of such "islands" (i.e. high places on levees), but since basins are not dependent on isolated islands, but rather on circuits of higher land (including dykes and 'islands') the name sounds somewhat odd as well. However, the word ⲙⲟⲟⲩⲉ "island", originally "new land" goes back etymologically to the Egyptian root *m3wi* "to be new". In Coptic, this root has barely survived, although there is a rare word ⲙⲟⲩⲓ of this meaning.[114] Arguably, ⲥⲓⲙⲟⲩ could go back to *š m3wi* "the new basin". This would, at the time of the creation of an artificial basin, have been an appropriate name for a locality.

This interpretation seems the most attractive to us, and it has the important consequence that the toponym consists of a noun followed by an adjective. In Demotic and Coptic, this adjective would be connected to the preceding noun by an indirect genitive (*š n m3w.t*; ⲥⲙ̄ⲙⲟⲩⲓ), which is not in keeping with what the writings of the toponym suggest. Therefore, the place name is arguably of pre-Demotic date. If the etymology holds, it presupposes that the toponym dates back to at least as early as the Third Intermediate Period.

113 I express my gratitude to Willy Clarysse, with whom I discussed this problem, and who was entirely of the same opinion. The same opinion is found in Peust, 2010, p. 54.

114 The only dictionary mentioning it is Westendorf 1965-1977, p. 88.

3. Tānūf. This is certainly not an Arabic place name. Drew-Bear relates it to the Greek settlement of Ιβιών Τανουπεως, which goes back to as early as A.D. 128.[115] Peust relates it to the Demotic female name *T3-Ỉnpw*.[116]

4. Daljā. The etymology is again unclear, but it does not sound like an Arabic name, and it is in fact attested in medieval Arabic, Coptic, and Greek sources going back to at least as early as A.D. 709 (Τέλκε; ⲧⲏⲗⲕⲉ).[117] Perhaps this name includes the feminine article ⲧ,[118] in which case an etymology with an earlier Egyptian ancestor of ⲏⲗⲕⲉ hould be sought for. The town is built on a large tell, so that it must be of very considerable age. This tell lies on what may be an earlier levee of the 'palaeo-Baḥr Yūsif' (see 4.1.2 (3)). Research carried out there by Verstraeten's team in collaboration with the Dayr al-Barshā project has revealed that agricultural fields in this area go back to the early third millennium B.C.[119]

5. Badramān. Despite its Arabic appearance, this toponym, referring to a settlement located on the Baḥr Yūsif close to al-Shaykh Shibaykā, has been argued to go back to Greek Πατριμο[...], Coptic ⲡⲁⲧⲣⲉⲙⲱⲛ (and variants), references to which go back to the sixth century A.D. It may be referred to even earlier, in the fourth century A.D., in a Greek text referring to an "Island of P[a] trim[...]".[120]

6. Tūkh. This name apparently goes back to ⲧⲱϩⲉ. This place name is widely attested throughout Egypt, and therefore no source can with certainty be related to the village of this name in the region here investigated. However, it is clear that we are facing an ancient indigenous name.[121]

7. *Nazlat al-Shaykh Ḥussayn.* Although this place name is not ancient, the settlement probably is. Jomard reports having seen there some limestone blocks that were between three and four metres long, and that the locals speak of the presence of a "Birbé" there.[122] As is well known, this word birba derives from the Egyptian word *r-pr* 'temple', and the Copts used this term for referring to pagan

115 DREW-BEAR, 1979, p. 129.

116 PEUST, 2010, p. 94.

117 DREW-BEAR, 1979, p. 274; TIMM, 1984, p. 502-504. Timm assumes the town existed already much earlier than this.

118 Thus also PEUST, 2010, p. 32.

119 MOHAMED, 2012, p. 107-110; 122-123; VERSTRAETEN et al., 2016.

120 DREW-BEAR, 1979, p. 242; TIMM, 1984, p. 277-278; PEUST, 2010, p. 17.

121 DREW-BEAR, 1979, p. 312-313; TIMM, 1992, p. 2865 ff.

122 JOMARD, 1821, p. 320.

temples. It is therefore likely that this settlement goes back to the pharaonic period.

8. Tandā. This village is first referred to in an Arabic document mentioning Arianus, the governor of Anṣinā, who lived in the fourth century A.D.[123]

9. Sinjirj. This toponym goes back to Σενκύρκις, a name attested at least since A.D. 136, and perhaps since 128.[124] Yoyotte has argued that the element –jirj, -κύρκις goes back to Eyptian *grg* "poser un filet de chasse, un piège, une nasse", while –n- would be the Egyptian indirect genitive and the first part of the name Coptic **ⲥⲉ** or **ⲥⲓ**, "seat, place".[125] The name would accordingly mean "la place où l'on chasse". This is now widely accepted,[126] but the reasoning that brings Yoyotte to his conclusion is not really convincing.

His argumentation is as follows. Egyptian possessed two roots *grg*, one with the meaning "to place a hunting net or trap" (**ϭⲱⲣϭ**), and a second one that can be translated as "to estabish, to found, to organize, to equip, to populate" (**ϭⲱⲣ̄ϭ**). The existence of marshlands throughout Egypt render likely that hunting and catching birds by traps would have been a widespread custom, and for this reason, according to Yoyotte, "il serait … étonnant que *grg* "chasser" n'ait laissé aucun vestige dans la toponymie d'époque pharaonique". However, "malheureusement, seules les transcriptions greques (et encore!) permettraient de reconnaître si une forme 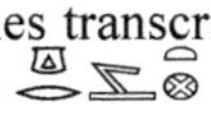 signifie "La Chasse" (*grg* = κυρκ-) plutôt que "La Fondation" (*grg/grgt* = κερκ-)".

This reasoning is hardly compelling. Yoyotte's use of the subjonctifs "serait" and "permettraient" and the exclamation "et encore!" suggest that he himself was only partly convinced by his hypothesis. While Yoyotte seems to take for granted that κυρκ- = Coptic **ϭⲱⲣϭ** and κερκ- = **ϭⲱⲣ̄ϭ**, he fails to pay attention to the facts that 1) the Coptic spellings are near identical (and entirely so if the line over *r* would be omitted, as would frequently happen); 2) in the examples he quotes the Greek forms κυρκ and κερκ are to an extent interchangeable (he explicitly quotes variants Σιν**κερκ**ις, Σιν**κυρκ**ις, etc. for the toponym we are here discussing, and 3) the form κερκ seems to be predominantly attested in the first position of a direct genitive, whereas κυρκ occurs mostly in cases like Σενκύρκις as the second element in an indirect genitive. It is quite possible that

123 Timm, 1992, p. 2498-2499

124 Drew-Bear, 1979, p. 242.

125 Černý, 1976, p. 145 relates this word to Egyptian *s.t* "place." The interpretation as "place" for the toponym Sinjirj can also be found in Crum, 1939, p. 316b; Peust, 2010, p. 87.

126 Yoyotte, 1962, p. 84; accepted by Drew-Bear, 1979, p. 242.

we are facing one and the same word in which the difference in pronunciation may be due to a difference in grammatical construction, to dialectical differences, or both. Since Yoyotte is unable to prove in a single case that the element *grg* refers to the verb "to hunt," it is for the time being better to assume that all toponyms with the element *grg/grg.t* refer to the ubiquitous word 'agricultural domain' or similar.[127]

This is all the more likely since several ancient toponyms designating settlements in the immediate vicinity of Sinjirj are of the type Κερκε + divine name, "settlement of god X". This is the case for Κερκεθοηρις "the agricultural domain of Thoeris" in Greek texts dating back to the mid-third century A.D,[128] as well as for Τκερκεθωθις, "the domain of Thoth".[129] If *grg.t* in this name means 'agricultural domain,' a further consequence is that the traditional interpretation of the element Sin- in Sinjirj must be reconsidered. As noted above this is usually understood as an indirect genitive preceded the Coptic word ⲥⲓ- "place, seat". However, the resulting meaning of the toponym as "place of the agricultural domain" sounds odd, since a domain is by definition a "place". ⲥⲓ- might therefore rather derive from *š* "basin". In toponyms of the type *Š-n-Ḥr* "basin of Horus", this name survives to the present day in forms like Shanhūr (referring to the village of that name just north of Luxor), but the Greek form Ψενυρις shows that *š n* could evolve into Sen-. The Arabic descendant of a village of this name in the Fayyūm is Sinnūris.[130] Accordingly, the place name Σινκερκις might well go back to "the basin of the agricultural establishment". The names Κερκεθοηρις and Τκερκεθωθις might refer to specific establishments within the basin. If there is such a causal relationship between the origin of the three toponyms, this could be chronologically relevant, since names of the type *Grg(.t)* + divine name are not attested before the first millennium B.C.[131] This reasoning would permit us to date the origin of the name Sinjirj to a date as early as this. It is however also conceivable that there is no causal connection, and that the name Sinjirj is in fact much older. Since domains called *grg.t* are frequently attested since the Old Kingdom, it might in fact be very old.

127 Erman, Grapow, 1931, p. 188,14-15; same etymology proposed by Peust, 2010, p. 87. For the nature of a *grg.t*, see Moreno Garcia, 2013, p. 96, with references to further literature.

128 Drew-Bear, 1979, p. 139. She translates this toponym as "le domaine aménagé pour Thoueris". This toponym must also contain the element *grg.t*.

129 Id., p. 271; Timm, 1992, p. 2567.

130 Peust, 2010, p. 87-88.

131 Yoyotte, 1962, p. 85.

10. al-Birka. This place name in Arabic means "the Lake" (see p. 288-289), a name probably derived from the location of the settlement close to the western desert edge, where a lake is effectively indicated on the Survey of Egypt maps. Although al-Birka is certainly an Arabic name, settlements lying in similar positions were already called *Brk.t* in the tenth century B.C., using a term of Semitic origin that is etymologically related to the modern Arabic word.[132] It is therefore theoretically possible that the name is not originally Arabic, but goes back to a far more ancient precursor. However, against this hypothesis, it should be pointed out that no toponym similar to al-Birka is known from the papyrological record.

11. Mallawī. This place name is generally believed to go back to Coptic ⲙⲁⲛⲗⲁⲩ, which is attested since AD 1296.[133] There are reports of finds of columns and other architectural elements even including a naos in the town, but nothing of it was found *in situ* or in controlled excavations. Yet it has been argued that these finds suggest the floruit of the town was in the Late Roman and early Islamic era.[134] Tombs dated to the Graeco-Roman period have also been reported, but this dating is of uncertain value.[135]

12. Al-Rairamūn. In this town, there is a very large tell, indicating a significant age for this settlement. Remains of Coptic buildings visible today in the streets of Rairamūn suggest the village goes back at least to the early Christian period. Because of the element –amūn, it has been argued that this toponym contains a reference to the god Amun, although this is not quite certain. D. Kessler has intensively discussed the indications for the identification of the ancient name of Rairamūn. It could be the *Nʿy-Wsr-mꜢʿ.t-Rʿ Mr.y-Ỉmn* mentioned in P. Wilbour II, § 89.[136] Although this cannot be considered certain, Kessler plausibly argues that the cemeteries of al-Ashmūnayn at al-Shaykh Saʿīd and Dayr al-Barshā should be linked to the town by an easily negotiable road or waterway, and he

132 See ANTOINE, 2016, p. 25.

133 DREW-BEAR, 1979, p. 165. The Coptic etymology proposed by CRUM 1939 ("place of things"), to which she refers, is hardly credible. ČERNÝ's alternative idea that it might refer to a "place of textile" (ⲗⲁⲩ) is more acceptable (1976, p. 346). See also TIMM, 1988, p. 1542-1543.

134 See the overviews in DREW-BEAR, 1979, p. 35-36; KESSLER, 1981, p. 93.

135 Found by Usīrīs Ghubriāl, see LECLANT, 1971, p. 234; ID., 1973, p. 405. However, the same author has attributed tombs in Dayr al-Barshā to the Graeco-Roman period that are more likely to date to the Late Period. This suggests his dating may not be accurate.

136 KESSLER, 1981, p. 92-103.

deems likely that Rairamūn "like today" would constitute the ideal place for a ferry to cross the Nile. This he takes as an argument to assume that the waterway between al-Ashmūnayn and Rairamūn is as ancient as the cemeteries are. This is a convincing line of argument. However, the shortest connection to Dayr al-Barshā would not be through Rairamūn, as Kessler suggests, but through Dayr al-Mallāk, two kilometers further north. This implies that the ferry harbor used for the trip to Dayr al-Barshā lay at the mouth of al-Sabakh. The Tirʿa al-Shaykh Hajāza, leading from al-Ashmūnayn to Rairamūn, would, however, be very suitable to travel to the Old Kingdom cemetery at al-Shaykh Saʿīd. In this case, the harbour at Rairamūn might go back in time as far as this, and Rairamūn might be equally ancient.

13. al-Ashmūnayn. Textual attestations of the town of al-Ashmūnayn go back in time to the early third millennium B.C., even though archaeological remains of that early date have not surfaced thus far.[137] Since late Old Kingdom settlement remains have, however, been found here, there is no reason to doubt that the town is very ancient.[138] According to Butzer and other authors, it then lay on a Nile levee. The Nile would only later have shifted its bed to the more easterly location where it is now. However, no evidence has been marshaled to prove this eastward migration of the Nile bed. We have argued above that no such shift took place, that the Tirʿa al-Shaykh Ḥajāza is an ancient Nile branch, and that its levee led to the emergence of the tells of the settlements at Rairamūn, perhaps Qulubba, and al-Ashmūnayn. In the near future, we hope to be able to carry out augerings in this area to put this hypothesis to the test.

14. Nawāi. The earliest papyrological sources mentioning Nawāi date to the sixth-seventh centuries A.D,[139] but already in the fourth century A.D. there was a monastery in this village.[140] Kessler suggests that a precursor of the name can be found in the toponym *N3-ḥr-ḥw*, mentioned in a papyrus from the time of Ramses II.[141] Jomard has seen limestone blocks there.[142]

15. Itlīdim. It is assumed that this modern place name goes back to Greek Τληθμις, which is attested in papyrological sources at least as early as the first century A.D.[143] It has been argued that this village is identical to the pharaonic

137 GOMAÀ, 1978, p. 189-190.

138 ROEDER, 1959, p. 48 and the literature there cited.

139 DREW-BEAR, 1979, p. 178-179.

140 TIMM, 1988, p. 1755.

141 KESSLER, 1981, p. 69-70; P. BM 10447,1 and 5 (see GARDINER, 1948, p. 59).

142 PEUST, 2010, p. 68.

143 DREW-BEAR, 1979, p. 302-303; TIMM, 1985, p. 1204-1205.

town of Nefrusy. However, Kessler stresses that this identification is uncertain.[144] He localizes Nefrusy in the centre of the floodplain due west of Banī Ḥasan,[145] far outside the area we are here investigating. Etymologically, the word may be related to Demotic *ltm*, referring to a tree.[146]

16. Maḥras. Evidence that this place might date back to the pharaonic era is rather unsecure.[147]

5.1.2 Interpretation of the Toponymy within the Context of Information on Sedimentation, Soil Composition, and Archaeological Indications

We have seen in 4.2 that E. Subias, I. Fiz, and R. Cuesta observed a dyke system between Ṣamalūṭ and the Fayyūm entrance that was much more regular than the one in the al-Ashmūnayn-Dairūṭ region. Based on remote sensing analysis, they also showed that, before the introduction of this system, a similarly irregular pattern of waterways had existed there as we have observed between al-Ashmūnayn and Dairūṭ based on the *Description de l'Egypte* maps. We accordingly drew the conclusion that the hydrology of the latter region that still existed in 1798 is more ancient than the one observed further north.[148]

To this can be added the observation that, between the Middle Ages and the early nineteenth century, only few innovations in the waterworks seem to have been carried out in Egypt.[149] It is not easy to assess the importance

144 KESSLER, 1981, p. 159-161.

145 KESSLER, 1981, p. 120-185.

146 VYCICHL, 1983, p. 214; PEUST, 2010, p. 55.

147 KESSLER, 1981, p. 72-73.

148 As noted in n. 105-106, GOMAÀ/MÜLLER-WOLLERMANN/SCHENKEL put forward the idea that the dykes in the region between Ṣamalūṭ and the Fayyūm entrance are spaced at distances that would reflect the ancient Egyptian *itrw* measure. By consequence, the dyke system in this region would date back to the New Kingdom. The idea that the system in the al-Ashmūnayn region is less regular, and therefore arguably less recent, would date the Ashmūnayn dyke pattern even further back in time. It goes without saying that this argument, which rests on the similarity between the distance of extant dykes and the ancient *itrw* measure, is not compelling, and we will not use it here.

149 GARBRECHT/JARITZ, 1990, p. 188-189. For the collapse of the irrigation system after medieval plagues, see BORSCH, 2004, p. 458-463. The effects of the Black Death on population size must have had a long-term impact. MIKHAIL, 2010, p. 580-582 gives an enlightening account of the chaotic way in which the central

of this consideration for the relative dating of the different systems of water management, but if it is true that the dyke systems remained generally unchanged since medieval times, the consequence is that the more modern system in the Minia region was itself realized probably no later than that.

Carrying this reasoning further, the implication would be that the phenomena observed in the al-Ashmūnayn-Dairūṭ region were earlier still, at least in conception. We do not have to stress that this reasoning is tentative, but in support of it one might refer to the basin system in the Delta, which, already in the thirteenth century A.D., operated on the basis of large, rectangular basins.[150] This could be taken as an argument to push the al-Ashmūnayn system back in time to a date well before the end of the Middle Ages.

It is likely that the two systems may reflect different degrees of state penetration in the organization of irrigation. In the Ottoman period, large irrigation works servicing extensive areas were called al-jusūr al-ṣulṭānīya "state dams", a term that applied for instance to the large dams at al-Lahūn and between Itsā and Shidmū. By contrast, al-jusūr al-baladīya "peasant dams", were built and maintained by local initiative and served local concerns only. The terminological difference between the two kinds of waterworks goes back to the Middle Ages.[151] It is likely that the former of the two systems operated in the al-Ashmūnayn region as well, although evidence for state dams is restricted to dyke 23, which is designated as "gesr Soultâni" in the map of the *Description de l'Egypte.* Arguably, this is part of a destroyed dyke system linking dyke 21 to dyke 23. If this hypothesis is correct, a large E-W state dam linking the Nile past al-Ashmūnayn to Ṭūna al-Jabal may have existed in the Middle Ages and the Ottoman period.[152]

---

state intervened in the maintenance of the dams in the Fayyūm still in the early eighteenth century.

150 Lecture by Borsch, held at the Mainz conference on the Nile. Unfortunately his paper could not be included in this volume.

151 Borsch, 2004, p. 458-460; Michel, 2005, p. 260; Mikhail, 2010, p. 588, n. 50. It seems not unlikely that the larger-scale system in the Minia region and the small-scale system still prevailing in the al-Ashmūnayn area reflect these two different systems, and accordingly a less deep penetration of state involvement in irrigation affairs in the latter area.

152 Kessler, 1981, Karte 1, indicates that al-jisr al-ṣulṭānī continues all the way from al-Ashmūnayn to the Nile. In his map, this is an entirely straight dyke, an interpretation that is not borne out by the evidence provided by the *Description de l'Egypte.* This observation also warns us that simplified reconstructions of dyke

Other aspects are the date of origin of some of the landscape features that shaped the hydrology of the region. The levees of the Baḥr Yūsif are a case in point. It has been claimed that in Antiquity this waterway did not branch off from the Nile at Dairūṭ, the argument being that medieval records suggest it began further south, near Manhā. This may well be correct, as we have referred on p. 277 above to waterways in the depression southwest of Dairūṭ leading to the Baḥr Yūsif, which may be the remains of this system.[153] However, the existence of a more southerly Nile branch connecting to the Baḥr Yūsif in no way implies that there was not also a branch starting near Dairūṭ. The very prominent levees between Dairūṭ and Daljā leave no doubt that this waterway must have a very considerable age. As we have seen in the preceding analysis, many waterways branched off from the Nile in the direction of the Baḥr Yūsif. We would therefore posit that the Dairūṭ branch may well be ancient.

We have seen in 4.1.1 that the Baḥr Yūsif is a Nile branch that emerged when the gradually descending landscape west of the Nile had already been formed. Initially there was a depression in the western part of the floodplain, of which the topographic maps made in the mid-twentieth century indicate an elevation in the +43 m realm (unnumbered basin southeast of Daljā; basins XVII, XX, XXVI). In this area, the elevation reached by the levee of the Baḥr Yūsif is higher than 45 m a.s.l. This means that Baḥr Yūsif has risen at least over 2 m since it emerged. Following the rule of thumb of an average of 1 mm of sedimentation per year, the levees around the Baḥr Yūsif would go back at least two millennia. However, since the basins beside the levees have in the meantime also been silting up, thereby covering the lower parts of the levees, the Baḥr Yūsif levees must be much older than two millennia.

In fact, geomorphological research carried out north of Daljā suggests that a precursor of Baḥr Yūsif or a similar channel originating from further south already existed around 1400 A.D. in the region that is nowadays covered with dunes (i.e. 2 km further to the west of the current-day channel of Baḥr Yūsif). Although no other palaeochannels could be clearly identified, the presence of floodbasin sediments in between dune ridges, dated at 4520-4440 cal BC, demonstrates that tributary channels of the Nile, such as Baḥr Yūsif, were able to transport sufficient amounts of water and sediment to the western desert edge at the time. Moreover, an area to the west of the Baḥr Yūsif seems already to

systems such as are often based on papyrological documents, can only convey a very schematic view of what the dyke pattern may have looked like in reality.

153 KESSLER, 1981, p. 26-27, referring also to the earlier literature; DREW-BEAR, 1979, p. 289-290.

have been under cultivation by the early Old Kingdom.[154] All of this suggests that the major hydrological elements of the landscape, the Nile and the Baḥr Yūsif, go back to a period at least early in the pharaonic era.

This is confirmed by some of the toponyms analysed above. With the exception of Mallawī, which is first attested in 1296 A.D., all investigated place names go back to at least the (late-) Roman period, suggesting at least this time depth for many of the settlements recorded by the *Description de l'Egypte.* Some settlements, however, are likely to be much older. Two villages both called Dairūṭ lie on the levees of the Baḥr Yūsif: one at its entrance and a second one (called "Daroût Achmoun" in map 14 of the *Description de l'Égypte*) west of al-Ashmūnayn. This toponym is etymologically connected to Egyptian *ỉw rd* "plant island", a name known since the Old Kingdom for agricultural domains (see p. 310-311). The term *ỉw* 'island' designated not only permanent islands, but also higher parts of the floodplain that were islands only during the inundation.[155] The two villages called Dairūṭ are located on the levees of the Baḥr Yūsif, and they may well have lain on parts of these levees that were such temporary islands.[156]

As designations of such places, Egyptian knew two names: *ỉw* "island" and *m3.wt* (Coptic ⲙⲟⲩⲉ), literal translation "new land". While it is true that the two terms were in simultaneous use in the Twentieth Dynasty expression *ỉw n m3.wt* "island of new land", *ỉw* was generally a more ancient name, which is attested since the Old Kingdom, and was gradually being supplanted by *m3.wt* from the Amarna period onwards.[157] This suggests that the various Dairūṭs are very old, and may go back to one of the Old Kingdom royal domains named *ỉw rd.*

All of this suggests that the Baḥr Yūsif has followed a course very similar to the current one since very early in pharaonic history. This is confirmed by the probably great age of the tell of Daljā, and the nearby agricultural fields dating back to the early third millennium B.C. between the same site and al-Shaykh Shibaykā, and which relate to the Baḥr Yūsif.

The Tirʿa al-Shaykh Ḥajāza and al-Sabakh have levees that today rise at least 1 m above the surrounding landscape, and that may be of comparable age. A chronological indication is that two major Old and Middle Kingdom cemeteries dependent on al-Ashmūnayn are located near al-Shaykh Saʿīd just north of al-Amarna, and near Dayr al-Barshā. In the decision to locate these burial grounds

154 Mohamed, 2012, p. 107-110; 122-123. See also p. 249-250 in this volume.

155 A similar ambiguity in meaning exists with the Arabic word *jazīra.*

156 Schenkel, 1978, p. 62-65.

157 Id., p. 64; Antoine, 2016, p. 25-27.

where they were, considerations of accessibility must have played an important role. Now the Tirʿa al-Shaykh Ḥajāza, connecting the very old town of al-Ashmūnayn with al-Rairamūn (with a likely time depth going back at least to the New Kingdom) goes in the direction of the Old Kingdom cemeteries at al-Shaykh Saʿīd, suggesting it must be ancient. For similar reasons, the al-Sabakh is ideally located as a transport way linking al-Ashmūnayn to Dayr al-Barshā, where vast cemeteries existed in the Old and Middle Kingdoms (see p. 316-317).[158]

The early date of at least the Tirʿa al-Shaykh Hajāza receives support from the interpretation of the landscape in fig. 10, which combines the plans already rendered in figs. 5 and 6, but adds information on the form of the levees. Here the grey areas surrounded by a green line indicate the levee belts built up alongside the major river systems (Nile and Baḥr Yūsif). The rose-coloured areas surrounded by a red line indicate areas where levees of a different kind built up. These are probably crevasse splays: they take their point of issue at the outside of a bend of the river and deposit alluvium in a wide area fanning out from where the main levee broke. The fan-shaped form of the channels in the Mallawī area are very typical for crevasse splay channels. It is quite apparent that the settlements of Qulubba and al-Ashmūnayn lie on a single levee system that is likely to have been in origin a crevasse splay. In fact, from this perspective the location of al-Ashmūnayn is easy to understand: it is surrounded by irrigable land, but lies itself on higher ground. The situation of the town, which is known to go back to the third millennium B.C., suggests that the levees of this crevasse splay go back to at least as early as that.[159]

Other settlements also merit being mentioned here. The name of Sinjirj has been argued above to go back to Egyptian *š n grg.t* "the basin of the agricultural domain" (or similar). This name must be pharaonic, and its agricultural meaning is interesting, because the eighteenth century village lay on a dyke, and therefore on the edge of a flood basin. This is an argument for assuming that the dyke on which Sinjirj lies is ancient, and the same may be the case for other dykes in the same area.

158 Currently, a project directed by the first author in collaboration with G. Verstraeten and W. Toonen is investigating the geomorphology of this region.

159 One reason why BUTZER assumed the Nile must originally have followed a course immediately past al-Ashmūnayn was his idea that early settlement emerged on levees. This has also been assumed by archaeologists working in other areas, like the Rhine Delta. However, as VAN DINTER/VAN ZIJVERDEN, 2010, have shown, crevasse splays are admirably suited for incipient settlement.

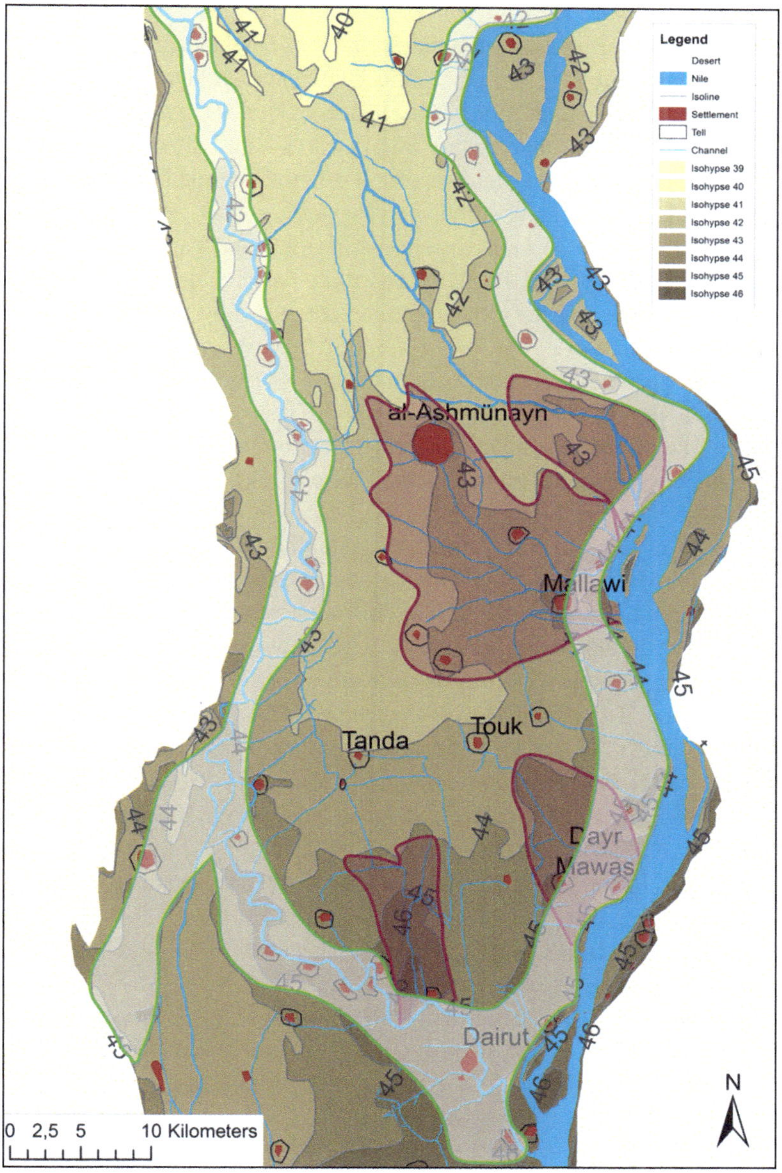

Figure 10. Levees and crevasse splays in the region investigated. Grey with green line: levee. Rose with red line: crevasse splay. Map Hanne Creylman, Gert Verstraeten and Georgia Long.

The village of Ismū may have similar roots. As the (possible) precursor of the name CIMOY suggests, this name may originally have meant "new basin", and may date back to at least as early as the Third Intermediate Period. It is highly interesting that this village still in the eighteenth century lay on the dyke enclosing basin XV. This strongly suggests that dykes 4 and 17 enclose a basin that may have been created during or before the Third Intermediate Period, a basin which, at the time of its creation, may have been baptized "new basin".The location of this basin, at great distance from the Baḥr Yūsif, would moreover suggest that it must always have been part of a basin chain. This offers support for the assumption that the dyke system encountered by the French army may be of very ancient date.

Finally, the obviously ancient architectural remains (vast blocks of ashlar masonry) at the village of Nazlat al-Shaykh Ḥussayn and the lore about the presence of a 'birbā' ('temple') there suggests the presence of a settlement at least going back to the Pharaonic period (see p. 313-314). This settlement in the eighteenth century lay directly beside a dyke.

A series of texts in the quarries of Hatnub, and dating back to the early Middle Kingdom, offers supplementary information. These texts describe a warlike situation in the reign of pharaoh Amenemhat I, in the course of which the town of al-Ashmūnayn was under siege. Several of the texts suggest that the population fled the city. In this connection one of the leaders of the town describes himself as "someone who acted as its rearguard in *Šdy.t-š3* when all people had fled".[160] The place name *Šdy.t-š3* has always been interpreted as a marsh-like wilderness that was hard to find access to by enemies unfamiliar with local conditions.[161] Egyptologists have never attempted to localize *Šdy.t-š3*, but it stands to reason that it must have designated the humid depression between the levees of the Nile and the Baḥr Yūsif. In the eighteenth century, the area south

160 ANTHES, 1928, Gr. 17,12-13. Similar expressions occur in Gr. 16,5; 23,5; 24,8; 25,7; 26,6. For the date of the texts, see WILLEMS, 2014, p. 79-87.

161 In Middle Kingdom texts the term *šdy.t* refers to a region rich in water and where fish live, although it is also said to refer to a type of land (ERMAN/GRAPOW, 1930, 567,11-14). *Š3* has a similar range of meaning: it refers to flooded land, or land where flowers grow. It is used in opposition to *sḫ.t*, the normal expression for agricultural fields (ID., 399,7-400,1). It is only in texts of the Graeco-Roman period that both terms are apparently used with reference to farmland. *Šdy.t-š3* was probably a direct genitive, but in view of the range of meaning of both components of the expression it is hard to translate. In any case it seems clear that a kind of marsh-like environment is being referred to.

of dyke 21 seems to have been under cultivation, but north of it, i.e. northwest of al-Ashmūnayn, there was an area without dykes, and which the waters of the Sabakh and the Baḥr Yūsif had free access to (basin XXVI and perhaps XXV). It stands to reason that the Hatnub texts are referring to a precursor of this area in the neighbourhood of al-Ashmūnayn.

The above indications suggest that the main hydrological arteries that were observed in the late eighteenth century by the French lay at more or less the same places as their precursors had since early in pharaonic history: the Nile, the Baḥr Yūsif, the Tirʿa al-Shaykh Ḥajāza, and the Sabakh, and a precursor of basin XXVI. Moreover, the settlements of Sinjirj and Ismū have names that may point to an originally agricultural background, which would fit in well with their location on dykes still extant in the eighteenth century. Since, as has been remarked by Gomaà, Müller-Wollermann, and Schenkel, maintenance of dykes is more likely to entail repairs of already existing earth works than the wholesale construction of new ones, it should come as no surprise that dykes might reach back far in time. The long history of agriculturally-based toponyms, and the persistence of the general layout of the landscape suggest that the pattern we have been discussing may be very old.[162]

This brings us to a last point. The area we have been discussing encompasses an extensive system of basins that starts where the Baḥr Yūsif branches off the Nile near Dairūṭ, and ends between Iṭlīdim and al-Ashmūnayn, in the humid basin XXVI, with its open connections both with the Nile and the Baḥr Yūsif. Both near Dairūṭ and south of Iṭlīdim, the flood basins within dykes and levees cover only a narrow strip along the Nile, whereas in the area in between, artificial irrigation basins extend westward almost as far as the western desert. Stated differently, the artificial irrigation basins rendered in the *Description de l'Egypte* do not constitute a continuous system, but only a local network of basins centred around al-Ashmūnayn.

There are again indications that this situation may have originated long before it was recorded by the French army. In fact, the areas where the irrigation basins are reduced to mere strips of land bordering on the Nile, seem to be

162 Accordingly, the assumption of a general eastward migration of the Nile (suggested, for instance, by Bunbury/Malouta, 2012, p. 119-122) is for this region rather unlikely. Note that Michel, 2005 has convincingly argued that, technically, the irrigation system has not changed much between the Graeco-Roman and Ottoman period. Over this long period of time, according to him, there is little or no evidence for the construction of artificial canals, the only major intervention in the landscape consisting in the construction of dams and the cleaning of channels.

located precisely where the northern and southern borders of the fifteenth Upper Egyptian nome were located already in the Middle Kingdom, and perhaps before.[163] It is unlikely that this correspondence is due to mere coincidence. We would suggest that the network of basins surrounding al-Ashmūnayn may have played a crucial role in defining the surface of the nome. Here it is worth recalling that the hieroglyph for *sp3.t* "nome" depicts a network of fields surrounded by small dykes (𓈈). We would suggest that the ancient fifteenth Upper Egyptian nome, and perhaps other nomes as well, may have crystallized around agriculturally coherent zones, which in Egypt implies the presence of irrigation basins.

The suggested relationship between the successful maintenance of agricultural basins and provincial administration does not necessarily imply a return to Wittfogel's model of oriental despotism, however (see n. 7). To him, a centralised irrigation authority lay at the roots of state administration. To us, it seems entirely acceptable that the origin of basin systems may have been based on local and informal systems of collaboration, which only in the course of the state formation process were partly integrated in more encompassing systems of administration. The form and nature of the basins in the al-Ashmūnayn area in fact suggest that human interference in the landscape was very restricted.

This is is confirmed by fig. 10. As explained above many channels in the region can be explained as naturally formed waterways, whereas it is in

163 This requires some explanation, as Helck 1974, p. 109; 205 placed the southern border between Dairūṭ and Dayr Mawās, and the northern one on the west bank even to the north of the latitude of Banī Ḥasan. His reasoning for this is 1) that the nome list in the *chapelle blanche* attributes a length of 3 *itrw* and 3 *ḫ3*, i.e. 33.069 km, to the Hare nome; 2) that the northern border of the nome must have passed north of Iṭlīdim and Balanṣūra, and 3) that because the southern border must have lain c. 33 km further south, Dairūṭ must have lain in the 14th Upper Egyptian nome, and not in the 15th. This whole reasoning rested on Helck's identification of Iṭlīdim and Balanṣūra with ancient Egyptian place names. In a thorough reevaluation of this toponymical evidence, however, Kessler, 1981, p. 120-185 showed convincingly that the northern border of the 15th nome cannot have passed north of Balanṣūra. He instead places it at the height of al-Shaykh Tīmai on the east bank, which corresponds roughly to the latitude of Iṭlīdim on the west bank. Because the northern border must accordingly be moved southwards for a considerable distance, and the location of the southern border can only be determined based on its 33-km distance to the northern border, Dairūṭ clearly must have been part of the 15th nome.

no case evident that a waterway is of man-made origin. This means that very little hydrological engineering was required to create the landscape we have discussed.[164]

5.2 Northern Middle Egypt and the Fayyūm

In previous sections we have shown that the natural process of sedimentation from both the main bed of the Nile and the Baḥr Yūsif led to the emergence of a shallow depression between the two waterways. At the end of the flood period, the dykes were opened, an occasion referred to in ancient texts as *wb3 š* "the opening of the basin'.[165] Only after the residual water had drained off from the basins in this way, it was possible to start plowing and sowing the fields.

The "opening of the field" took place throughout Egypt after the flood period, and it must have generated a major water displacement in the direction of the Fayyūm though the depression between the levees of the Nile and the Baḥr Yūsif, as well as through these waterways themselves. Although the depression has occasionally been referred to in the literature, the vast impact it must have had on the hydrology of all of northern Upper Egypt, on the spread of agriculturally useful fields, and on the demography of the region, has not been addressed before.[166] It seems likely that the magnitude of this factor explains the scale of the dyke system in the Fayyūm, of which the functioning in the eighteenth century has been discussed above (4.3.1-4). In this section we will attempt to show that it goes back much farther in time.

The sluice system at the entrance to the Fayyūm, between the modern villages of al-Lahūn and Hawwārat al-ʿAdlān, is repeatedly mentioned in medieval accounts dating back to the eleventh to thirteenth centuries A.D. The sluice currently in use was adapted in the nineteenth century, but in essence the

164 Of course this does not rule out that existing natural channels may have been maintained. This confirms the results of MICHEL, 2015, which was based on papyrological and Ottoman administrative documents.

165 For an overview of the pertinent literature, see EGBERTS, 1995, p. 321-322, n. 10.

166 This is not the place for a calculation of the water volume concerned. Although losses in water volume during the flood period have to be reckoned with due to infiltration and evaporation, the impact of the former factor must in the deeper areas have been significantly below average due to the relative closeness of the groundwater table. Moreover, as these deep areas were flooded far longer than the higher ones, the groundwater level itself must have been higher than average.

present structure dates back to 1245 A.D.[167] The archaeological study of the dams at the Fayyūm entrance and near the al-Mala'a reservoir, and the stone-lined walls north of the Baḥr Yūsif near Hawwārat al-Maqṭā, has shown that these structures are at least partly of Roman date.[168] An interesting point is that the Jisr al-Shaykh Jād-Allāh (i.e. the al-Lahūn dyke) has a Roman period stone casing. Since this casing is likely to have been applied in response to flood damage, the presently visible dyke must have been preceded by an older clay dam. The Jisr al-Bahlawān, the partner dyke of the Jisr al-Shaykh Jād Allāh, never received such a coating. The clay dams therefore seem to be earlier structures, perhaps of pre-Roman date.[169] The enormous expansion of agricultural activity in the Fayyūm can only be explained by the implementation of large-scale systems of water management from the Ptolemaic period onwards. Moreover, Herodot offers an account of the irrigation system in the Fayyūm that is in many regards unreliable, but that at least presupposes the presence of mechanisms to control the amount of water flowing into the depression. Since this would have been impossible without major waterworks being in place, some dykes and sluices must have existed at the entrance to the Fayyūm at least by Herodot's day, i.e. around 445 B.C.[170]

According to Classical lore, irrigation processes were, however, already undertaken in the Fayyūm in the Middle Kingdom. These works are attributed to a king Lamares, who is generally believed to be Amenemhat III.[171] In fact very little evidence remains of the long intervening period, except an Eighteenth Dynasty statue of a Sobkhotep, who claims the title of *ḥ3.ty-ʿ n š-rs.y š-mḥ.ty* "Lord of the southern lake and of the northern lake".[172] It has been argued that in this text, "the northern lake" refers to Lake Fayyūm, while "southern lake" designates the area of the al-Mala'a reservoir, which would accordingly reach back in time at least as far as this.[173] In fact, the term "southern lake"

167 GARBRECHT/JARITZ, 1990, p. 141; 147.

168 The oldest dam in the Mala'a area is Roman: ID., 1990, p. 133.

169 ID., p. 109.

170 ID., p. 140-141. RÖMER, 2016, p. 136 takes seriously Herodot's statement that surplus water could flow back from the Fayyūm depression into the Nile valley floodplain. We agree to the work of the authors cited in the preceding notes that this is rather unlikely.

171 RIAD, 1958, p. 203-206, with references to the pertinent Classical literature. See also RÖMER, 2016, p. 173-177.

172 HELCK, 1956, p. 1588,14-15.

173 GARBRECHT/JARITZ, 1990, p. 31.

already occurs much earlier, being attested several times during the early Old Kingdom.[174] The earliest references occur in the inscriptions of Metjen of the early Fourth Dynasty and in a text from the reign of king Khufu recently found in the Wādī al-Jarf.[175] Recent geomorphological reconstructions suggest this 'lake' already existed since the late eighth millennium B.C.[176]

There are in fact good reasons to assume that Amenemhat III (and his predecessors) pursued a policy of developing the Fayyūm. In this connection, the remarkable location of the pyramids of Senwosret II (near al-Lahūn) and Amenemhat III himself (near Hawwārat al-Maqṭā) in the Fayyūm has been duly noted, as has the presence of two colossal statues of Amenemhat III on the Middle Kingdom shore of Lake Fayyūm near Biahmū.[177] While all of this clearly bears witness to an unusual interest for the Fayyūm, it is possible to be far more specific about what was happening there than has been realized.

A first point of note is that the pyramids of Senwosret II and Amenemhat III are not just located in de Fayyūm region, but at very conspicuous places there. The pyramid of Senwosret II lies at the end of the al-Lahūn dam, and the pyramid of Amenemhat III lies close to Hawwārat al-Maqṭā. Like the location of the al-Lahūn pyramid, this is an important place in the irrigation landscape of the Fayyūm. On the one hand, the Bats drain, the most significant drainage system leading to the Fayyūm lake, begins here. In the Roman period, an important sluice system was built here serving to control the amount of water flowing off in this direction. But owing to the natural landscape, there must have been a dam system here earlier as well, because otherwise it would have been difficult to control the hydrology of the Fayyūm at all. The extensive erosion areas visible here must date back to periods before the construction of the dams, or to periods when they were broken, and this dramatically illustrates the impact of uncontrolled flooding in this area.

---

174 ĆWIEK, 1997, p. 17-22.

175 SETHE, 1933, p. 3,12; TALLET/MAROUARD/LAISNEY, 2012, p. 413 and p. 441, fig. 22. Here, *š rs.y* is erroneously translated as "Fayyūm".

176 HASSAN/TASSIE, 2006, p. 39.

177 Other remains have also been found in Madīnat Fayyūm and at Abjīj (the structure at the latter site even going back to as early as Senwosret I). For an overview of the Middle Kingdom evidence, see TALLET, 2005, p. 101-109. Other publications pointing out the significant location of the pyramids of Senwosret II and Amenemhat III are GOMAÀ/MÜLLER-WOLLERMANN/SCHENKEL, 1991, p. 27-28 and SCHENKEL, 1978, p. 65-67.

Further south, but still close to Hawwārat al-Maqṭā, lies the medieval sluice system of ʿIzbat Muṣṭafā al-Jindī, which served to control water flowing into the al-Mala'a reservoir. The question is, of course, whether this reservoir was already operational in the Middle Kingdom, but if it was, the location of the Hawwāra pyramid in front of two important water works is most conspicuous.

We have already noted evidence for an Eighteenth Dynasty official and others dated to the Old Kingdom who may have been in charge of the al-Mala'a reservoir, but there is more compelling evidence that the reservoir may have been used in the Middle Kingdom. Here one preliminary remark must be made. Throughout this article, repeated mention has been made of the al-Mala'a reservoir, but it has not yet been pointed out that at its southwestern end, the al-Gharaq channel constitutes the connection to the more westerly al-Gharaq depression, which, if sufficient water was available, could also be flooded. It is around the southern banks of the al-Mala'a reservoir and the Gharaq depression that two major settlements evolved in the Ptolemaic period: Tebtynis and Madīnat Mādī (see fig. 9). However, the latter settlement certainly goes back to the Middle Kingdom, witness a temple built there under Amenemhat III and Amenemhat IV. Interestingly this temple is dedicated to Renenutet, an agricultural goddess, and Sobek, a deity linked to the Nile flood.[178] Somewhat further east, between Tebtunis and Madīnat Mādī, lies the extensive Middle Kingdom cemetery of Kawm al-Khalwā, interestingly on a promontory overlooking both the Mala'a reservoir and the Gharaq depression. The imposing tombs here resemble those of nomarchs further south in Upper Egypt.[179] The location of all the monuments discussed thus far lies so conspicuously close to places that played a crucial role in the Ptolemaic and Roman irrigation regime of the Fayyūm that it is hard to escape the impression that this infrastructure had Middle Kingdom precursors at exactly the same spots. That this should be so is not surprising, considering the constraints imposed by the natural landscape.

The most important indication as regards the functioning of the hydrological system is, however, provided by the toponymy of settlements in the region.

Here, pride of place goes to the village of al-Lahūn. It is well known that this name goes back to *R-ḥn.t*, a toponym that is attested from the Middle Kingdom onwards.[180] Its meaning has been intensively studied. Since *ḥn.t* indicates

178 This point is duly remarked by Tallet, 2005, p. 105, who does not, however, comment on the geographical proximity of the reservoirs.

179 Bresciani, 1997–1998, p. 9–48. The tombs are briefly discussed also by Tallet, 2005, p. 105-107.

180 Erman/Grapow, 1929, p. 398,3.

a waterway, lake, or similar, there is general agreement that it conveys some such meaning as "the opening/beginning of the *ḥn.t*-water". Early researchers like Brugsch assumed that *ḥn.t* means "canal", a reading accepted by Erman and Grapow.[181] In a detailed reassessment of all the then available evidence, Gardiner and Bell dismissed this interpretation, arguing instead that a *ḥn.t* is a lake, and that, in the Fayyūm, it could only designate the "Lake of Moeris", a designation they assumed refers to Lake Fayyūm.[182] However, this reading of the evidence seems to be far too restricted. In the Book of the Fayyūm, the entrance area near al-Lahūn is explicitly called *ḥn.t*. The same word is also used for watery areas further west. Beinlich has argued that the word designates the entire humid area ("Feuchtgebiet") in the Fayyūm, but not the lake itself (which would be named *š*). He seems to restrict this interpretation somewhat to the more moor-like parts of the hydrology, excluding the main waterway, the Baḥr Yūsif, which according to him is designated in the text as *Mr-wr* "Great Waterway".[183] However, none of the passages cited by Beinlich rule out that the Baḥr Yūsif forms part of the *ḥn.t*. In fact, the eastern part of the *ḥn.t* as depicted in the Book of the Fayyūm includes a large channel which flows past an area called "Pyramid land", *i.e.* Hawwāra.[184] This general area is called *ḥn.t n.t Mr wr* "the flood area of the Great Waterway".[185] In a detailed analysis of the available Demotic and Greek papyrological (and earlier) evidence, Vandorpe has moreover argued that, in the Fayyūm, this element of the topography is only referred to in the singular, and that these texts locate settlements either to the north or to the south of it. Since these settlements lie on opposite sides of the Baḥr Yūsif within the Fayyūm, she argues that the *ḥn.t* must correspond to the Baḥr Yūsif, and that it therefore means "canal".[186] This argumentation seems entirely convincing to me, except in one regard: a "canal" is, as Gardiner already pointed out in the article cited above, a man-made waterway, which the Baḥr Yūsif certainly is not. The translation "channel" seems better.

181 Id., p. 105, 1-5

182 Gardiner/Bell, 1943, p. 37-50; followed by Černý, 1976, p. 346. According to Garbrecht and Jaritz, Lake Moeris does not designate Lake Fayyūm, but the al-Mala'a reservoir. Gardiner and Bell were clearly not aware of the existence this reservoir, and did not consider it in their interpretation.

183 Beinlich, 1991, p. 289-293.

184 For the interpretation of this area, see Beinlich, 1991, p. 79.

185 Id., p. 138 (line 1). For this interpretation of the term *ḥn.t*, cf. Yoyotte, 1987-1988, p. 146-147.

186 Vandorpe, 2004, p. 61-78.

In view of all of this, the toponym *R-ḥn.t* "al-Lahūn", i.e. "mouth of the channel", must refer to a settlement at the head of the part of the Baḥr Yūsif leading into the Fayyūm. Its location is on the Jisr al-Shaykh Jād-Allah, and opposite Hawwārat al-ʿAdlān on the Jisr al-Bahlawān. The Baḥr Yūsif and the sluice systems controlling its access into the Fayyūm lie right between the two settlements. The "mouth" referred to in the toponym is undoubtedly the opening between the two dams (or precursors of those). There is accordingly no room for doubt that the name al-Lahūn refers to this constellation of waterworks. And since *R-ḥn.t* is attested since the Middle Kingdom, the waterworks must go back at least as far as that.[187]

This impression is strengthened by the location, date, and toponymy of other pharaonic settlements in the region. Several rather important ones cluster in the area of the Fayyūm entrance. We have already mentioned Hawwārat al-ʿAdlān, at the northeastern end of the Jisr al-Bahlawān. Another important settlement is the site of Ghurāb at the southern end of the same dyke, where a royal palace was located in the New Kingdom. According to early results of recent corings undertaken here, a subsidiary branch of the Baḥr Yūsif may have passed Ghurāb in the New Kingdom.[188] And since the name of al-Lahūn only makes sense under the assumption that a precursor of the Jisr al-Bahlawān existed already before the New Kingdom, it stands to reason to assume there was a second sluice system in this dam near Ghurāb.

The etymology of the name Hawwāra, a toponym attested twice in the region here being discussed, is likewise of importance. It has been shown that this name goes back to Egyptian *Ḥw.t-wr.t*, a designation of ancient date that was since the Old Kingdom used to describe a legal institution.[189] It seems striking that one settlement with this name lies immediately opposite al-Lahūn, beside the sluice system through which the Baḥr Yūsif entered the Fayyūm, while the second, Hawwārat al-Maqṭā, lies beside the Roman sluices controlling the influx of water into the Bats drain.

---

187 HASSAN/TASSIE, 2006, p. 40 state that the jisr al-Shaykh Jād-Allāh "dates to the Middle Kingdom" (an interpretation accepted elsewhere by RÖMER, 2016, p. 177. However, no evidence to this effect is indicated in the publication.

188 BUNBURY, 2012, p. 52-54.

189 For the identification, see GAUTHIER, 1927, p. 59; GARDINER/BELL, 1943, p. 43. HABACHI, 1977, has claimed that the name of Hawwāra rather has an Arabic origin, as "a designation of isolated places". However, the sources discussed by Gardiner, which were apparently unknown to Habachi, leave no room for doubt that the proposed etymology is correct.

Taking a look at the plan of fig. 8, it is clear that all the settlements discussed are strategically located on the dyke systems A-B and D-E. This figure also indicates the former presence of a dyke B-C, which ran from the village of al-Lahūn to the Jabal Abū Ṣīr. Of this dyke, nothing now remains, but as explained above it must have played an important role in managing the distribution of water in the Nile Valley. It has not been remarked before that, precisely where dyke B-C reached the Jabal Abū Ṣīr, there is another prominent site: the vast Middle Kingdom cemetery of al-Ḥarāja.[190] It is usually assumed that this cemetery belongs to the Middle Kingdom site of Kahun. This interpretation is, however, not convincing, as Kahun lies on the desert edge, and it cannot have been difficult to find suitable cemetery ground here. The possibility that there may have been an important settlement close to al-Ḥarāja should not be discarded too lightly, therefore. However, even if the cemetery of Kahun town really lay at al-Ḥarāja, the fact that the town was separated from it by a depression where the water speed must have been rather high during the flood, suggests that there must have been a land route between the two. Perhaps, dyke B-C dates back to as early in time as this.

Taken together, indications that the dyke system in the Fayyūm may go back to as early as the Middle Kingdom are certainly not restricted to the remarkable location of two pyramids, but to a whole array of indications for human activity at rather conspicuous places. The name of al-Lahūn, attested since the Middle Kingdom, in our view leaves no room for doubt that precursors of the Roman dyke system here must date back to at least the Twelfth Dynasty. Also, the constellation of important settlements and cemeteries only makes sense if dykes A-E in the entrance area of the Fayyūm already existed then.

If dykes of sufficient magnitude to control the flood could be built with the means available in the Twelfth Dynasty, then it should be no cause for surprise that a second dyke system was built between Itsā and Shidmū, and this in fact explains the location of a settlement at Madīnat Mādī and of a major cemetery at Kawm al-Khalwā.

In view of the major efforts that must have been made in this region, it cannot fail to surprise that the archaeological record for irrigation works in the Fayyūm immediately stops again after the Twelfth Dynasty. In fact, archaeological evidence is in general extremely poor then. It is impossible at the present juncture to explain this satisfactorily. However, there is an ongoing debate over how the extremely high flood records at Semna should be interpreted. These

190 ENGELBACH, 1923.

records only start under Amenemhat III, continuing through the first decades of the Thirteenth Dynasty.[191] If these records really reflect catastrophic floods, this must have had a great impact on the freshly built dams at the Fayyūm entrance. One might speculate that the damage caused was of such magnitude that further attempts to bring the Fayyūm under cultivation were forestalled for many hundreds of years.

6. Summary and Conclusions

The aim of this article was to show how historical maps can be used to reconstruct the preindustrial Egyptian irrigation landscape in the area between Dairūṭ al-Sharīf and the Fayyūm depression. It made use primarily of the maps of the *Description de l'Egypte*, produced in 1798-1799, which provide detailed insight in the hydrology of this region before the landscape was fundamentally changed by the large scale modernisations of the irrigation system implemented in the nineteenth century. These maps were georeferenced with maps of the early twentieth century which provide information on natural relief. In this way, a very detailed reconstruction was proposed of the eighteenth century hydrology in the region between Dayrūṭ al-Sharīf and an area to the north of al-Ashmūnayn. After a detailed critique of the (sometimes erroneous or unclear information provided by the) ancient maps, an analysis of the irrigation basins was presented. The reconstruction revealed a highly complex system, with basins of rather irregular shapes. Some of these functioned as units in their own right, while others constituted chains of basins, some of which were fed directly from the Nile, and yet others received their water from the Baḥr Yūsif or other Nile branches. An aspect not hitherto paid attention to in Egyptological studies of the irrigation landscape is that, owing to the form of the floodplain, only a relatively small part of the floodwater could drain back into the Nile after the flood season, because it remained trapped between the levees of the Nile and the Baḥr Yūsif.

North of al-Ashmūnayn, and south of the entrance to the Fayyūm, the *Description de l'Egypte* shows a far more regular system of dykes. However, recent remote sensing investigations have shown that prior to this, an equally irregular system of waterways existed there as the one still encountered around al-Ashmūnayn in the Napoleonic age. This suggests that the system in the latter area represents a less developed irrigation system than the one recorded further

191 SEIDLMAYER, 2001, p. 73-80.

north. The de Vaugondy map, published in 1753, but based on information collected around 1720, and the d'Anville map published in 1765, show that, at that time, the low area between the Nile and the Baḥr Yūsif was in fact a wide, humid area, through which residual flood water flowed north to the mouth of the Fayyūm depression.

The existence of this important hydrological system, combined with natural factors, must have led to large accumulations of water in front of the Fayyūm entrance long after the end of the flood season. Under natural conditions, part of this water could drain off further north to the Nile Delta, while another part of it would be transported by the Baḥr Yūsif into the Fayyūm depression. The vast dyke systems at the entrance of this depression, near the town of al-Lahūn, must have served to monitor the amount of water being allowed into the Fayyūm. Internal dyke and sluice systems within the Fayyūm moreover served to store water for later distribution after the flood. For this, the al-Mala'a reservoir existed, southeast of Madīnat al-Fayyūm. We assume that it could be replenished with water until long after the flood season, and that, accordingly, the al-Mala'a basin may have provided a far larger area within the Fayyūm with water than was hitherto thought.

Particularly for the al-Fayyūm area, there is well-documented evidence showing that the dyke systems discussed must have been functional already in Medieval, Byzantine, and Roman times. The hydrology of Middle Egypt as elaborated in this article shows, however, that the water trapped in the depression between the Nile and the Baḥr Yūsif must already at that early date have led to a substantially higher water supply to the al-Fayyūm region than was hitherto suspected, and this very fact may explain the scale of the waterworks in the Fayyūm region. The amount of the water reaching this area was, however, dependent on the hydrology of Middle Egypt.

Since these conditions are mainly contingent on natural conditions, it stands to reason that in Antiquity, the hydrological system must have been similar in some essential regards to that of the late eighteenth century. For this reason, an investigation was undertaken to determine how old this system was. Historical toponymy suggests that many of the settlements recorded by the *Description de l'Egypte* have names that can be traced to villages and towns known from the Graeco-Roman period. In several cases, place names are demonstrably much earlier, and their etymology occasionally relates them to hydrological conditions that resemble those highlighted by the maps of the *Description de l'Egypte* quite closely. Moreover, the scant geomorphological evidence available, as well as the spread of archaeological sites (mostly in the Fayyūm) and textual data, provide strong evidence that some of the essential elements of this system may have roots as early as the Old or Middle Kingdoms.

New hypotheses developed in this study concern 1) the longevity of pharaonic hydrological patterns, many of which may have survived into the Middle Ages and the Napoleonic era; 2) the existence of a vast wet zone between the Nile and the Baḥr Yūsif south of the Fayyūm, which must have had an enormous impact on the functioning of the Fayyūm and on the necessity to build large scale water works here from an early date. 3) This moreover has implications for calculations of the carrying capacity of land for sustaining a population. In current estimates, the wet zone, which must have covered a vast area of land, is never taken into consideration. Since it is unlikely to have been suitable for the cultivation of cereals as much as for habitation, estimates of population size may have to be lowered significantly. In fact, even in the eighteenth century, the wet zone seems to have been only thinly inhabited.

Acknowledgements

The research here reported upon was made possible by a Johannes-Gutenberg research fellowship at the Johannes-Gutenberg-Universität Mainz and by several project grants: the APLADYN project funded by the STEREO II-program of the Belgian Science Policy-project SR/00/132, the project Spatial analysis of the cemeteries of Dayr al-Barshā (Middle Egypt). A Study of the cultural and environmental context of an archaeological site (FWO project G.0277.06), Settlement patterns in the greater Dayr al-Barshā region. An archaeological investigation of population dynamics in the northeastern Hare nome (FWO project G.0A94.14), and From Khemenu to Tjerty. Towards a physical and social geography of a region in pharaonic Middle Egypt (KU Leuven project OT/13/042). The authors also express their gratitude to Wouter Claes (Royal Museums of Arts and History), for making available scans of the 1926 Survey of Egypt maps, and to Georgia Long for finalizing the drawing of fig. 10.

Bibliography

ALLEAUME, GUISLAINE, Les systèmes hydrauliques de l'Égypte prémoderne. Essay d'histoire du paysage,' in: Itinéraires d'Égypte. Mélanges offerts au père Maurice Martin s.j., ed. by CHRISTIAN DÉCOBERT (Bibliothèque d'Étude 107), Le Caire 1992, p. 301-322.

ANTES, JOHN, Observations on the Manners and Customs of the Egyptians, the Overflowing of the Nile and its Effects; with Remarks on the Plague, and Other Subjects, Written during a Residence of Twelve Years in Cairo and its Vicinity, London, 1800.

ANTHES, RUDOLF, Die Felseninschriften von Hatnub, Leipzig 1928.

ANTOINE, JEAN-CHRISTOPHE, The Wilbour Papyrus Revisited: the Land and its Localisation. An Analysis of the Places of Measurement, in: Studien zur altägyptischen Kultur 40 (2011), p. 9-27.

ID., Modelling the Nile Agricultural Floodplain in Middle Egypt. from the Wilbour Papyrus and Tenth Century B.C. Landregisters, in: JAN-M. DAHMS, HARCO WILLEMS (eds.), The Nile. Natural and Cultural Landscape. Proceedings of the International Symposium held at the Johannes-Gutenberg Universität Mainz, 22 – 23 February 2013, Bielefeld 2016.

ARROWSMITH, AARON, New and Elegant General Atlas, Comprising all the New Discoveries, to the Present Time. Containing Sixty-three Maps. Intended to Accompany the New Improved Edition of Morse's Geography, but Equally Well Calculated to Be Used with his Gazetteer, or any Other Geographical Work, Boston 1812.

BALL, JOHN, Contributions to the Geography of Egypt, Cairo 1938.

BEINLICH, HORST, Das Buch vom Fayum. Zum religiösen Eigenverständnis einer ägyptischen Landschaft (Ägyptologische Abhandlungen 51), Wiesbaden 1991.

BIETAK, MANFRED, Tell el-Dabʿa II. Der Fundort im Rahmen einer archäologisch-geographischen Untersuchung über das ägyptische Ostdelta (ÖAW Denkschriften IV), Wien 1975.

BONNEAU, DANIELLE, Le régime administratif de l'eau du Nil dans l'Égypte grecque, gréco-romaine et byzantine (Probleme der Ägyptologie 8), Leiden 1993.

BORSCH, STUART J., Environment and Population. The Collapse of Large Irrigation Systems Reconsidered, in: Comparative Studies in Society and History 46 (2004), p. 458-463.

BOWMAN, ALAN K., ROGAN, EUGEN (eds.), Agriculture in Egypt from Pharaonic to Modern Times (Proceedings of the British Academy 96), Oxford 1999.

BRESCIANI, EDDA, Khelua, l'indagine e le scoperte', in: Egitto e Vicino Oriente 20–21 (1997–1998), p. 9–48.

BRIDGE, JOHN, Rivers and Floodplains. Forms, Processes, and Sedimentary Record, Oxford 2003.

BROWN, ANTHONY G., Alluvial Geoarchaeology. Floodplain Archaeology and Environmental Change, Cambridge 1997.

BUNBURY, JUDITH, Geoarchaeology, in: IAN SHAW, The Gurob Harem Palace Project, Spring 2012, in: Journal of Egyptian Archaeology 98 (2012), p. 52-54.

BUNBURY, JUDITH/MALOUTA, MYRTO, The Geology and Papyrology of Hermopolis and Antinoopolis, in: eTopoi 3 (2012), p. 119-122.

BUTZER, KARL, Early Hydraulic Civilization in Egypt. A Study of Cultural Ecology, Chicago/London 1976.

ČERNÝ, JAROSLAV, Coptic Etymological Dictionary, Cambridge 1976.

COOKSON-HILLS, CLAIRE .J., Engineering the Nile. Irrigation and the British Empire in Egypt, 1882-1914, PhD Kingston 2013.

CRUM, WALTER EWING, A Coptic Dictionary, Oxford 1939.

ĆWIEK, ANDRZEJ, Fayum in the Old Kingdom, in: Göttinger Miszellen 160 (1997), p. 17-22.

DALY, MARTIN W., The British Occupation; 1882-1922, in: DALY, MARTIN W. (ed.), The Cambridge History of Egypt II, Cambridge 1998, p. 239-251.

VAN DINTER, MARIEKE/VAN ZIJVERDEN, WILKO K., Settlement and Land Use on Crevasse Splay Deposits; Geoarchaeological Research in the Rhine-Meuse Delta, the Netherlands, in: Netherlands Journal of Geosciences – Geologie en Mijnbouw 89, No. 1 (2010), p. 19-32.

DREW-BEAR, M., Le nome hermopolite. Toponymes et sites (American Studies in Papyrology 21), Ann Arbor 1979.

EGBERTS, ARNO, In Quest of Meaning. A Study of the Ancient Egyptian Rites of Consecrating the Meret-Chests and Driving the Calves I (Egyptologische Uitgaven VIII,1), Leiden 1995.

ENGELBACH, REGINALD, Harageh (British School of Archaeology in Egypt 28), London 1923.

ERMAN, ADOLF/ GRAPOW, HERRMANN, Wörterbuch der ägyptischen Sprache III-V, Berlin, 1929–1931.

EYRE, CHRIS, The Water Regime of Orchards and Plantations in Pharaonic Egypt, in: Journal of Egyptian Archaeology 80 (1994), p. 57-80.

FERGUSON, NIALL, Empire. How Britain Made the Modern World, London 2003.

GARBRECHT, GÜNTHER/JARITZ, HORST, Untersuchung antiker Anlagen zur Wasserspeicherung im Fayum/Ägypten (Leichtweiss-Institut für Wasserbau der Technischen Universität Braunschweig. Forschungsvorhaben Ga 183/28-1), Braunschweig/Kairo 1990.

ID., Neue Ergebnisse zu altägyptischen Wasserbauten im Fayum, in: Antike Welt 23 (1992), p. 238-254.

GARDINER, ALAN HENDERSON, Ancient Egyptian Onomastica, 3 vols, Oxford 1947.

ID., Ramesside Administrative Documents, Oxford 1948.

GARDINER, ALAN HENDERSON/BELL, HAROLD IDRIS, The Name of Lake Moeris, in: Journal of Egyptian Archeology 29 (1943), p. 37-50.

GAUTHIER, HENRI, Dictionnaire des noms géographiques contenus dans les textes hiéroglyphiques IV, Le Caire 1927.

GOMAÀ, FAROUK, Ägyptische Siedlungen nach Texten des Alten Reiches (TAVO B 19), Wiesbaden 1978.

GOMAÀ, FAROUK/MÜLLER-WOLLERMANN, RENATE/SCHENKEL, WOLFGANG, Mittelägypten zwischen Samalūṭ und dem Gabal Abū Ṣīr. Beiträge zur historischen Topographie der pharaonischen Zeit (TAVO B 69), Wiesbaden 1991.

GRIESHABER, FRANK, Lexikographie einer Landschaft. Beiträge zur historischen Topographie Oberägyptens zwischen Theben und Gabal as-Silsila anhand demotischer und griechischer Quellen (GOF IV,45), Wiesbaden 2004.

GRIMAL, NICOLAS, Histoire de l'Égypte ancienne, no place 1988.

HABACHI, LABIB, s.v. Hawara, Lexikon der Ägyptologie II, Wiesbaden 1977, col. 1073.

HASSAN, FEKRI, The Dynamics of a Riverine Civilization: a Geoarchaeological Perspective on the Nile Valley, in: World Archaeology 29 (1997), p. 51-74.

HASSAN, FEKRI/TASSIE, GEOFF, Modelling Environmental and Settlement Change in the Fayum, in: Egyptian Archaeology 29 (2006), p. 37-40.

HELCK, WOLFGANG, Die altägyptischen Gaue (TAVO B 5), Wiesbaden 1974.

ID., Urkunden der 18. Dynastie, Heft 18, Berlin 1956.

HURST, HAROLD EDWIN, The Nile. A General Account of the River and the Utilization of its Waters, 2nd ed., London 1957.

JACQUET-GORDON, HELEN K., Les noms des domaines funéraires sous l'Ancien Empire Égyptien (Bibliothèque d'étude 34), Le Caire, 1962.

JEFFREYS, DAVID/TAVARES, ANA, The Historic Landscape of Early Dynastic Memphis, in: Mitteilugen des Deutschen Archäologischen Instituts Abt. Kairo 50 (1994), p. 143-173.

JOMARD, EDME FRANÇOIS, Description de l'Égypte ou recueil des observations et des recherches qui ont été faites en Égypte pendant l'expédition de l'armée française, Paris 1821.

KEES, HERMANN, Das alte Ägypten. Eine kleine Landeskunde, Berlin 1955.

KEMP, BARRY, Settlement and Landscape in the Amarna Area in the Late Roman Period, in: JANE FAIERS, Late Roman Pottery at Amarna and Related Studies (EES Excavation Memoir 72), London 2005, p. 11-56.

KESSLER, DIETER, Historische Topographie der Region zwischen Mallawi und Samalut (Beihefte TAVO B, 30), Wiesbaden 1981.

LECLANT, JEAN, Fouilles et travaux en Egypte et au Soudan, in: Orientalia 40 (1971), p. 224-266.

ID., Fouilles et travaux en Egypte et au Soudan, in: Orientalia 40 (1971), p. 393-440.

LEHNER, MARK, The Fractal House of Pharaoh. Ancient Egypt as a Complex Adaptive System, a Trial Formulation, in: KOHLER, TIMOTHY A./GEORGE J. GUMERMAN (eds.), Dynamics in Human and Primate Societies. Agent-based Modeling of Social and Spatial Processes, New York/Oxford 2000, p. 275-353.

LORAND, DAVID, À la recherche d'Itj-Taouy/el-Licht. À propos des descriptions et cartes du site au XIXe siècle, in: Talking along the Nile. Ippolito Rosellini, Travellers and Scholars of the 19th Century in Egypt, ed. by MARILINA BETRÒ/GIANLUCA MINIACI, Pisa 2013, p. 137-150.

LUTFI AL-SAYYID MARSOT, A., Egypt in the Reign of Muhammad Ali, Cambridge 1984.

LYONS, HENRY GEORGE, The Physiography of the River Nile and Its Basin,Cairo 1906.

MASPERO, JEAN, Fouilles exécutées à Baouît (Mémoires publiées par les membres de l'Institut Français d'Archéologie Orientale du Caire 59), Le Caire 1931.

MENU, BERNADETTE (ed.), Les problèmes institutionnels de l'eau en Égypte ancienne et dans l'antiquité méditerranéenne. Colloque AIDEA Vogüé 1992 (Bibliothèeque d'étude 110), Le Caire 1994.

MICHEL, NICOLAS, Travaux aux digues dans la vallée du Nil aux époques papyrologique et ottoman. Une comparaison, in: CRIPEL 25 (2005), p. 253-276.

MIKHAIL, ALAN, An Irrigated Empire. The View from Ottoman Fayyum, in: International Journal of Middle East Studies 42 (2010), p. 569-590.

VAN MINNEN, PETER, Une nouvelle liste de toponymes du nome Hermopolite, in: Zeitschrift für Papyrologie und Epigraphik 101 (1994), p. 83-86.

MOELLER, NADINE, The First Intermediate Period. A Time of Climate Change?, in: Ägypten & Levante 15 (2005), p. 153-167.

MOHAMED, IHAB N.L., Evolution of the South-Rayan Dune-Field (Central Egypt) and its Interaction with the Nile Fluvial System, Leuven 2012.

PEUST, CARSTEN, Die Toponyme vorarabischen Ursprungs im modernen Ägypten (GM Beihefte 8), Göttingen 2010.

MORENO GARCIA, JUAN CARLOS, The Territorial Administration of the Kingdom in the 3rd Millennium B.C., in: Ancient Egyptian Administration, ed. by

JUAN CARLOS MORENO GARCIA (Handbuch der Orientalistik I, 104), Leiden/Boston 2013, p. 85-151.

RIAD, HENRI, Le culte d'Amenemhat III au Fayoum à l'époque ptolémaïque, in: Annales du Service des Antiquités de l'Égypte 55 (1958), p. 203-206.

RIPAUD, <LOUIS MADELEINE>, Report of the Commission of Arts to the First Consul Bonaparte, on the Antiquities of Upper Egypt, and the Present State of all the Temples, Palaces, Obelisks, Statues, Tombs, Pyramids, &c. of Philœ, Syene, Thebes, Tentyris, Latopolis, Memphis, Heliopolis, &c. &c. from the Cataracts of the Nile to Cairo, with an Accurate Description of the Pictures with which they Are Decorated, and the Conjectures that May Be Drawn from them, Respecting the Divinities to whom they Were Consecrated, London, 1800.

ROEDER, GÜNTHER, Hermopolis 1929-1930, Hildesheim 1959.

ROEMER, CORNELIA, The Nile in the Fayum – Strategies of Dominating and Using the Water Resources of the River in the Oasis in the Middle Kingdom and the Graeco-Roman Period, in: JAN-M. DAHMS/HARCO WILLEMS (eds.), The Nile. Natural Landscape – Cultural Landscape. Proceedings of the International Symposium held at the Johannes-Gutenberg Universität Mainz, 22 – 23 February 2013, Bielefeld 2016.

RUF, THIERRY, Questions sur le droit et les institutions de l'eau dans l'Égypte ancienne, in: MENU 1994, p. 281-293.

SCHENKEL, WOLFGANG, Die Bewässerungsrevolution im alten Ägypten (SDAIK 6), Mainz 1978.

SEIDLMAYER, STEPHAN J., Historische und moderne Nilstände. Untersuchungen zu den Pegelablesungen des Nils von der Frühzeit bis in die Gegenwart (Achet A1), Berlin 2001.

SETHE, KURT, Urkunden des Alten Reichs, Leipzig 1933.

SHAFEI, ALI, Lake Moeris and Lahûn Mi-Wer and Ro-Hûn. The Great Nile Control Project Executed by the Ancient Egyptians, in: Bulletin de la Société de Géographie d'Égypte 33 (1957), p. 187-217.

STRATHERN, PAUL, Napoleon in Egypt, New York 2007.

SUBIAS, EVA/FIZ, IGNACIO/CUESTA, R., The Middle Nile Valley: Elements in an Approach to the Structuring of the Landscape from the Greco-Roman Era to the Nineteenth Century, in: Quaternary International 312 (2013), p. 27-44.

TALLET, PIERRE, Sésostris III et la fin de la XIIe dynastie, no place 2005.

TALLET, PIERRE/MAROUARD, GREGORY/LAISNEY, DAMIEN, Un port de la IVe dynastie au Ouadi el-Jarf (Mer Rouge), in: Bulletin de l'Institut Français d'Archéologie Orientale du Caire 112 (2012), p. 399-446.

TIMM, STEFAN, Das christlich-koptische Ägypten in arabischer Zeit I-II (TAVO B41/1-2), Wiesbaden 1984.

ID., Das christlich-koptische Ägypten in arabischer Zeit III (TAVO B41/3), Wiesbaden 1985.

ID., Das christlich-koptische Ägypten in arabischer Zeit IV (TAVO B41/4), Wiesbaden 1988.

ID., Das christlich-koptische Ägypten in arabischer Zeit V (TAVO B41/5), Wiesbaden 1991.

ID., Das christlich-koptische Ägypten in arabischer Zeit VI (TAVO B41/6), Wiesbaden 1992.

VANDORPE, KATELIJN, The Henet of Moeris and the Ancient Administrative Division of the Fayum in Two Parts, in: Archiv für Papyrusforschung 50 (2004), p. 61-78.

VERSTRAETEN, GERT/MOHAMED, IHAB/NOTEBAERT, BASTIAAN/ WILLEMS, HARCO The Dynamic Nature of the Transition from the Nile Floodplain to the Desert in Central Egypt since the Mid-Holocene, in: JAN-M. DAHMS/HARCO WILLEMS (eds.), The Nile. Natural and Cultural Landscape. Proceedings of the International Symposium held at the Johannes-Gutenberg Universität Mainz, 22 – 23 February 2013, Bielefeld 2016.

VYCICHL, WERNER, Dictionnaire étymologique de la langue copte, Leuven 1983.

WARD, STEVE, s.v. Yazoo, in: ANDREW S. GOUDIE (ed.), Encyclopedia of Geomorphology, London 2004, p. 1121.

WESTENDORF, WOLFHART, Koptisches Handwörterbuch, Heidelberg 1965-1977.

WHITTOW, JOHN, The Penguin Dictionary of Physical Geography, 2nd edition, London 2000.

WILLCOCKS, WILLIAM, Egyptian Irrigation, London 1889.

ID., Egyptian Irrigation, 2nd ed., London 1899.

WILLCOCKS, WILLIAM/ CRAIG, J.I., Egyptian Irrigation, 3rd ed., London/New York 1913.

WILLEMS, HARCO, The Physical and Cultic Landscape of the Northern Nile Delta according to Pyramid Texts Utterance 625, in: "Parcourir l'éternité." Hommages à Jean Yoyotte II, ed. by CHRISTIANE ZIVIE-COCHE/IVAN GUERMEUR (ed.), (Bibliothèque de l'École des Hautes Études. Sciences religieuses 156), Turnhout 2012, p. 1097-1107.

ID., Nomarchs and Local Potentates: the Provincial Administration in the Middle Kingdom, in: Ancient Egyptian Administration, ed. by JUAN CARLOS MORENO GARCIA (Handbuch der Orientalistik Section 1, Ancient Near East 104), Leiden/Boston, 2013, p. 341-392.

Id., Historical and Archaeological Aspects of Egyptian Funerary Culture. Religious Ideas and Ritual Practice in Middle kingdom Elite Cemeteries (CHANE 73), Leiden/Boston 2014.

Wilson, John A., The Burden of Egypt, Chicago 1951.

Wittfogel, Karl, Oriental Despotism. A Comparative Study of Total Power, New Haven 1957.

Yoyotte, Jean, Études géographiques II. Les localités méridionales de la région memphite et le "Pehou d'Héracléopolis", in: Revue d'Égyptologie 14 (1962), p. 75-111.

Id., Hérodote et le "Livre du Fayoum." La crue du Nil recyclée, in: Revue de la Société Ernest Renan N.S. 37 (1987-1988), p.53-66, republished in Ivan Guermeur (ed.), Histoire, géographie et religion de l'Égypte ancienne. Opera selecta (Orientalia Lovaniensia Analecta 224), Leuven/Paris/Walpole 2013, p. 135-148.

Zibelius, Karola, Ägyptische Siedlungen nach Texten des Alten Reiches (TAVO B 19), Wiesbaden 1978.

Landscapes of the Bashmur

Settlements and Monasteries in the Northern Egyptian Delta from the Seventh to the Ninth Century

PENELOPE WILSON

1. Introduction

The protective Delta marshes are a powerful motif of Egyptian religious and historical stories. From Isis attempting to keep Horus safe from Seth, Psamtek hiding from the Assyrians, the *boukoloi* of the Mendesian nome[1] and the Bashmurites using the trackless marshes from which to resist the Arab tax collectors,[2] the marshes of the Delta have traditionally offered a safe haven from persecution of different kinds. After the Bashmurite insurgency of 831 A.D., settlements and churches in the area were destroyed to remove the possibility of further revolts. Little is known, however, of the extent or nature of settlement in this area from the seventh to the ninth century or of the way of life in the remote areas at the apex of the Delta. Recent survey work, however, has begun to record information about the ancient sites which lie north of, inside and along

1 BLOUIN, 2014, p. 285–295.

2 EVETTS, 1906–1915, III, p. 157, p. 487–494; KENNEDY, 1998, p. 83–84; MIKHAIL, 2014, p. 118–127.

the southern fringe of Lake Burullus.[3] This paper describes some of those sites[4] in order to put the archaeological material obtained so far into the wider context of the landscapes of the north Delta and the political and social upheavals of the two centuries after the Arab invasion of Egypt. This preliminary discussion has the objective of demonstrating the potential of the archaeology of the area to explore the relationships between the towns and the environment, the settled and the wilderness and, despite a lack of papyrological or textual evidence, to illuminate the material culture and history of the north Delta.

2. The Geographical Setting

Heliodorus' novel *Aithiopika* from the fourth century A.D. evokes a Roman impression of the north of Egypt: "What shores are to seas, swamps are to lakes" (Book I, 5),[5] also the abode of brigands and pirates. This literary motif evokes the unknown, the strange and uncivilized world in the northern extremes of Egypt, and paints a desolate picture of the natural environment of the Nile mouths as they reached the Mediterranean Sea. The Nile Delta apex, however, has been a changing and volatile environment in geological terms. The northern Delta in the first millennium A.D. was characterized by a band of beach and coastal dune to the very north, brackish lagoons and marsh and wetland to the south before the floodplain and levee landscape of the central and southern Delta.[6] In the Roman period, from 30 B.C. to the fourth century A.D., the sea levels of the Eastern

3 The work of the Egypt Exploration Society's Delta Survey has been conducted in the Kafr el-Sheikh province by Jeffrey Spencer and Patricia Spencer in the 1990s and Penelope Wilson since 2000. EES Delta Survey website: http://www.delta-survey.ees.ac.uk/ and Delta Survey work http://www.dur.ac.uk/penelope.wilson/Delta/Survey.html.

4 I would like to thank Harco Willems and Jan-Michael Dahms for their invitation to take part in the conference and their patience. I also acknowledge the support of Jeffrey Spencer and the Egypt Exploration Society's Delta Survey project, Roger Dickinson for assistance with the map and Dimitrios Grigoropoulos, Rebecca Bradshaw, Fatma Keshk and Hind Ramadan, as well as the Ministry of State for Antiquities in Egypt and Kafr el-Sheikh office under Gamal Selim and Dr Mohamed abd el Rifaat.

5 Lamb, 1961, p. 5.

6 Stanley/Warne, 1993.

Mediterranean were around 1.5 m lower than at present.[7] Dated archaeological material on the Italian, Greek and Levantine coasts would have stood at or above the level of the sea, whereas now the sites are under or almost submerged by water.[8] As well as lower sea-levels, the lagoons of Burullus and Menzala were probably not as extensive as they were by the nineteenth century[9] and the northern belt consisted of a series of smaller lagoons or lakes, closed in by spits of land, caused by the various cuspate subdeltas. Agricultural possibilities would therefore have been good from the Ptolemaic to Roman periods.[10] With increasing subsidence behind the beach barrier, however, caused by various factors including earthquakes, the swamps and lagoons have spread since the fifth to tenth centuries.[11] The present Delta lakes were most likely created in their current form in 961 A.D. by a marine transgression.[12] In Lake Burullus, at the very north of Egypt, the islands that exist inside the Burullus lagoon now may have been standing within swampy, salty land for most of the year and a shallow freshwater lagoon at the time of the inundation. Although the waters of the lagoon are now brackish, that is a mixture of salt water and fresh water,[13] the annual inundation would have caused the levels of salinity to have changed during the year.[14] Marine and freshwater fish abounded in the lake and provided sustenance for local towns and larger cities at the end of the nineteenth century. Migratory and resident birds, saline and freshwater plants were abundant in the lagoonal areas,[15] with perhaps, in ancient times wild pigs and hippopotami. The natural resources were thus attractive, and required little husbandry. The lake was joined to the sea by an opening or *boghaze* of Arabic texts, less dangerous than the main river branch mouths, but still only navigable by vessels of shallow draught.[16]

The Sebennytic Nile branch was the major waterway flowing north from Sebennytos (Samanud), to the east of the Burullus lagoon and debouching near modern El-Borg.[17] There were other distributaries to the west, changing

7 STANLEY/WARNE, 1993, p. 632.

8 LAMBECK et al, 2010, p. 81–88; FLEMMING, 1992.

9 TOUSSOUN, 1922, pl. 1 map of Bois Aymé from the Description de l'Égypte.

10 BUTZER, 1959, p. 63.

11 SESTINI, 1992, p. 568.

12 KAMEL, 1926, Pt.4, p. 892 after Makhzumi; FRIHY, 1992.

13 RAMDANI et al, 2001, p. 291–292.

14 BIRKS/BIRKS, 2001, p. 474–476.

15 DUMONT/EL-SHABRAWY, 2007.

16 COOPER, 2009, p. 114.

17 ARBOUILLE/STANLEY, 1991, p. 60–64.

over time and noted by different classical authors and geographers. Such waterways included the Pneptimi and Dioclus "false mouths" on the northern shore[18] the "Saitic" branch, perhaps debouching near Agnou (Ikhnu) or modern Mastarua.[19] The Bolbitine branch may have only reached the coast from the late first millennium B.C., having been canalized in its lower reaches and starting from the Canopic Branch.[20] The Bolbitine-Rosetta river branch and promontory only formed from about 2000 years ago,[21] resulting in the foundation of Rashid (Rosetta) as late as 870 A.D.[22] and a shift of the balance of water in the Delta to the Rashid and Dumyat branches with the silting up of the Canopic and Sebennytic branches by around seventh to the ninth century A.D. In addition, transverse canals such as the Butic canal, as recorded by Ptolemy and others also operated[23] and may have assisted both communication and drainage.

Within this changing and diverse landscape there are many north Delta archaeological sites, whose creation, development and abandonment do not fit the overall model of Ptolemaic and early Roman "boom" followed by decline into the Late Antique period,[24] instead they are part of a recently recognized pattern of Late Antique reconfiguration, wealth and continuation or displacement into the Islamic period.[25] The ideal scenario would be to take one site at a time and understand the individual internal dynamics of particular situations,[26] but in the northern Delta, however, the situation is still at an early stage, as this paper will show.

3. Historical and Archaeological Setting

David Hogarth visited the northern province of Kafr el-Sheikh in 1895 and 1903, reporting to the Hellenic and Egypt Exploration Societies on thirty one Late Roman, Byzantine and early Arab sites in the "Nile fens".[27] He identified Kom

18 As in Ptolemy, Toussoun, 1922, p. 43–52.

19 Arbouille/Stanley, 1991, p. 60, fig. 10.

20 Toussoun, 1922, p. 27.

21 Chen et al, 1998, p. 551.

22 Timm, 1984–1992, V, p. 2198–2203.

23 Ball, 1942, p. 128–130; Blouin, 2014, p. 32–33.

24 Based on the Fayum papyri, Rostovtzeff, 1926; van Minnen, 1995.

25 Pollard, 1998; Keenan, 2005.

26 Rathbone, 1997; Hartung, 2009 (for Buto); Blouin, 2014 (for Mendes).

27 By this time the area was semi-drained, but still not densely inhabited, Hogarth, 1904; 1910, p. 91–107.

el-Khanziri as Pachnemounis from an inscription found there and discussed the identification of sites, provinces and Nile branches. Further survey work by Pascale Ballet and Thomas von der Way[28] and then by the Egypt Exploration Society has shown that the archaeological prospects for the northern Delta are good, in the sense that there are many extant sites worth scientific study. The sites have been under threat since the end of the nineteenth century, however, as many have been levelled or built over (see figure 1). Out of 153 "sites" logged by the EES Delta Survey database, that is places designated with the name "kom" or "tell"[29] in the areas of Kafr el-Sheikh, Hamul, Biyala and Sidi Salem, 29 % have been levelled, 22 % have been built upon, 28 % mostly in the Hamul/ Biyala area have not been researched, leaving only 21 % with useful scientific data.[30] Most of these 33 sites are in the area between Buto (Tell Farain) and Lake Burullus. Some of the sites are multi-period sites and of those 22 % have Ptolemaic-Roman material (third century B.C. to second century A.D.), 61 % Late Roman material (third to seventh century A.D.) and 17 % Islamic-medieval material (eighth to eleventh century A.D.). While the figures are rather crude and the data is incomplete, the information can be compared with the similarly weak amount of data from historical and textual sources, in order to show how much further work there is to do in the area but also the potential for looking at environmental and socio-economic interactions, particularly in the Late Roman and early Islamic periods.[31]

28 Ballet/von der Way, 1993.

29 The name may refer to naturally high places, such as mud hills or levees, but they are included here as places with the potential to find ancient sites of some kind. There is often too little information until a survey visit is made to a "kom" or "tell" to be certain that it was an older settlement.

30 The sites were collected from the EES database and allocated a status according to whether they had been investigated. After counting, the numbers in each category (leveled, built over, investigated, not investigated), the percentages were calculated.

31 Mikhail, 2014.

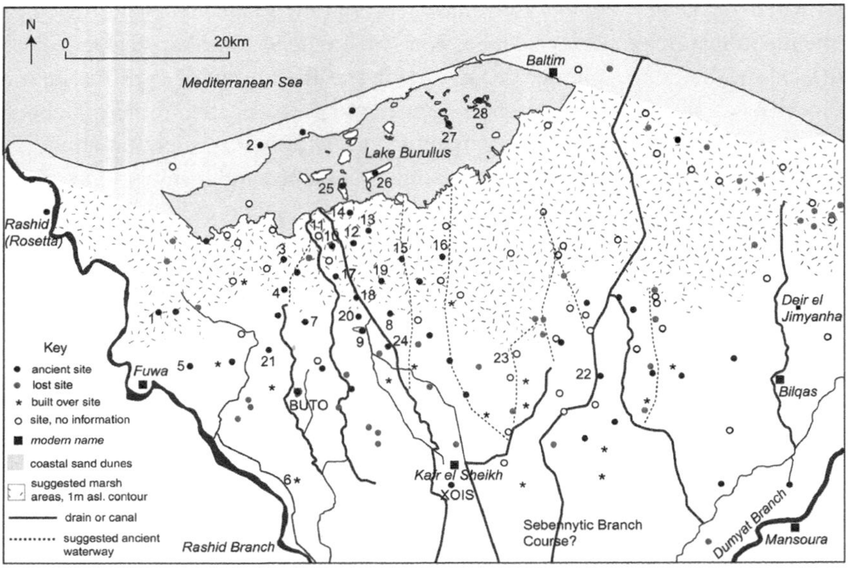

Figure 1. Map of the northern area of Egypt, with Late Antique sites and places named in the text. After Survey of Egypt maps (1997) by Roger Dickinson and Penelope Wilson.

Bashmur Map Key

1 MUTUBIS (Kom el-Ahmar)	2 Mastaraua	3 Tell el-Aluwe
4 Kom el-Arab	5 KOPRET	6 Shabas Shuhada
7 Kom Sheikh Ibrahim	8 PHRAGONIS (Kom el-Khawalid)	9 Kom Sidi Selim
10 Tell Foqaa	11 Kom el-Meiteh el-Bahri	12 Kom el-Ahmar el-Ain
13 Tell el-Khubeiza	14 Tell el-Retabi	15 Nashawein
16 PACHNAMOUNIS (Kom el-Khanziri)	17 Haddadi	18 Kom Bunduq
19 Kom Khirbeh	20 Kom el-Misk	21 Kom Abu Ismail
22 Kom el-Tawil	23 Umm el-Gafar	24 Tida, Tell el-Daba
25 Gezira Dakhla	26 Gezira Kom el-Akhdar	27 Singar
28 Mehgareh		

It is possible that an agrarian development of the north Delta began in the Ptolemaic period similar to the settlement "boom" in the west Delta and Fayum during the Ptolemaic period,[32] but there is a limited amount of evidence due, in part, to the relative lack of archaeological work in the Delta in general and a specific lack of documentary material. Where papyri have survived, as at Thmouis in the eastern Delta they have been able to provide precise details about specific periods, in this case the second century A.D.[33] Ptolemaic levels are most likely to be buried under Delta sites and sediments, but during survey work around Buto,[34] limited sondages were made in some places and pottery was found that could be dated to the Ptolemaic period, for example at Kom el-Gir. Recent work at Kom el-Gir, north-west of Buto, has confirmed the Ptolemaic origins and Roman development of this site.[35] Similarly, excavations for the Egypt Exploration Society by Jeffrey Spencer at Kom Daba found standing remains and foundations from a tower-house, with Ptolemaic pottery at or under the foundations.[36] Surface survey from the EES Delta Survey, has only identified Ptolemaic material at a limited number of sites such as Kom el-Khawalid (Phragonis),[37] Kom el-Arab and Tell Mutubis,[38] while Ptolemaic coin hoards, town-houses and industrial quarters, as well as bath-houses have been identified at the large, urban centres of Buto and Xois (Sakha).[39]

The establishment of new provinces such as Metelis, Phtenetu, Cabasa and Helearchia from the Ptolemaic and Roman periods, suggest that the administration was effectively dealing with "new" settlements and agricultural lands as they came "on-stream". The exact location of these areas is, however, not certain. The Phthenetic and Cabasite nomes are first mentioned in the *Natural History* of Pliny and then described more fully in the *Geography* of Ptolemy.[40] Buto was the capital of the former and Cabasa of the latter, being somewhere north of Buto, perhaps to form a focus for the intensification of settlement due to the increased level of agricultural exploitation and to manage the changed waterway system. The Cabasa element may be represented in the modern toponym Shabas which

32 Butzer, 1976, p. 95–96.

33 Blouin, 2014.

34 Ballet/von der Way, 1993.

35 Schiestl, 2013.

36 Spencer, 2011.

37 Hogarth, 1904, p. 4–5; Timm, 1984–1992: II, p. 940–944.

38 Wilson/Grigoropoulos, 2009, p. 253–255, 473–475 and 454–462.

39 Christiansen, 2004, p. 46–54; Ballet et al, 2011; Trümper, 2009.

40 Ball, 1942, p. 72, 85.

was used of at least four towns south of Buto: [41] Shabas el-Shuhada, Shabas el-Umayyir, Shabas el-Mehl and, perhaps, Shabas Sanhur. Emile Amélineau and John Ball favoured Shabas el-Shuhada as Cabasa,[42] because it is in the centre of this group and there is indeed a 3 m high mound underneath the modern cemetery at the site with stratified burials and Roman-Late Roman pottery.[43] Campbell Edgar[44] also noted the Shabas element in a group of sites centred on the modern Tell el-Qabrit, a group which may also be a candidate for the new capital of Cabasa. This kind of problem in identifying main cities, even administrative centres, known from textual records, with remains on the ground exemplifies the wider problem of Delta archaeology. Similarly, the north-east Delta region seems to have been known as Helearchia,[45] which Hogarth was inclined to place near Abu Madi south-east of Baltim.[46] It is possible, however, that this was the earlier name for Bashmur, which was in the Christian period, the whole region north-east of Fuwwa and north of Dirkirnis – that is the area south of Lake Burullus.[47] The town of Pisharot (Coptic) or al-Bashrud (Arabic) is known from the third to the ninth century in tax lists and may have been located somewhere near a town called Sidi Ghazi, south of large tell called Kom Umm Jafar, 15 km south of Kafr el-Sheikh.[48] It is likely that some terms – especially for areas – were used rather vaguely and that strict equivalences between the different periods cannot be made. The mapping of toponyms to known sites is difficult to say the least, with remnants of different periods reflected in the modern landscape.[49]

4. The Late Antique Sites: Monasteries and Ports

The Late Antique period from the fourth to the seventh century A.D. represents a changing political and economic focus from Rome to Constantinople to Fustat (Cairo). During this period documentary evidence shows that agricultural land

41 Timm, 1984–1992, V, p. 2218–2222.

42 Amelineau, 1893, p. 419–421; Ball, 1942, p. 109, 122, 164, 178.

43 See http://www.dur.ac.uk/Penelope.Wilson/DeltaSurvey/

44 Edgar, 1911.

45 Maspero/Wiet, 1919, p. 29–30.

46 Hogarth, 1904, p. 13.

47 Timm, 1984–1992, I, p. 354–356.

48 Daressy, 1927.

49 Engsheden, 2008.

was owned by villagers and city dwellers alike, there were large "estates" such as that of Aurelius Appianus in the Fayum, and, by the sixth century A.D. the property holdings of the church and monasteries were substantial.[50] In this period, wheat was shipped out to Constantinople and other major Christian centres in the Levant such as Antioch, when trans-shipment took place through the northern lagoonal area and ports on Burullus. An account of boats dating to the fourth century A.D. mentions the home ports of boats in the following nomes of the north delta: Metelite, Prosopite, Phtenote, Lower Diospolite, Elearchia and Nilopolite.[51] The northern waterways would thus have aided the movement of goods in different directions from those of the Roman and Ptolemaic periods. In addition, the bishoprics in the north Delta were certainly active with involvement in significant economic activities. At Parallos, in the area of Baltim, a Bishop Nonnos is known from the year 339 A.D. and his successors up to the late sixth century A.D..[52] Forty-four bishoprics are recorded in the Delta out of around one hundred in the whole of Egypt, including Libya and the Pentapolis of Cyrenaica, perhaps not too surprising in view of the relative proximity to Alexandria.[53] Similarly, a Christian elite can be identified in cities and towns, who were responsible for church building and administration amongst other activities.[54]

In the countryside, however, a small amount of archaeological evidence implies a striking community of agricultural settlements, perhaps the source of wealth for some of the elite. Late Antique Period dated material predominates on those sites where pottery survey has been carried out in the Delta,[55] but the phenomenon has also been noted in Middle Egypt around Amarna[56] and now, also, in the Fayum.[57] Although the survey collections in the Delta included material from ditches around the sites, in order to obtain pottery that may have come from deeper strata, it is possible, however, that this material could also have been spread out from the central part of the sites and covered earlier

50 Bagnall, 1993, p.148–153, p. 289–293.

51 P. Oxy. XXIV 2415. Lobel et al., 1957, p. 176–179; dating after Sijpesteijn/Worp, 2011; Bagnall, 1998, p. 37.

52 Worp, 1994, p. 304.

53 Wipszycka, 1983, p. 183–186.

54 Mikhail, 2014, p. 37–50.

55 Wilson/Grigoropoulos, 2009, p. 276–281; Rowland/Wilson, 2006; Ballet/von der Way, 1993; Trampier, 2009.

56 Kemp, 2005; Parcak, 2005.

57 Kirby/Rathbone, 1996; Keenan, 2003.

material. The cultural units used in the survey combine the first century B.C. to the second century A.D. as a distinct Early Roman unit of material culture, reflecting the joint Hellenistic culture of the eastern Mediterranean, while Roman is second to the third century A.D. and Late Roman/Late Antique is the fourth to seventh century A.D. Often, however, cultural markers such as "Christian" or "Byzantine"[58] lag behind absolute chronology of periods and tend to move at slower rates. In effect, material from the sites could have been in use and reuse over many years before deposition so the survey methods used are necessarily only crude markers – a beginning. The following description and discussion highlights future directions for investigation of the north Delta arising from four selected sites.

5. Fives Sites: Form and Function

The site of Tell Mutubis (see figure 1, no. 1) (Kom el-Ahmar)[59] has been more intensively surveyed than others in the North Delta Survey in order to understand the development of the town from its origins to a period of floruit in the Late Antique Period and abandonment around the ninth to tenth century. The site, now surrounded by agricultural land, covers an area of around 25 ha, with a central mound rising to a height of 12 m above the field level. The mound height can be compared to the original mound at Sakha (Xois), which was estimated at 80 feet (c. 24 m).[60] Tell Mutubis is strategically located east of the Bolbitine (Rosetta) branch and west of the marsh zone of Lake Burullus. The town could conceivably have controlled access across the Delta and, in the Roman period, have formed a link to the Butic canal. The pottery found at the site overwhelmingly dated to the Late Roman period[61] and very little discernible material of earlier periods could be detected, except for one possible diagnostic sherd dated to the Ptolemaic period, a fragment of *tholos*-type bath-house and some Ptolemaic coins found in the 1970s.[62] Most of the glass also dated to the

58 On the problems of nomenclature Mikhail, 2014, p. 1–4.

59 SCA register number 090175.

60 By Percy Newberry in a letter to Gardner in 1947, quoted in Reeves/Taylor, 1992, p. 105.

61 Identifications by Aude Simony, Mikaël Pesenti and Penelope Wilson, for the Mutubis Project.

62 I am grateful to Dr. Ayman Wahby for this information from Cairo Museum.

fourth century onward and a few possible examples from the third century.[63] The geological investigation of the palaeo-landscape suggests that the site was founded on land emerging after a change in the fluvial regime, perhaps no earlier than the Late Ptolemaic to Early Roman period, depending upon the date of the pottery from the earliest features.[64] The latest material suggests that the site was already in decline in the seventh century at the time of the Arab conquest and perhaps was abandoned soon after that time. The abundance of Late Roman 1 type amphora fragments, African, Egyptian and Cypriot Red Slip Ware, and Roman cooking pots suggests a domestic occupation and consumer culture, which fits within the general pattern of the Late Antique period in North Africa as a whole.[65] Buildings are visible over the tell surface and in satellite imagery and one such building with limestone tile floors, columns on brick piers and plaster mouldings suggests some kind of wealthier structure. Large monumental walls made of mud-brick with fired-brick courses at the edge of current mound may suggest a more fortified set of structures at the site.

A different type of site is exemplified by Tell el-Khubeiza (see figure 1, no. 13)[66] situated amongst modern fish-farms. The site consists of one oval mound around 8.5 m above the fish-farms, with an outlying, lower and more circular mound to the west (see figure 2). The whole site is around 660 m in length from east to west and about 285 m, at its maximum extent, from north to south, thus covering around 18 ha. The colour of the main mound was red because the surface was covered in degraded pottery and red brick, some of which had been vitrified, as well as glass and some fragments of corroded bronze material. The surface varied from being hard underfoot to soft in places, the difference perhaps representing walls and internal features of buried structures. The western side of the mound seemed to have been cut away and large amounts of red brick, including square bricks 21 by 21 cm in size, and larger pieces of pottery from cooking vessels and amphorae, including Late Roman 1 types, lay at the bottom of the embankment on this side. Nearby, two kilns inside a brick surround were clearly visible and the ground surface around them was covered in slag. The pottery from the site showed a good range of Late Roman material, including painted wares, Cypriot and Egyptian Red Slip finewares, as well as amphorae with a bright orange fabric and white-cream wash over the surface. Corroded bronze coins and glass fragments also lay on the surface, the latter

63 Information from Daniela Rosenow for the Mutubis Project.

64 Information from Benjamin Pennington for the Mutubis Project.

65 Dossey, 2010, p. 62–97.

66 Supreme Council of Antiquities register number 090120.

in large numbers in places. There were three fragments of red granite lying on the surface of the mound, including two parts of grindstones, suggesting the presence of food processing facilities of some kind at the site. Building plans were clearly visible on the mound surface. The remoteness of this mound and perhaps the marshy environment that once surrounded it may be significant in determining the nature of the settlement here.

Figure 2. View up to the top of Tell el-Khubeiza (photograph by Penelope Wilson).

A third site, Tell Nashawein,[67] (see figure 1, no. 15) was not too far to the west but was of a somewhat different character (see figure 3). The Arabic dual form of the name is confirmed by the topography of the site, as it has two main mounds, either side of an approximately 100 m wide band running from north to south through the middle of the site. The nature of the band running through the site is not clear, and the field patterns to the north and south are not very helpful in this regard, but the area seems to consist of hard, compact silt with only a few fragments of pottery lying on the surface. The flat areas may be the remnant of a waterway which ran through the area and which was "decommissioned" when the new irrigation system was completed or may have run dry of its own accord sometime in antiquity. The site at Nashawein is large, covering a maximum of around 1000 m in length by around 550 m in width and with a maximum elevation of the eastern mound of up to 10 m.

67 SCA register number 090144.

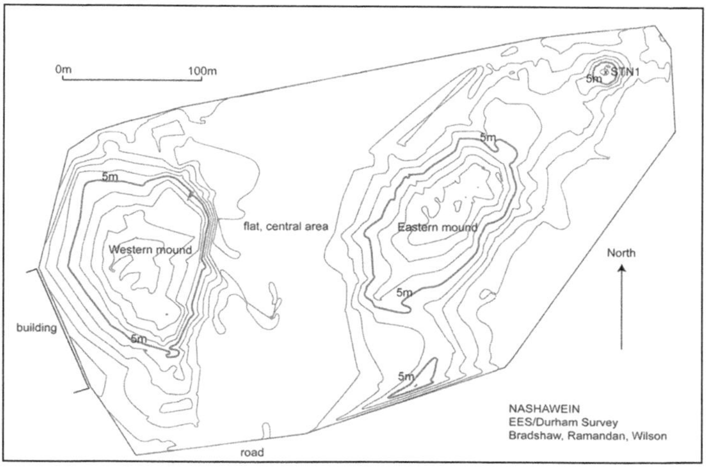

Figure 3. Plan of Nashawein (Bradshaw, Hind and Wilson see WILSON*, 2014 p. 43).*

The western mound shows building plan traces on the surface, either as pale bands running along the ground or in the growth of small plants where water has collected usually in the inner part of a room of a building. The difference between the higher, more solid "walls" of the buildings and the softer, vegetation-filled room depressions can be easily seen and felt underfoot. A large rectangular building on the southern part of this mound, running from north-west to south-east, was clearly visible to the naked eye as well as on satellite images. The main eastern mound has a small mound further to its east and the two are separated by a depression between them. The southern sides of the eastern mounds were gently sloping, with water-cut gullies running through the sides. Building plans are also visible on the mound surfaces.

The mounds were covered in pottery and fired brick, including some vitrified brick fragments. There were also pieces of glass in specific places. The pottery studied dated mostly to the Late Roman period and included examples of the Late Roman 1 type of amphorae, cooking vessels with ledged rims, pie crust dishes and some with painted decoration on the inside, as well as ledged "bitronconique" AE3 amphora spikes. They all date to within the Late Roman period and perhaps Early Arab period for the painted pie crust dishes (fourth to ninth century).

Figure 4. View of Singar island (photograph by Penelope Wilson).

Tell Singar[68] (see figure 1, no. 27) now lies within Lake Burullus itself and is only accessible by boat (see figure 4). The archaeological zone was defined as a pottery and brick covered area on the south side of the island, which was bare of scrub and bushes. By contrast, the northern part of the island had an area of marshes and reed beds, along with harder, sandier surfaces and some vegetation. As a result it is used in modern times for cattle pasture. The pottery-covered area seemed to be very low-lying and covered an area of approximately 90 m by 140 m at its maximum extent. The maximum height of the tell was barely one metre above the level of the water. As the site had previously been occupied by fishermen in the not too distant past, the remains of their dwelling places were clearly visible on the island. In many places, hollows had been dug out of the island, leaving earth piled up alongside the hollows and the pottery and material from the holes lying beside them. The pottery dated from the Late Roman period through to the Islamic period and there were some red slip finewares along with the glazed wares and glass. Although the area was not very extensive the material seemed to occur in dense strata across the site and these were around one to two metres deep in places. The edges of the site were bordered by reed beds. Meinardus noted that Singar was known from the third until the thirteenth cen-

68 SCA register number 090150.

tury, had once been important as a safe haven for relics and had been accessible on foot.[69] Other sites comparable to Singar lie within Lake Burullus itself, such as those on the islands of Kom Dakhlah, el-Akhdar and Mehgarah. They may have been connected to the marsh area by causeways or dykes, standing proud of low-lying water and thus been similar to the more famous site of Tinnis in Lake Menzaleh to the east. [70]

Figure 5. Red granite column at Mastarua, length 1.80m (Photograph by Penelope Wilson).

A fifth site at Mastarua (see figure 1, no. 2) lies on the northernmost sea-shore of Egypt. The site is situated amongst the coastal dunes where the scrubby grass shows undulations in the local topography. The differential growth of grass shows rectangular building traces buried underneath the sand, also visible from satellites, and green clumps of grass may represent the interior of buildings and individual rooms. Pottery of Late Antique and Islamic date litters the surface and a red granite column was seen during an exploratory visit in 2012 (see figure 5).[71] The presence of a local coastguard station at the site shows its continuing strategic importance, perhaps having been a monitoring station for shipping. In this case, the site most likely can be identified with a settlement from the Late Antique to Medieval period called Nastarawa, which is mentioned often in the

69 MEINARDUS, 1963–1966.

70 BUTLER, 1998, p. 350–356.

71 http://www.dur.ac.uk/penelope.wilson/Delta/Survey.html

Arabic Itineraries of travellers in the north.[72] For example, Idrisi (1154 A.D.) notes that Nastarawa al-Beheira Bashmur was between El-Mahagum (location unknown) and el-Burullus, while others suggest that it may have been an island approachable only by causeways, although it seems to be near the sea-shore (Abu el-Fida).[73] The importance of this site and of Burullus, along with other coastal sites such as Rashid (Rosetta), Tinnis and Dumyat (Damietta) is indicated when, under the Patriarch Kosmas (851-858 A.D.), walls were built there against Byzantine attacks.[74]

6. Discussion

The five sites described above could be used to construct a narrative framework for understanding the landscape changes which affected human settlement from the Ptolemaic through to the Islamic period in the northern Delta. Tell Mutubis, not yet identified with any place name from antiquity, seems to be a key strategic place for the management of the newly created Bolbitine canal to the sea and the western basin of Burullus lagoon. Through the Roman period, it became perhaps the seat of a bishop, but after the Arab conquest and the development of the new *boghaze* port of Rashid, it became less viable and was abandoned. In the case of Tell el-Khubeiza, the site could be connected with a group of four monasteries that Maqrizi noted were situated "in the region of the salt marshes near Lake al-Burullus" – two of them at Bilqas[75] and Dayr Jimyanah and Dayr al-Maghtis further north.[76] Tell Nashawein (see figure 1, no. 15) could have been the harbourage for the large settlements serving the central basin of Burullus, that was Kom el-Khanziri (see figure 1, no. 16) or Pachnemounis, the capital of the Lower Sebennyitic province. Such "lake" edge sites may have been once high-lying settlements on waterways exploiting the natural resources of the area, some continuing into the Islamic period, as for example Tell Foqaa (see figure 1, no. 10).[77] Tell Foqaa is one of a group of sites along the south edge of the marshes/Lake Burullus, including Meiteh el-Bahri (see figure 1, no. 11),[78] Kom

72 Amélineau, 1893, p. 275–276; Timm, 1984–1992, IV, p. 1739–1742.

73 Guest, 1912, p. 960–961.

74 Timm, 1984–1994, I, p. 450–451.

75 Ramzi, 1955, Vol. 2, p. 27.

76 Timm, 1984–1992, II, p. 731–732.

77 Wilson/Grigoropoulos, 2009, p. 233–236.

78 Known as Miyetain in 1903, Hogarth, 1904, p. 15, pl. 1.

el-Ahmar el-Ain (see figure 1, no. 12), Tell el-Retabi (see figure 1, no. 14) and Tell el-Khubeiza (see figure 1, no. 13) which are covered in fineware pottery and have granite grindstones on them. The sites could be service settlements (ports, warehouses), with nearby "satellite" villas, farmsteads or monasteries dotted along the lake shore, taking advantage of the lagoonal resources, but reliant upon the harborage. The sea incursion of the tenth century A.D. would have made life in this area almost impossible, so the Lower Sebennytic floodplain was abandoned. Sites now lying inside the lake such as Singar (see figure 1, no. 27), Dakhla (see figure 1, no. 25) Kom el-Akhdar (see figure 1, no. 26), Mehgareh (see figure 1, no. 28) and others could have had similar functions as the "lake-side" sites but at high-water would have made been places for low-draught boats to have moored before heading out to larger ships at sea.

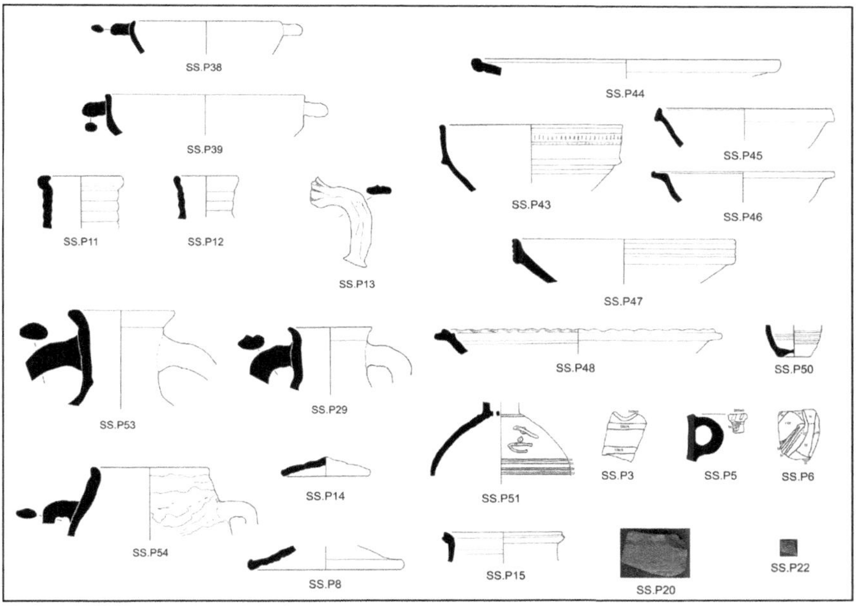

Figure 6. Pottery collected during survey work at Sidi Selim (D. Grigoropoulos).

Mastarua is the latest dated site in the medieval period of this small group, sea-ward and lakeward looking and only a short connection away to the south to Tell Foqaa and thence into the waterway network onward to al-Fustat or west to Alexandria. Other sites in the EES survey showed some continuation into the ninth and tenth century A.D., namely Sidi Selim (see figure 1, no. 9), Abu Ismail (see figure 1, no. 21) and Foqaa (see figure 1, no. 10). The dating depended upon the appearance of specific types of material in the ceramic collections, namely glazed wares in kaolin fabrics (see figure 6). In addition, Late Antique, "Coptic", pottery such as white slipped forms and glass vessels, continue from the seventh century into the eighth and ninth century A.D., so that they are evidence of an underlying material culture continuity not of chronology.[79] It seems that, given the early Arab ceramics, there was some form of occupation well after the Arab expansion and into the Fatimid period, when the rural ar-Rif of the Delta was important for its produce and particularly flax and linen into the tenth century A.D.[80]

Ultimately, the changing northern environment also contributed to the decline in settlement. The Sebennytic branch had silted up by the early medieval period, perhaps with some earlier gradual problems such as sea incursions into the area, and it is possible that this led to the eventual real decline of the area as the focus of attention for the main arteries of transport switched to the Rosetta and Damietta branches of the river.[81] Both the Canopic and Sebennytic river branches had been the life blood of the areas through they flowed. So long as there were viable waterways, the towns beside or close to them offered transshipment points through the network of Delta waterways, ensuring fast movement of goods within Egypt and to ports for export. The settlements also offered high ground above the flood plain, upon which towns were situated purposely to sit out the inundation. Additional resources such as fish, reeds and papyrus could have ensured that the inundation months were not unproductive, so that there was the possibility of all-year-round exploitation of the delta environment. The end of the Byzantine boom may have been a gradual phenomenon, noticeable only two centuries into the Islamic period, as is also the case in other parts of the Islamic world.[82] The Islamic period also coincided with the increased unsustainability of the high mound settlements, when supplying them with water became too difficult and people moved down onto cleaner, less crowded sites on new levees. Such displacements of settlements from older mounds has also been noted in Upper Egypt at Tell Edfu[83] as well as Tebtunis-Tutun in the Fayum[84]

79 Grigoropoulos reflecting comments on the difficulty of dating late Antique to Early Arab pottery in general, in WILSON/GRIGOROPOULOS, 2009, p. 282–286.

80 SANDERS, 1998, p. 161–165.

81 ARBOUILLE/STANLEY, 1990: p. 59–63; STANLEY/WARNE, 1993.

82 KING et al, 1994.

83 GASCOIGNE, 2005.

84 KEENAN, 2003, p. 132–137.

and may be a general phenomenon for the early Islamic period. In the north, populations of old settlements were being displaced to or replaced in new, but nearby places, perhaps in the cases of Sidi Selim, Metoubas and Tell el-Daba-Tida.

7. Conclusion

A stark photograph of the monastery of Saint Damiana around 1903, shows the building standing in a salt-mud flat, with nothing else around it.[85] Flat expanses of water with labyrinthine reed beds of modern Lake Burullus (see figure 7),

Figure 7. The reed beds of Burullus, with a hidden entrance-track to an island in the lake (Photograph by Penelope Wilson).

hiding the remnants of ancient sites evoke a desolate marsh environment prone to human interference and natural stress. Much survives worth archaeological investigation but the prioritisation of that work is difficult due to the nature of the sites involved and modern agricultural and urban pressures. Classification of sites and zones of sites as urban, self-sufficient "monasteries"/farmsteads, military/customs outposts, industrial/warehousing, small scale farms and fisheries shows the diversity of production and activity of the north. The work discussed here presents a small, perhaps imperfect dataset, but set against the framing story

85 HOGARTH, 1910, facing p. 100.

of the natural fluvial landscape and from the wider Eastern Mediterranean and North African perspective, it is worth teasing out the strong regional character and cultural context of life in the unique, north Egyptian Delta.

Acknowledgements

I would like to thank Harco Willems and Jan-Michael Dahms for their invitation to take part in the conference and their patience. I also acknowledge the support of Jeffrey Spencer and the Egypt Exploration Society's Delta Survey project, Roger Dickinson for assistance with the maps and Dimitrios Grigoropoulos, Rebecca Bradshaw, Fatma Keshk and Hind Ramadan, as well as the Ministry of State for Antiquities in Egypt and Kafr el-Sheikh office under Gamal Selim and Dr Mohamed abd el Rifaat.

Bibliography

AMÉLINEAU, EMILE, La Géographie de l'Égypte à l'époque Copte, Paris 1893.

ARBOUILLE, DIDIER/STANLEY DANIEL J., Late Quaternary Evolution of the Burullus lagoon region, north-central delta, Egypt, in: Marine Geology 99 (1991), p. 45-66.

BAGNALL, ROGER, Egypt in Late Antiquity, New Jersey 1998.

ID., Egypt in the Byzantine World 300-700, Cambridge 2007.

BALL, JOHN, Egypt in the Classical Geographers, Cairo 1942.

BALLET, PASCALE/VON DER WAY, THOMAS, Exploration archéologique de Bouto et de sa région. Epoques romaine et byzantines, Mitteilungen des deutschen archäologischen Instituts, Abt. Kairo 49 (1993), p. 1-22.

BALLET, PASCALE ET AL., Et la Bouto tardive?, in: Bulletin de l'institut français d'archéologie orientale 111 (2011), p. 75-93.

BIRKS, HILARY/BIRKS, HARRY J. B., Recent ecosystem dynamics in North African Lakes in the CASSARINA Project, in: Aquatic Ecology 35 (2001), p. 461-478.

BLOUIN, KATHERINE, Triangular Landscapes. Environment, Society and the State in the Nile Delta under Roman Rule, Oxford 2014.

BUTLER, ALFRED J., The Arab Conquest of Egypt and the Last Thirty Years of the Roman Dominion. Oxford 1998.

BUTZER, KARL, Environment and human ecology in Egypt, in: Bulletin de la societé de géographie d'Égypte 32 (1959), p. 43-87.

ID., Early Hydraulic Civilization in Egypt, Chicago, 1976.

CHEN, ZHONGYAN/WARNE, ANDREW/STANLEY, DANIEL J., Late Quaternary Evolution of the Northwestern Nile Delta between the Rosetta Promontory and Alexandria, Egypt, in: Journal of Coastal Research 8, No. 3, Summer (1992), p. 527-561.

CHRISTIANSEN, ERIK, Coinage in Roman Egypt. The Hoard Evidence, Aarhus. 2004.

COOPER, JOHN P., The Medieval Nile. Route, navigation and landscape in Islamic Egypt, PhD Southampton University 2008.

DARESSY, GEORGES, Recherches géographiques II. Eléarchia, Annales du Service des Antiquités de l'Égypte 26 (1927), p. 259-272.

DOSSEY, LESLEY, Peasant and Empire in Christian North Africa, Berkeley/Los Angeles/London 2010.

DUMONT, HENRI J./EL-SHABRAWY, GAMAL, Lake Borollus of the Nile Delta. A Short History and an Uncertain Future, in: Ambio 36, No. 8, Dec. 2007, p. 677-682.

EDGAR, CAMPBELL C., Notes from the Delta, Annales du Service des Antiquités de l'Égypte 11 (1911), p. 87-97.

ENGSHEDEN, ÅKE, A View on the Toponyms in the Governorate of Kafr el-Sheikh, in: Altägyptische Weltsichten. Akten des Symposiums zur historischen Topographie und Toponymie Altägyptens vom 12.-14. Mai 2006 in München (Ägypten und Altes Testament 68), ed. by FARIED ADROM/KATRIN SCHLÜTER/ ARNULF SCHLÜTER, Wiesbaden 2008, p. 35-49.

EVETTS, BASIL, History of the Patriarchs of the Coptic Church of Alexandria I-IV. Patrologia Orientalis Vol. I.2, I.4, V.1 and X.5, Paris 1906-1915, Accessed online at: http://www.tertullian.org/fathers/severus_hermopolis_hist_alex_patr_01_part1.htm

FLEMMING, NICHOLAS C., Predictions of Relative Sea-Level Change in the Mediterranean Based on Archaeological, Historical and Tide-Gauge Data, in: Climate Change and the Mediterranean, ed. by LUDUMIR LEFTIC/JOHN D. MILLIMAN/GIULIANO SESTINI, London et al. 1992, p. 247-281.

FRIHY, OMAR, Holocene Delta changes at the Nile Delta Coastal Zone of Egypt, in Geo Journal 26, No. 3, March (1992), p. 389-394.

GASCOIGNE, ALISON, Dislocation and Continuity in early Islamic Provincial Urban Centres. The Example of Tell Edfu, Mitteilungen des deutschen archäologischen Instituts, Abt. Kairo 61 (2005), p. 153-189.

GUEST, ARTHUR RH., The Delta in the Middle Ages, in: Journal of the Royal Asiatic Society (1912), p. 941-980.

HARTUNG, ULRICH et. al., Tell el-Fara'in – Buto 10. Vorbericht, Mitteilungen des Deutschen Archäologischen Instituts, Abt. Kairo 65 (2009), p. 83-190.

HOGARTH, DAVID, Three North Delta Nomes, in: Journal of Hellenic Studies 24 (1904), p. 1-19.

ID., Accidents of an Antiquary's Life, Cambridge 1910.

KAMEL, YOUSEF, Monumenta cartographica, Africa et Aegyptus Epoque Arabe, Hague 1926.

KEENAN, JAMES G., Deserted Villages: From the Ancient to the Medieval Fayyum, Bulletin of the American Society of Papyrologists 40 (2003), p. 119-139.

KEMP, BARRY, Settlement and Landscape in the Amarna Area in the Late Roman Period, in: Late Roman Pottery at Amarna and related studies (EES Excavation memoir 72), ed. by JANE FAIERS, London 2005, p. 11-56.

KENNEDY, HUGH, Egypt as a province in the Islamic caliphate, 641-868, in: The Cambridge History of Egypt, Vol. I, Islamic Egypt, 640-1517, ed. by CARL F. PETRY, Cambridge 1999, p. 62-85.

KING, GEOFFREY R.D./CAMERON, AVERIL, The Byzantine and Early Islamic Near East II. Land Use and Settlement Patterns (Studies in Late Antiquity and Early Islam I), Princeton, New Jersey 1994.

KIRBY, CHRISTOPHER/RATHBONE, DOMINIC, Kom Talit. The Rise and Fall of a Greek Town in the Faiyum, in: Egyptian Archaeology 8 (1996), p. 29-31.

LAMB, WALTER, Heliodorus, Ethiopian Story, London 1961.

LAMBECK, KURT et al., Paleoenvironmental Records, Geophysical Modeling, and Reconstruction of Sea-Level Trends and Variability on Centennial and Longer Timescales, in: Understanding Sea-Level Rise and Variability, ed. by JOHN A. CHURCH et al., Chichester 2010, p. 61-121.

LOBEL, EDGAR et al., The Oxyryhnchus Papyri Part XXIV, London 1957.

MASPERO, JEAN/WIET, GASTON, Matériaux pour servir à la géographie de l'Égypte (Mémoires publiés par les membres de l'Institut français d'archéologie orientale 36), Cairo 1919.

MEINARDUS, OTTO, Singar. An Historical and Geographical Study. Bulletin de la Société d'Archéologie Copte XVII-XVIII (1963-1966), p. 175-179.

MIKHAIL, MAGED, From Byzantine to Islamic Egypt. Religion, Identity and Politics after the Arab Conquest, London/New York 2014.

PARCAK, SARAH, Settlement Pattern Studies in the Nile's Floodplain. Satellite Imagery Analysis and Ground Survey in Middle Egypt and the Delta, PhD Cambridge University 2005.

POLLARD, NIGEL, The Chronology and Economic Condition of Late Roman Karanis. An Archaeological Reassessment, in: Journal of the American Research Center in Egypt 35 (1998), p. 147-162.

RAMDANI, MOHAMED ET AL., North African wetland lakes. Characterization of nine sites included in the CASSARINA project, in: Aquatic Ecology 35 (2001), p. 281-302.

RATHBONE, DOMINIC, Surface Survey and the Settlement History of the Ancient Fayum, in: Archeologia e papyri nel Fayyum. Storia della ricerca, problemi e prospettive. Atti del convegno internazionale, Siracusa, 24-25 Maggio 1996, ed. by ANNA DI NATALE, Syracuse 1997, p. 7-19.

REEVES, NICHOLAS/TAYLOR JOHN H., Howard Carter before Tutankhamun, London 1992.

ROWLAND, JOANNE/WILSON, PENELOPE, The Delta Survey 2004-05, in: Journal of Egyptian Archaeology 92 (2006), p. 1-13.

ROSTOVTZEFF, MICHAEL, Social and Economic History of the Roman Empire, Oxford 1926.

SANDERS, PAULA, The Fatimid State, 969-1171, in: The Cambridge History of Egypt, Vol. I, Islamic Egypt, 640-1517, ed. by CARL F. PETRY, Cambridge 1998, p. 151-174.

SCHIESTL, ROBERT, Kom el-Gir in the western Delta, in: Egyptian Archaeology 42, Spring (2013), p. 28-29.

SESTINI, GIULIANO, Implications of Climatic Changes for the Nile Delta, in: Climate Change and the Mediterranean, ed. by LUDUMIR LEFTIC/JOHN D. MILLIMAN/GIULIANO SESTINI, London et al. 1992, p. 535-601.

SIJPESTEIJN, PIETER J./WORP, KLAAS A., A Transportation Archive from Fourth-Century Oxyrhynchus (P. Mich XX), in: American Studies in Papyrology 49, Durham, North Carolina 2011, p. 175-180.

SPENCER, JEFFREY, The EES Delta Survey in Spring 2011, in: Egyptian Archaeology 39 (2011), p. 3-5.

STANLEY, DANIEL J./WARNE, ANDREW G., Nile Delta. Recent Geological Evolution and Human Impact, in: Science New Series 260, No. 5108, Apr. 30 (1993), p. 628-634.

TIMM, STEPHAN, Das christliche-koptische Ägypten in arabischer Zeit. Eine Sammlung christlicher Stätten in Ägypten in arabischer Zeit, 6 Volumes (Beihefte zum Tübinger Atlas des Vorderen Orients 41), Wiesbaden 1984-1992.

TOUSSON, OMAR, Mémoire sur les anciennes branches du Nil. Époque ancienne. Époque arabe (Mémoires de l'"institut de l'Égypte IV), Cairo 1922.

TRAMPIER, JOSHUA, The Dynamic Landscape of the Western Nile Delta from the New Kingdom to the late Roman Periods, PhD Dissertation, UMI: 3419703, University of Chicago 2010.

TRÜMPER, MONIKA, Complex Public Bath Buildings of the Hellenistic Period. A Case Study in Regional Difference, in: Le bain collectif en Égypte (Institut français d'archéologie orientale Études urbaines 7), ed. by MARIE-FRANÇOISE BOUSSAC/THIBAUD FOURNET/BÉRANGÈRE REDON, Cairo 2009, p. 139-179.

VAN MINNEN, PETER, The other cities in later Roman Egypt, in: Egypt in the Byzantine World, 300-700, ed. by ROGER BAGNALL, Cambridge 2007, p. 207-225.

WILSON PENELOPE, Living the High Life, in: Egypt in the first Millennium AD. Perspectives from new fieldwork, ed. by ELISABETH O'CONNELL, Leuven 2014, p.43-58.

WILSON, PENELOPE/GRIGOROPOULOS, DIMITRIOS, The West Nile Delta Regional Survey, Beheira and Kafr el-Sheikh Provinces, London 2009.

WIPSZYCKA, EWA, La chiesa nell'Egito del IV secolo. Le strutture ecclesiastiche, in: Miscellanea Historiae Ecclesiasticae VI, Congrès de Varsovie 1978, Warsaw/Louvain/Brussels 1983, p.1 82-201.

WORP, KLAAS A., A Checklist of Bishops in Byzantine Egypt (AD 325-750), in: Zeitschrift für Papyrologie und Epigraphik 100 (1994), p. 283-318

Authors

JEAN-CHRISTOPHE ANTOINE, Department of Neurological Sciences, University of Saint-Etienne

MANFRED BIETAK, Austrian Academy of Sciences, Institute of Oriental and European Archaeology, Department "Egypt and the Levant"

MANSOUR BORAIK, General Director of the Antiquities from Giza to Sohag and of the Western Deserts, Ministry of State Antiquities

JUDITH BUNBURY, Department of Earth Sciences, University of Cambridge

HANNE CREYLMAN, Department of Archaeology, Art history and Musicology, KU Leuven

JAN-MICHAEL DAHMS, holds a PhD in Egyptology from the Ruprecht-Karls-University Heidelberg.

VÉRONIQUE DE LAET, Center for Archaeological Sciences, Department Earth and Environmental Sciences, Division of Geography and Tourism, KU Leuven

LUC GABOLDE, Director of research at the CNRS, UMR 5140, "Archéologie des sociétés méditerranéennes", Montpellier-Lattes

PEDRO GONÇALVES, Department of Archaeology, University of Cambridge

ANGUS GRAHAM, Wallenberg Academy Fellow, Department of Archaeology and Ancient History, Uppsala University; Honorary Research Associate, Institute of Archaeology, University College London

IHAB MOHAMED, Department Earth and Environmental Sciences, Division of Geography and Tourism, KU Leuven

BASTIAAN NOTEBAERT, Fund for Scientific Research Flanders, Department Earth and Environmental Sciences, Division of Geography and Tourism, KU Leuven

BENJAMIN PENNINGTON, Geography and Environment, University of Southampton.

FÉLIX RELATS MONTSERRAT, PhD Assistant, UMR 8167 "Orient et Méditerranée – Composante Mondes Pharaoniques", LABEX Resmed – Fondation Thiers, Université de Paris-Sorbonne

CORNELIA RÖMER, DAAD/DAI Cairo

SANDRA SANDRI, Institut für Altertumswissenschaften, Ägyptologie, Johannes Gutenberg-Universität Mainz

ANA TAVARES, Department of Archaeology, University of Cambridge

JOSHUA TRAMPIER, Research Affiliate, Center for Ancient Middle Eastern Landscapes (CAMEL), University of Chicago

GERT VERSTRAETEN, Center for Archaeological Sciences, Department Earth and Environmental Sciences, Division of Geography and Tourism, KU Leuven

HARCO WILLEMS, Center for Archaeological Sciences, Near Eastern Studies, KU Leuven

PENELOPE WILSON, Senior Lecturer, Department of Archaeology, Durham University